Joint 12th International Conference: Two-Phase Systems for Space and Ground Applications and 2nd International School of Young Scientists “Interfacial Phenomena and Heat Transfer” 2017

Journal of Physics: Conference Series Volume 925

Novosibirsk, Russia
11 - 16 September 2017

Editors:

Oleg Kabov
Yuriy Lyulin
Dmitry Zaitsev

ISBN: 978-1-5108-5272-3
ISSN: 1742-6588

Printed from e-media with permission by:

Curran Associates, Inc.
57 Morehouse Lane
Red Hook, NY 12571

Some format issues inherent in the e-media version may also appear in this print version.

Printed by Curran Associates, Inc. (2017)

For permission requests, please contact the Institute of Physics
at the address below.

Institute of Physics
Dirac House, Temple Back
Bristol BS1 6BE UK

Phone: 44 1 17 929 7481
Fax: 44 1 17 920 0979

techtracking@iop.org

Additional copies of this publication are available from:

Curran Associates, Inc.
57 Morehouse Lane
Red Hook, NY 12571 USA
Phone: 845-758-0400
Fax: 845-758-2633
Email: curran@proceedings.com
Web: www.proceedings.com

Table of contents

Volume 925

Joint 12th International Conference: "Two-Phase Systems for Space and Ground Applications" and 2nd International School of Young Scientists "Interfacial Phenomena and Heat Transfer"11–16 September 2017, Novosibirsk, Russian Federation

Accepted papers received: 1 November 2017
Published online: 9 November 2017

Preface

Papers

IOP Conf. Series: Journal of Physics: Conf. Series **925** (2017) 011001 doi:10.1088/1742-6596/925/1/011001

Preface

Joint 12-th International conference «Two-Phase Systems for Space and Ground Applications» and 2nd International School of Young Scientists «Interfacial Phenomena and Heat Transfer» were held at Kutateladze Institute of Thermophysics SB RAS, Academgorodok, Novosibirsk, Russia, during 11-16 of September 2017. The third largest city in Russia, often referred to as "the capital of Siberia", Novosibirsk is home to a number of research centres and institutes, including the Kutateladze Institute of Thermophysics. The long tradition of research in both fluid mechanics and heat transfer makes this institute a natural venue for the Conference and School.

The Conference and School provide a platform for researchers to exchange information and identify research needs in this important area encompassing several engineering, mechanical and physical disciplines. A wide range of topics in the rapidly developing and highly interdisciplinary field of interfacial phenomena and heat transfer were covered, including boiling, shear-driven films, droplet evaporation, contact line phenomena, thermocapillary flows, two-phase flows in microchannels and minichannels, and effect of gravity on various processes.

The Conference and School were conducted under the financial support of the Russian Science Foundation (№ 15-19-30038).

The editors would like to express their sincere appreciations and thanks to all the committee members of the Conference and School for their enormous efforts, to all the authors for their research contribution and to all expert referees for peer review of the articles. Thanks also to IOP Conference Series for producing the volume with articles.

Editors,

Prof. Oleg Kabov

Dr. Yuriy Lyulin

Dr. Dmitry Zaitsev

For more information about the Conference and School please visit the websites:

http://www.itp.nsc.ru/htl/spaceconf-2017/

http://www.itp.nsc.ru/htl//school-conference-2017/

IOP Conf. Series: Journal of Physics: Conf. Series **925** (2017) 011002 doi:10.1088/1742-6596/925/1/011002

Peer review statement

All papers published in this volume of *Journal of Physics: Conference Series* have been peer reviewed through processes administered by the proceedings Editors. Reviews were conducted by expert referees to the professional and scientific standards expected of a proceedings journal published by IOP Publishing.

IOP Conf. Series: Journal of Physics: Conf. Series **925** (2017) 012001 doi:10.1088/1742-6596/925/1/012001

Convective fluid flows in a horizontal channel with evaporation: analytical and experimental investigations

Y V Lyulin[1] and E V Rezanova[2]

[1] Kutateladze Institute of Thermophysics SB RAS, 1, Ac. Lavrentyieva ave., 630090 Novosibirsk, Russia
[2] Altai State University, Faculty of Mathematics and Information Technologies , pr. Lenina, 61, 656049 Barnaul, Russia

E-mail: katerezanova@mail.ru

Abstract. Heat- and mass transfer processes in a two-layer system of the liquid and gas are studied with respect to evaporation at interface. The stationary convective flows of two immiscible viscous incompressible fluids filling an infinite channel and being under action of the transverse gravitation field are studied analytically. Mathematical modeling of the flows is carried out with the help of the Navier-Stokes equations in Boussinesq approximation. The Dufour and Soret effects are taken into consideration in the gas-vapor phase. In the two-dimensional case the exact solutions of special type are constructed under condition of a given specific gas flow rate. Comparison of the analytical results with results of the physical experiments with the "liquid-gas" system like "ethanol-air" are presented.

1. Introduction

Investigations of the nonisothermal flows of fluids, complicated by the interfaces, are stimulated by necessity of explanation of results of the scientific and technological experiments aimed at the study the vortex fluid structures in the open cavities and in the moving layers, of the experiments directed to the study the film motion under action of the gas flows [1]. The study of exact solutions of a special type allows one to consider the influence of various physical factors on fluid flows [2-8]. Much attention is paid to investigation of flows in a two-layer system accompanied by evaporation. One of the first examples of the solutions of the problems with mass transfer at the interface has been obtained in [4]. A number of works are devoted to the investigation of convective flows with a thermocapillary interface under the action of a co-current gas flow and longitudinal temperature gradients (see [5] and [6-8])]. In [7] the results of the comparison of analytical data obtained using the exact solutions with experimental data are shown. The results of study of the stability of the two-layer flows with evaporation on the thermocapillary interface are presented in [9-11] (see also [12]).

In the present paper the analytical and experimental investigations of the liquid flows accompanied by a gas flux with evaporation at a thermocapillary interface are performed. In the upper layer a process of vapor diffusion is taken into account. To characterize the flow topology the examples of the velocity profiles and temperature distributions for the system "ethanol and air" are presented.

2. Mathematical model of the process

Let two immiscible viscous incompressible fluids (the liquid and the gas-vapor mixture) fill the infinite horizontal layers of depth equal to l and h, correspondingly. The upper and lower walls of the channel are the fixed impermeable boundaries. A common gas-liquid thermocapillary interface, remaining to be undeformed one, is given here by $y = 0$. The coordinate system is chosen so that the gravity acceleration vector is directed opposite to the Oy axis. The system of differential equations based on the Oberbeck-Boussinesq approximation of the Navier-Stokes equations [13] is used to model the flows of the liquid and gas and to find the unknown functions. Concentration of the vapor understood as a passive impurity in the gas layer is determined with the help of the diffusion equation. The Soret and Dufour are taken into consideration in the mathematical model [13, 14].

Let the solution of the system of equations be the solution of the Ostroumov-Birikh type [2, 3] in the form:

$$u_i = u_i(y),\ \upsilon_i = 0,\ p_i' = p_i'(x, y),\ T_i = (a_i^1 + a_i^2 y)x + \vartheta_i(y),\ C = (b_1 + b_2 y)x + \varphi(y).$$

Here u_i, υ_i are the velocity components, T_i is the temperature, p_i' is the modified pressure and C is the vapor concentration. The coefficients a_j^i and b_j (j = 0, 1) (the longitudinal gradients of temperature and concentration) are determined with the help of the boundary conditions for concentration and temperature. The functions u_i are defined as fourth-degree polynomials, ϑ_i and φ are the polynomials of the seventh degree, p' is the eighth degree polynomial on the transverse coordinate y (see [7]).

Let us formulate the boundary conditions for the unknown functions. On the thermocapillary interface the kinematic and dynamic conditions must be satisfied [14]. The kinematic condition is fulfilled automatically, the dynamic condition is written taking into account the thermocapillary effect. The conditions of the velocity and temperature continuity are set at the interface. The heat transfer condition and the mass balance equation at the interface should be written with respect to the Dufour and Soret effects. The saturated vapor concentration will be determined with the help of the relation that is a result of the Clapeyron-Clausius equation and Mendeleev-Clapeyron equation for ideal gas [6, 7, 14].

On the lower and upper solid impermeable walls of the flow domain the no-slip conditions should be fulfilled. The boundaries $y = -l$ and $y = h$ are heated linearly with respect to the longitudinal coordinate:

$$T_1|_{y=-l} = A_1 x + \vartheta^-, \qquad T_2|_{y=h} = A_2 x + \vartheta^+,$$

here A_1, A_2, ϑ^-, ϑ^+ are constant. The vapor concentration on the upper boundary is assumed to be equal to zero.

The gas flow rate in the upper layer is given and defined by the relation:

$$Q = \int_0^h \rho_2 u_2(y)dy,$$

ϱ_2 is the density of the gas.

Note that the condition of continuity of temperature at the interface $y = 0$ entails the equality of the temperature gradients $a_1^1 = a_2^1 = A$. Conditions on free and solid boundaries determine the relationship among longitudinal temperature gradients A, A_1 and A_2. One of the longitudinal gradients is chosen arbitrarily.

3. Results of analytical calculations. Comparison with physical experiments

Consider the two-layer system of liquid and gas such as "ethanol - air". The values of the physicochemical parameters can be found in [15]. We suppose that the Dufour coefficient δ is equal to 10^{-3} (see [10]); the Soret coefficient takes the value $-5\cdot10^{-3}$ (it corresponds to the normal thermocapillary effect, see [14]). The concentration coefficient of density γ is equal to -0.62.

IOP Conf. Series: Journal of Physics: Conf. Series **925** (2017) 012001 doi:10.1088/1742-6596/925/1/012001

Figure 1 shows the velocity profiles and temperature distributions (presented in Celsius degrees) for different values of the gas flow rate in the upper layer of the system, obtained with the help of the analytical formulas. The longitudinal temperature gradient on the interface A is taken equal to -20 K/m, the height of the gas layer of h is equal to $h = 0.3 \cdot 10^{-2}$ m, and the height of the liquid $l = 0.3 \cdot 10^{-2}$ m. Calculations show that the increase of the gas flow rate entails quantitative changes of the velocity (an increase in the liquid velocity), the qualitative character of the velocity profiles is conserved. The temperature distribution in the system quantitatively changes with increasing values of Q. The maximum value of temperature also grows.

Figure 2 presents the velocity profiles and the distributions of temperature by the values of the temperature gradients on the interface equal to 20 K/m. In the case, when the longitudinal temperature gradient changes the sign to the opposite one, the velocity values are changed weakly. The character of the temperature distribution in the system changes significantly, the zone of the greatest cooling of the liquid moves in the negative direction of the Ox axis.

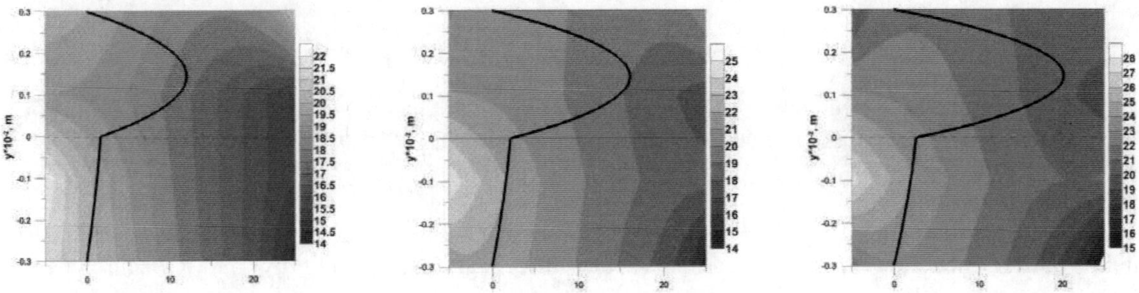

a b c

Figure 1. Velocity profiles and temperature distribution in the system "ethanol-air" at different values of the gas flow rate (a – $Q = 3.375 \cdot 10^{-4}$ kg/(m s), b – $Q = 4.5 \cdot 10^{-4}$ kg/(m s), c – $Q = 5.625 \cdot 10^{-4}$ kg/(m s)). Here $h = 0.3 \cdot 10^{-2}$ m, $l = 0.3 \cdot 10^{-2}$ m, A = -20 K/m

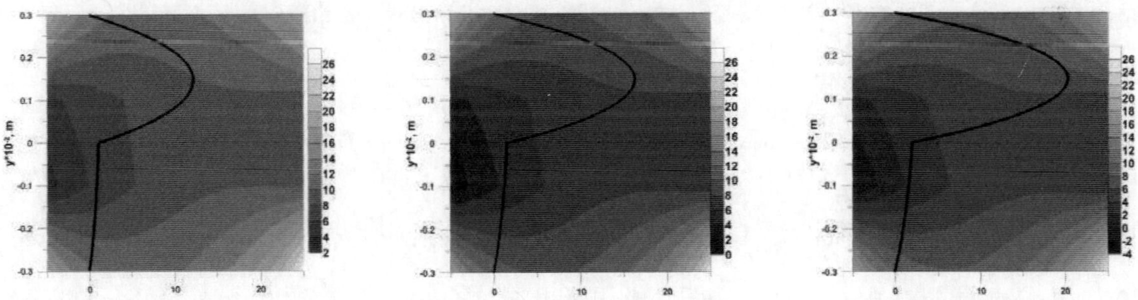

a b c

Figure 2. Velocity profiles and temperature distribution in the system "ethanol-air" at different values of the gas flow rate (a – $Q = 3.375 \cdot 10^{-4}$ kg/(m s), b – $Q = 4.5 \cdot 10^{-4}$ kg/(m s), c – $Q = 5.625 \cdot 10^{-4}$ kg/(m s)). Here $h = 0.3 \cdot 10^{-2}$ m, $l = 0.3 \cdot 10^{-2}$ m, A = 20 K/m

Figure 3 presents the experimental and analytical results demonstrating the values of the evaporative mass flow rate at the interface by the different values of the longitudinal temperature gradients on the lower solid boundary. The experimental data are presented in one system of measurements. By comparison of the experimental and analytical results of investigations of the dependence of the evaporative mass flow rate M on the gas flow rate it has been found that the results can have a similar quantitative and qualitative character. The rate of evaporation increases with increasing of the gas flow rate Q, but according to the experimental data this dependence of M on Q

has a nonlinear form. Figure 3 shows a rather good correlation of the analytical and experimental results for the value range of Q between 0.0003 and 0.0005 (kg/(m s) in the case of presented values of the longitudinal temperature gradient A.

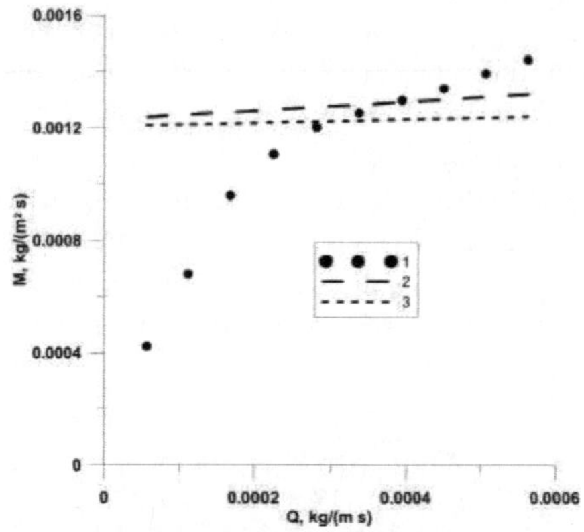

Figure 3. Dependence of the evaporation rate on the gas flow rate. Here $h = l = 0.3\cdot10^{-2}$ m, (1 – experimental data; 2 – calculations A= -20 K/m; 3 – A = -50 K/m)

Acknowledgments
This work was supported by the Russian Foundation of Basic Research (project 17-08-00291: mathematical modeling and presentation of the analytical results) and by the Russian Science Foundation (project 15-19-20049: experimental part of the work; analysis of the experimental results; comparison with analytical results).

References

[1] Lyulin Yu V and Kabov O A 2014 *Int. J. Heat and Mass Transfer* **70** 599-609
[2] Birikh R V 1969 *J. Appl. Mech. Tech. Phys* **3** 69-72
[3] Ostroumov G A 1952 *Free convection in the conditions of the internal problem* (M. - Leningrad: Gos. izd-vo tehniko-teoreticheskoy literaturi) 286 (in Russian)
[4] Shliomis M I and Yakushin V I 1972 *Uchenye zapiski Permskogo Gosuniversiteta* **4** 129-140
[5] Goncharova O N and Kabov O A 2010 *Int. J. of Heat and Mass Transfer* **53** 2795-2807
[6] Goncharova O N and Rezanova E V 2014 *J. of Appl. Mech. and Tech. Phys.* **55** 247-257
[7] Goncharova O N, Rezanova E V, Lyulin Yu V and Kabov O A 2015 *Thermophys. and Aeromech.* **5** 631-637
[8] Goncharova O N, Kabov O A and Pukhnachov V V 2012 *Int. J. of Heat and Mass Transfer* **55** 715-725
[9] Bekezhanova V B, Goncharova O N, Rezanova E V and Shefer I A 2017 *Fluid Dynamics* **52** 189–200
[10] Rezanova E V and Shefer I A 2017 *Journal of Applied and Industrial Mathematics* **11** 274–283
[11] Bekezhanova V B and Goncharova O N 2016 Stability of the exact solutions describing the two-layer flows with evaporation at interface *Fluid Dynamics Research Journal* **48** 061408
[12] Bekezhanova V B 2012 *Fluid Dynamics* **47** 702–708
[13] Andreev V K, Gaponenko Yu A, Goncharova O N and Pukhnachov V V 2012 *Mathematical models of convection* (de Gruyter Studies in Mathematical Physics, Berlin/Boston: De Gruyter) 417
[14] Prigogine I and Dufay R 1954 *Chemical thermodynamics* (Longmans, Green, London, New York)
[15] Ravdael A A and Ponomareva A M *Quick Reference physico-chemical quantities* 1998 (SPb, Spetc. lit.) 232 (in Russian)

Experimental determination of the hydraulic resistance coefficient at the microchannel inlet

V M Aniskin

S.A. Khristianovich Institute of Theoretical and Applied Mechanics SB RAS, Russia

E-mail: aniskin@itam.nsc.ru

Abstract. The paper presents the results of the hydraulic resistance coefficient measurements at the inlet of microchannel. The Reynolds number ranged from 900 to 2300. The channel with inlet diameter of 3.5 mm is abruptly narrowed to a diameter of 0.9 mm. We measured the pressure drop in the area of sharp change in diameter. Measured hydraulic resistance coefficient was compared with the previously obtained coefficients for microchannels of the same geometry. It is shown that for the studied interconnected channel geometry no differences were found between micro and minichannels.

1. Introduction

The advancement of technology made it possible to produce miniature fluidic devices. When developing and designing microfluidic devices, the question arises about the applicability of known fluid dynamics laws applicable for macrochannels to fluid flow in channels of micron and submicron sizes. Over the past 15-20 years, many research groups presented their results on pressure drop and hydraulic resistance coefficient for laminar and turbulent fluid flow in microchannels as well as studies on heat transfer in microchannels.

The application of microfluidic systems for cooling of microelectronic components is quite attractive and promising due to their large efficiency. Along with cooling systems, microfluidic systems find use in biomedical applications. Optimizing the design of microfluidic systems requires a clear understanding of the fluid transport mechanisms in both laminar and turbulent fluid flow.

When designing microfluidic devices, it is important to correctly account the hydraulic losses in all the areas of the hydrodynamic tract. When dealing with straight microchannels we know how to determine the hydraulic resistance value, the issue of hydraulic losses in coupling areas of different size channels still remains unclear. An additional difficulty in this problem is caused by the lack of information available in handbook on hydraulic resistance relevant to Reynolds numbers or geometric configurations of coupling of different channels that are typically used in practice. The complexity of fabrication of the microchannels with orifices in the wall makes it difficult carrying out experiments to determine the hydraulic resistance coefficients at the microchannel inlet.

Literature reviews dedicated to the hydraulic resistance in the microchannels are given in many papers [1-4], though the number of works on measurement of the pressure inside the microchannels is quite small [5-7].

Considerable interest in the context of design and the overall resistance of the microchannel is focused to the inlet region of the channel, namely the resistance of the inlet section. The study of this important issue is complicated by the extremely small size of the investigated area, which is responsible for pressure drop.

In our previous works [8, 9] we studied the fluid flow in microchannels with diameters from 70 to 100 microns. The channels were provided with the orifices made in the wall to measure the pressure along the channels. One of the orifices was located near the inlet to the channel at a distance of about one diameter of the channel. This made it possible to determine the hydraulic resistance coefficient at the inlet of the microchannel. The coupling geometry of two channels is shown in figure 1. Here D=1700 μm, while d is equal to 70 [8] and 100 [9] μm.

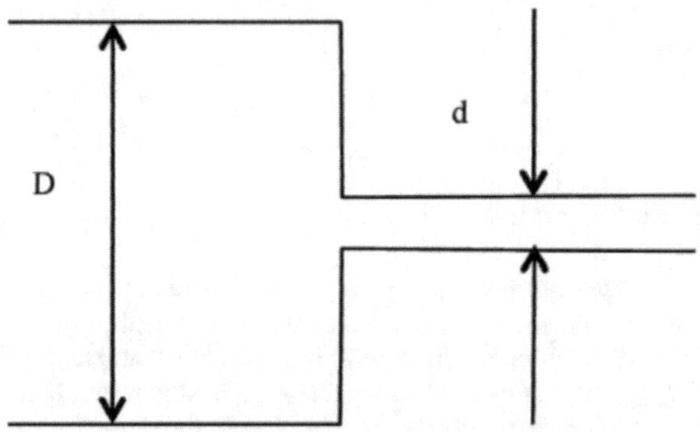

Figure 1. The coupling geometry of the channels

Hydraulic resistance coefficient of the inlet section of the microchannels in [8, 9] was defined as:

$$f = \frac{2\Delta P}{\rho V^2},$$

where ρ и V – are the density and the average fluid velocity, ΔP – is the pressure drop at the channel inlet, defined as the difference between the manifold pressure and the pressure in the first orifice of the microchannel.

Unfortunately, the handbook of hydraulic resistance [10] lacks data for a given geometric configuration and experimentally achieved Reynolds numbers. Accordingly, the aim of this work was to obtain information about the hydraulic resistance coefficient of the channel inlet with the same geometry but at larger scale.

2. Experiment

To conduct experiments, we have fabricated experimental setup consisting of two channels with diameters of 3.5 and 0.9 mm. The coupling geometry of the channels corresponded to that shown in figure 1. To measure pressure drop in the channels with a diameter of 3.5 and 0.9 mm we made orifices with a diameter of 300 and 100 μm, respectively. These orifices were located at a distance of 400 μm on either side of coupled channels.

The schematic diagram of experiment is shown in figure 2. Distilled deionized water (1) was used as a working fluid. Through the filter (2) water was fed into the channels (4) by means of Gilson-305 piston pump (3). Fluid flow rate was controlled by weighing system (5). The pressure drop was recorded by Honeywell PC24 differential pressure sensor at 30 psi. The Reynolds numbers ranged from 900 to 2300.

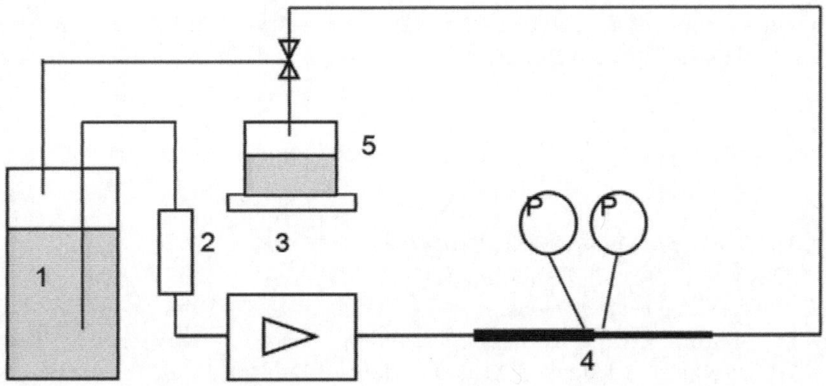

Figure 2. Schematic diagram of the experiment

Figure 3 shows the experimental results compared with the results of [8, 9]. We see good agreement of the data that indicates the absence of differences between hydraulic resistance coefficients obtained in macro and micro scale channels for a given coupling geometry within the studied range of Reynolds numbers.

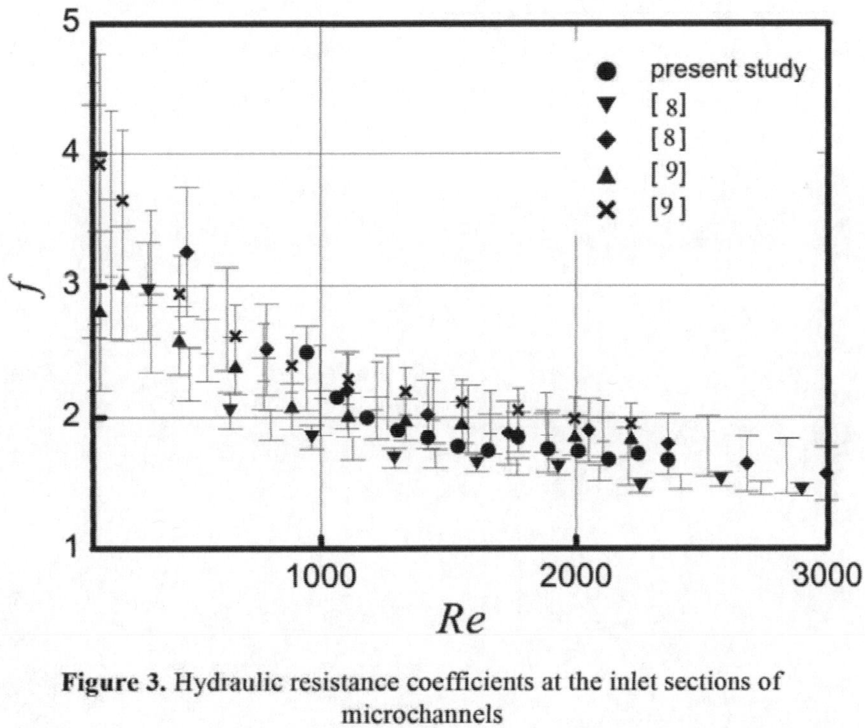

Figure 3. Hydraulic resistance coefficients at the inlet sections of microchannels

Conclusion

The results of the hydraulic resistance coefficient measurements at the inlet of microchannel is presented. The Reynolds number ranged from 900 to 2300. The channel with inlet diameter of 3.5 mm is abruptly narrowed to a diameter of 0.9 mm. The pressure drop in the area of sharp change in diameter is measured. Measured hydraulic resistance coefficient was compared with the previously

IOP Conf. Series: Journal of Physics: Conf. Series **925** (2017) 012002 doi:10.1088/1742-6596/925/1/012002

obtained coefficients for microchannels of the same geometry. It is shown that for the studied interconnected channel geometry no differences were found between micro and minichannels.

Acknowledgments

The work was financially supported by the Russian Science Foundation (project No. 16-19-10675).

References

[1] Obot N 2003 *Microscale Thermophys. Engineering* 155–173
[2] Morini G L 2004 *Int. Journal of Thermal Sciences* 631–651
[3] Hetsroni G 2005 *Int. Journal of Heat and Mass Transfer* 1982–1998
[4] Steinke M E and Kandlikar S G 2006 *Int. Journal of Thermal Sciences* 1073–1083
[5] Kohl M J, Abdel-Khalik S I, Jeter S M and Sadowski D L 2005 *Int. Journal of Heat and Mass Transfer* **48** 1518–33
[6] Costaschuk D, Elsnab J, Petersen S, Klewicki J C, Ameel T 2007 *Experiments in Fluids* **43** 907–916
[7] Baviere R and Ayela F 2004 *Meas. Sci. Technol* **15** 377–383
[8] Aniskin V M, Adamenko K V and Maslov A A 2012 *Nanosystems: Physics, Chemistry, Mathematics*, **3** 37-46
[9] Aniskin V M, Adamenko K V and Maslov A A 2010 *Siberian Journal of Physics* **5** 63-70
[10] Idelchik I E 1992 *The handbook of hydraulic resistance* (ed M O Steinberg(M: Mashinostroenie))

Model and on-orbit study of the International space station contamination processes by jets of its orientation thrusters

V N Yarygin[1], Yu I Gerasimov[2], A N Krylov[2], V G Prikhodko[1], A Yu Skorovarov[2] and I V Yarygin[1]

[1] Kutateladze Institute of Thermophysics SB RAS, Lavrentiev Ave 1, Novosibirsk, 630090, Russia
[2] Korolev Rocket and Space Corporation ENERGIA, Lenin Str. 4a, Korolev, Moscow area, 141070, Russia

E-mail: yarygin@itp.nsc.ru

Abstract. The main objective of this paper is to describe the current state of research for the problem of the International Space Station contamination by plumes of its orientation thrusters. Results of experiments carried out at the Institute of Thermophysics SB RAS modeling space vehicles orientation thruster's plumes are presented and experimental setup is discussed. A novel approach to reduction of contamination by thrusters with the help of special gas-dynamic protective devices mounted at the exit part of the nozzle is suggested. The description and results of on-orbit experiment at the International Space Station are given. Results show good agreement for model and on-orbit experiments validating our approach.

1. Introduction

Space stations are complex systems that have many components and a variety of on-board devices. On the one hand these devices are necessary for operation and on the other one some of them can cause undesirable effects. A case in point is liquid-propellant rockets of low thrust which are used for control of space station orientation. Their operation can lead to such negative phenomena as contamination of the station external surfaces by jets.

Currently liquid-propellant rocket engines of low thrust, employing self-igniting components – dinitrogen tetroxide (amyl) and unsymmetrical dimethylhydrazine (UDMH) are used as orientation thrusters (OT) of the International Space Station (ISS). During operation of the thruster its combustion chamber is subjected to high temperatures (about 3000 K) and high pressures (about $(6\text{-}10)\cdot10^5$ Pa). To protect chamber walls from harmful effect of high temperatures liquid film of one of propellant components is fed on the chamber walls. Operation of these thrusters is accompanied by ejection of burnt and unburnt propellant fractions (molecular clusters and droplets) into space. Results of both model and on-orbit experiments show that incomplete combustion products (ICP) scatter almost in all the directions: from 0 to 180° relative to the jet axis. This is caused by special character of gas and liquid outflow into vacuum. The flow of gas at angles over 90° is known as backflow. Such flows lead to contamination of the external surface of the ISS and outside equipment. This in turn poses a risk of toxic ICP penetration into the living space of the ISS. Therefore minimization of contamination is important and challenging problem.

Arguably experimental studies carried out in 1983-1988 at the Hamburg technical university [1-

2] were one of the first attempts to investigate contamination effects of thruster plumes. Monopropellant and bipropellant thrusters (thrust from 5 to 66 N) were tested in a vacuum chamber, and extensive data on the flow structure of the plume including the droplet phase were obtained. Experimental data includes droplets sizes and scattering angles. It was shown that scattering angles can well exceed 90°. However the problem of OT plumes negative impact minimization on space vehicles construction elements was not considered in these studies.

During 1993-1996 experimental studies of ICP ejection from the Russian bipropellant thruster 11D428A-16 were carried out in the vacuum chamber of the Keldysh Research Center [3]. Pulse duration in these experiments was 25-50 ms. Since build-up time of the thruster is about 35 ms, the results obtained included only ejection of droplet fraction of ICP. At the same time considerable ejection of droplets from a nozzle into peripheral part of a jet was demonstrated. Dependence of angular distribution of droplets mass in near-axial and peripheral zones of a jet up to polar angle equal to 90° was obtained in those studies.

For the first time the problem of spacecraft contamination by jets of OT attracted attention at the space station MIR. Contamination of different areas near thrusters was observed during on-orbit experiment "Dvicon" carried out in 1998. However, the attention was focused on this problem only at the International Space Station.

In this paper we study the problem of spacecraft and space station contamination by jets of OT. First we describe model experiments carried out at the Institute of Thermophysics SB RAS in which approaches to ISS contamination minimization with gas-dynamic protecting devices were evaluated. Then we describe on-orbit experiments on the efficiency of gas-dynamic protecting devices carried out at the ISS by Rocket and Space Corporation ENERGIA.

2. Model experiments

The objective of the model experiments was to study the processes of backflow contamination of spacecraft surfaces by thruster jets with a fuel film used for cooling the inner surface of the nozzle, and to suggest approach for minimizing contamination. Special attention was paid to behavior of droplets into which the liquid is decomposed at the initial stage of droplet motion and to the interaction of droplets with a freely expanding gas jet.

Model experiments were carried out at the VIKING facility of Vacuum gas-dynamic complex of the Institute of Thermophysics SB RAS [4]. This experimental facility with a 150 m3 working chamber provides wide opportunities for conducting experiments in pulse and continuous modes.
Since propellant components used in OT (unsymmetrical dimethylhydrazine and dinitrogen tetroxide) are toxic, possibility for investigation of the problem of contamination by real OT in the vacuum chambers is limited. In this case model investigations can provide valuable information on the problem of space station contamination. In model experiments correct choice of nozzle geometry is of outermost importance. In our work we used integral approach to modeling that can reproduce typical angle of jet divergence [5]. Supersonic nozzle with geometrical Mach number M=3 (throat diameter – 10 mm, exit cross-section diameter – 20 mm) was used as a model nozzle, ethanol – as working liquid (physical properties of ethanol and UDMH, such as viscosity, saturated vapor pressure, surface tension, heat of evaporation are rather close), and purified air – as working gas.

Study of the problem considered requires diagnostics of near-wall liquid film parameters inside the nozzle, first of all its thickness. For these measurements we used technique of capacity-type probes [6], mounted near the nozzle edge (Fig. 1). Since reliability of our data is highly dependent on precision of our measurements in our experiments we paid a lot of attention to calibration of capacity-type probes. We employed plugs made of dielectric. They were tightly inserted into the nozzle and had the gap of the given depth over the probe filled with liquid (Fig. 2). This method allowed us to obtain calibration curves for each probe taking into account its individual sensitivity and position concerning nozzle surface.

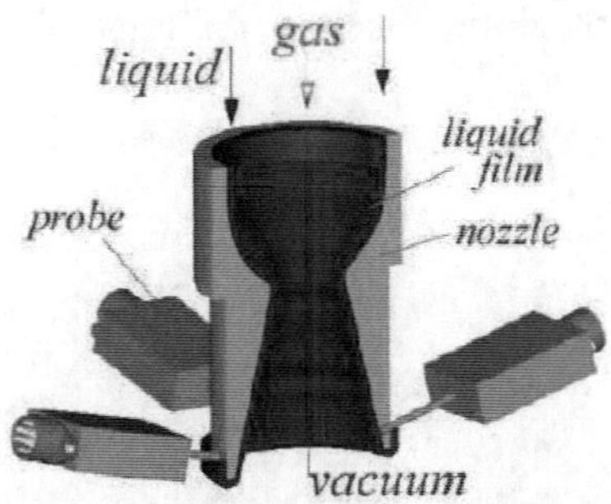

Figure 1. Schema of nozzle with probes

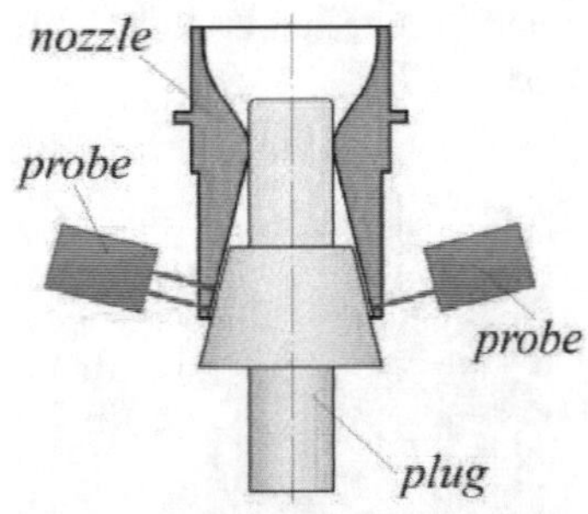

Figure 2. Calibration of probes

Typical results of ethanol film thickness measurement depending on Reynolds number of co-current gas flow are given in Fig. 3. One can see that thickness strongly depends on parameters of gas flow – film thickness decreases from 300 to 10 microns with increase of gas Reynolds number from 10^4 up to $4{\cdot}10^5$. Liquid film velocity was measured with the help of capacity technique. Velocity also depends on parameters of gas flow and is about 1 m/s. It was shown that detachment of droplets from film surface inside nozzle takes place.

Measurements of liquid film parameters in a nozzle helped us to understand mechanism of the film disintegration at the nozzle edge. In order to study gas-droplet flow which is formed behind a nozzle exit cross-section under disintegration of near-wall liquid film we measured angular distribution of droplet phase. The main challenge in carrying out these measurements was fast evaporation of droplets in vacuum.

To overcome this challenge we added special dye into model liquid. We chose a non sublimating in vacuum dye. In this approach the quantity of liquid phase was determined by the amount of remaining dye (the dry residue) on the sensor after liquid evaporation. Droplet phase angular distribution was measured with the help of spectrophotometry technique. The basic idea of this technique consists in using methods of spectral photometry for determining the amount of dye (solid residue) remaining on the sensors, mounted around the nozzle (Fig. 4), after evaporation of droplets incoming onto the sensor.

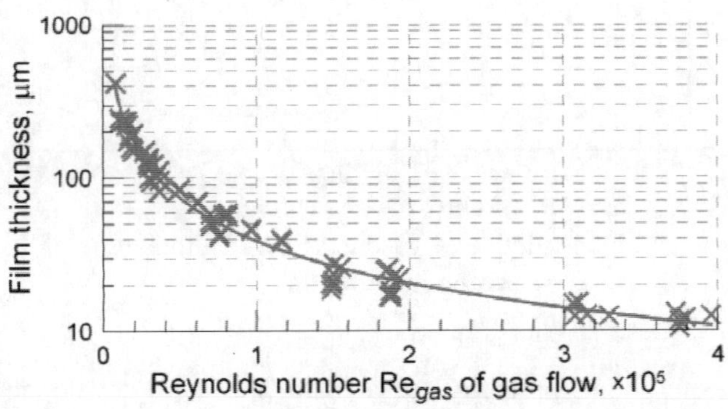

Figure 3. Dependence of film thickness on Reynolds number

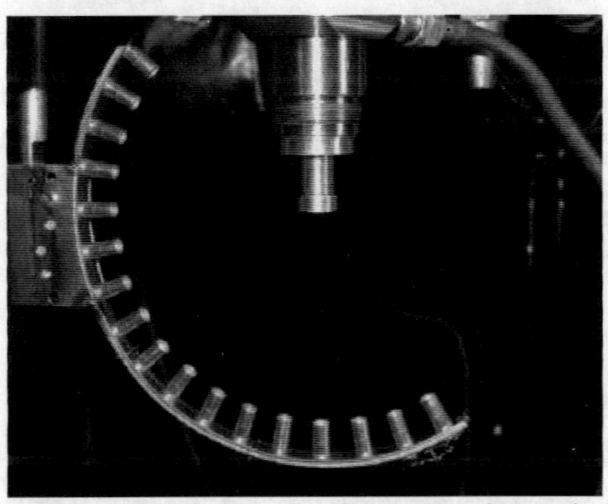

Figure 4. Diagnostic ring around nozzle

Experimental studies carried out allowed us to establish the detailed structure of gas-droplet flow which is formed under near-wall liquid film ejection with co-current gas flow from a supersonic nozzle into vacuum (Fig. 5). Appearance of two typical areas of droplet phase flow behind the nozzle exit – the central one and the peripheral one is shown (Fig. 6). The central *1* area is formed by droplets detaching from the film surface inside the nozzle and their subsequent fragmentation and acceleration by co-current gas flow. The peripheral one *3* is formed by disintegration of near-wall liquid film at the nozzle exit edge.

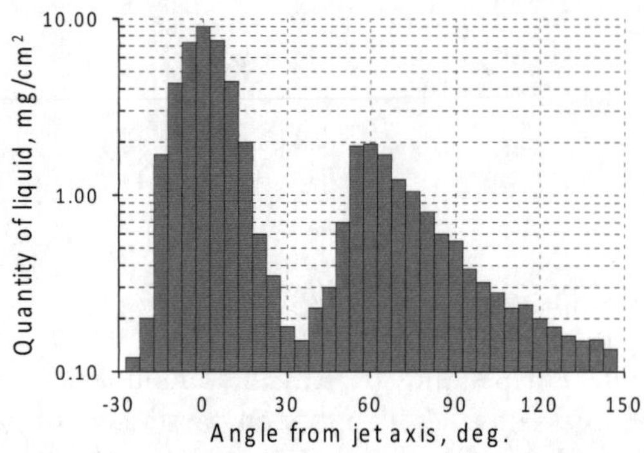

Figure 5: Droplet phase angular distribution

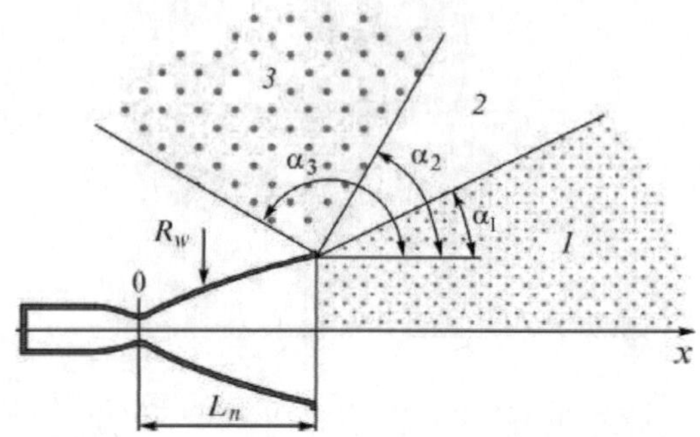

Figure 6: General structure of droplet phase

One of the ways to reduce back flows of droplets is to use special protective devices (screens) mounted on the exit part of a nozzle (Fig. 7). Schematic and constructive solutions of gas-dynamic protective devices (GDP) can vary and depend on a number of factors. The basic requirements for such devices – small weight, reliable operation under conditions of outer space, absence of influence on thrust characteristics of the rocket thruster. Besides, it is necessary to consider constructive restrictions on nozzles of thrusters, their configuration on the space vehicle and also space vehicle life-span. We show [7] that the correct choice of GPD is crucially important since incorrect one may cause increase in backflows instead of reduction (Fig. 8).

Figure 7. Gas-dynamic protective device (screen) on the model nozzle

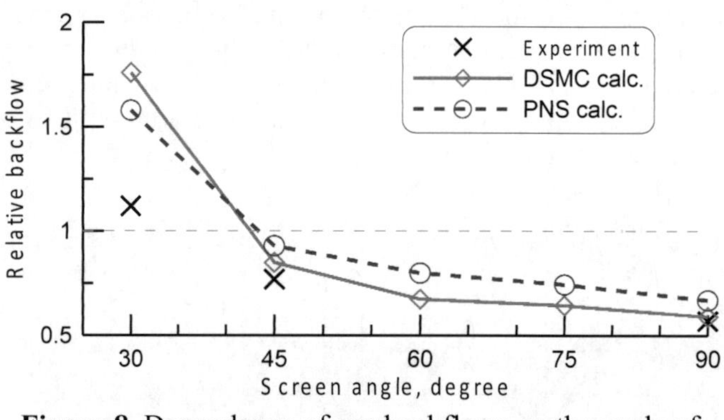

Figure 8. Dependence of gas backflows on the angle of screen installation.

3. On-orbit experiments

Based on the promising results of our model experiments protective devices for ISS were developed and manufactured by Korolev Rocket and Space Corporation ENERGIA. In 2002 they were delivered at ISS and mounted at ZVEZDA Service Module. By that time first stage of on-orbit experiment "Kromka" was completed. Its objective was to evaluate contamination of ISS surface without GPD to use as a reference for the next stages of experiment. These stages took place in 2002-2006 and allowed to come to important conclusions.

The main goal of on-orbit experiments was to evaluate GPD efficiency under long-term exposure of control plates in a zone of ICP ejection from orientation thrusters. Control plates were equipped with samples of different materials and covered with different coatings (Fig. 9). Preliminary results of on-orbit experiment were obtained by control plates photographs analysis made by crew through ISS viewports or during spacewalk sessions. Final results of on-orbit experiment were obtained after control plates had been returned from orbit to Earth and analyzed at the laboratory [8].

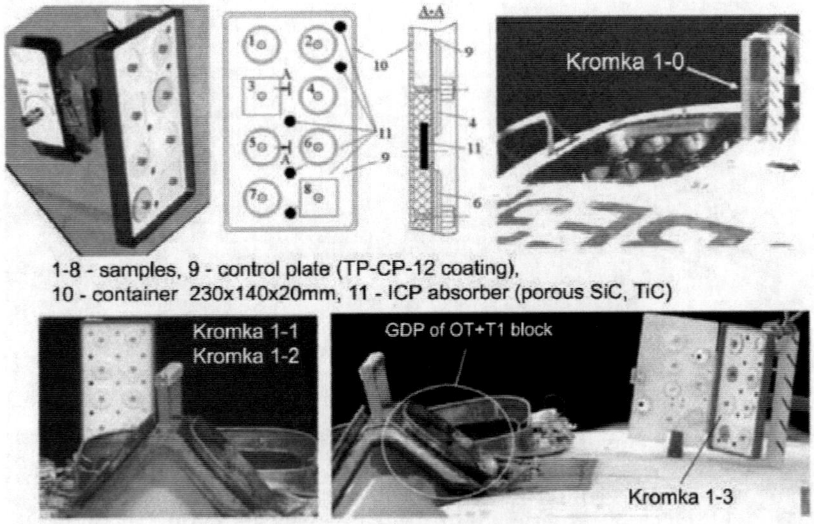

Figure 9. On-orbit experiment «Kromka»

Exposure of control plates was carried out before (stage "Kromka 1-0") and after GDP installation at OT blocks (stages "Kromka 1-1", "Kromka 1-2", "Kromka 1-3") [9]. Results of on-orbit experiment confirmed efficiency of GDP for jet peripheral part shielding from ejection of the ICP droplet fraction (Fig. 10).

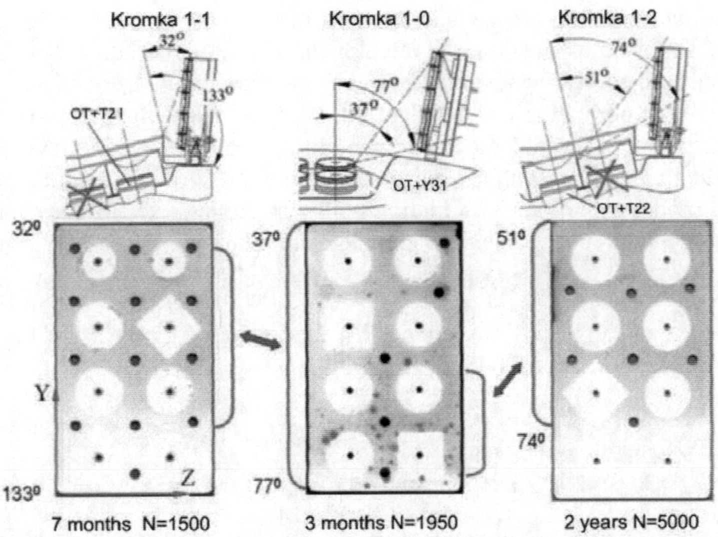

Figure 10. On-orbit experiment «Kromka»: contamination of control plates

An important result of on-orbit experiment was angular distributions of ICP ejected from OT of the ISS Service Module without GDP and after its installation (Fig. 11). Distributions were obtained with the help of control plates densitometry and quantitative estimations after chemical analyses of fresh ($M_{cont(fresh)}$ – in ICP container) and dry (M_{dry} – on control plate) ICP contaminants [9]. These results demonstrate quantitative characteristic of ICP efficiency, namely the fact that in a peripheral part of the flow at θ = 90° the intensity of ICP contaminants decreased almost by 3 orders of magnitude [9]. This fact is in good agreement with results of model experiments we carried out earlier.

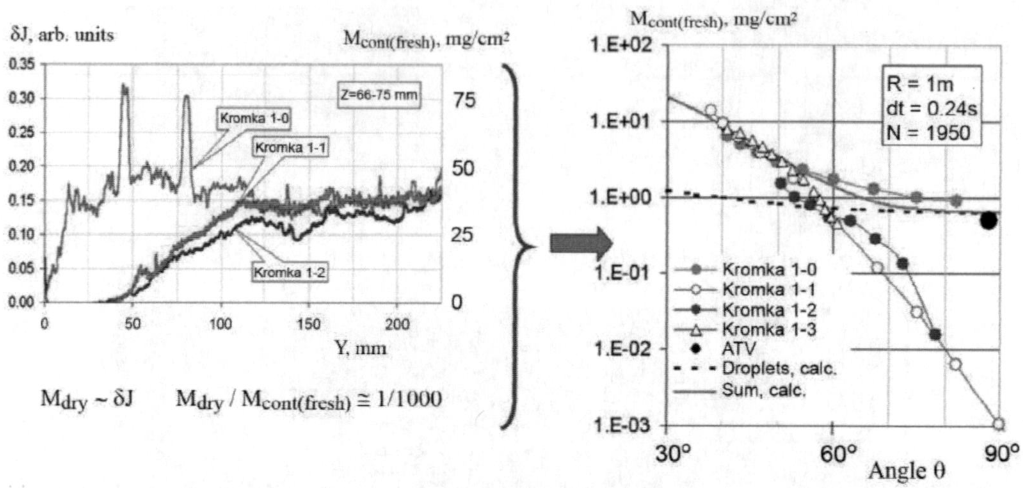

Figure 11. On-orbit experiment «Kromka»: distribution on contaminants

4. Conclusion

Our experiments have shown that operation of liquid rocket engines of low thrust that are used as orientation thrusters of spacecrafts is accompanied by ejection of incomplete combustion products. These products which scatter practically in all the directions from 0 to 180° are the main cause of both external surface of spacecraft and technical equipment contamination.

A novel approach for protection of surfaces and scientific equipment from contaminants is suggested. Design of accumulative type gas-dynamic protection devices mounted near blocks of thrusters nozzles for variety of spacecrafts is developed. The model studies evaluating efficiency of different schemes of the developed protective devices were carried out.

Results of on-orbit experiments that validate our designs and demonstrate remarkable shielding characteristics (reduction by several orders of magnitude) of protective devices installed at the International space station are presented. These results are in excellent qualitative agreement with results of our earlier model experiments validating both experimental techniques and supporting GDP as a valid approach for solving the problem of spacecraft contamination by orientation thrusters.

Acknowledgment

Experimental part of the model study was financially supported by the Russian Foundation for Basic Research, scientific project No.16-08-00436.

References

[1] H. Trinks 1987 *AIAA Paper* 87-1607
[2] H. Trinks, I. Kaelsch 1987 *AIAA Paper* 87-1603
[3] S. Rebrov, Y. Gerasimov 2001 *AIAA Paper* 2001-2818
[4] V.G. Prikhodko, G.A. Khramov, V.N. Yarygin 1996 *Instrum. Exp. Tech.* **39** 309-311
[5] V. Yarygin, Yu. Gerasimov, A. Krylov, V. Prikhodko, and I. Yarygin 2011 *Microgravity Sci. Technol.* **23** 15-23
[6] A.F. Serov, S.V. Kotov, A.D. Nazarov et al. 1997 *Instrum. Exp. Tech.* **40** 136-139
[7] V.N. Yarygin, V.G. Prikhodko, P.A. Skovorodko, I.V. Yarygin 2016 *Thermophys. Aeromech.* **23** 937-940
[8] Yu.I. Gerasimov, A.N. Krylov, S.P. Sokolova et al. 2003 *Thermophys. Aeromech.* **10** 555-565
[9] Yu.I. Gerasimov, V.N. Yarygim 2016 *Phys.-Chem. Kinetics Gas Dynam.* **17**(4) (in russian)

IOP Conf. Series: Journal of Physics: Conf. Series **925** (2017) 012004 doi:10.1088/1742-6596/925/1/012004

Drying droplet deposited on poor wetting substrate: beyond the lubrication approximation

P Lebedev-Stepanov[1,2], S Efimov[1] and A Kobelev[2]

[1] Photochemistry Center of FSRC Crystallography & Photonics RAS Novatorov St., 7a/1, Moscow, 119421, Russia

[2] National Research Nuclear University MEPhI (Moscow Engineering Physics Institute), 31 Kashirskoeshosse, 115409, Moscow, Russia

E-mail: petrls@yandex.ru

Abstract. Evaporating sessile droplet of aqueous solution deposited on hydrophobic surface is an urgent object of theoretical modeling (evaporation dynamics, microfluidics inside the drop, particle dynamics in evaporating drop, etc) and applied researches (printing technologies, nanoparticle ensemble self-assembly processes, hydrophobic coatings, etc). Although self-assembly investigation in evaporating droplet of colloidal solution on smooth surfaces with quite acute contact angles has been widely studied recently for liquids of different properties, nanoparticles ensemble self-assembly processes in droplet deposited on hydrophobic and superhydrophobic surfaces has not received much attention up to date. This work includes the analysis of application of existing droplet evaporation models, the boundary conditions for the hydrodynamic flows on the drop surface, as well as the nanoparticle dynamics in the volume of aqueous solution droplet deposited on hydrophobic flat surface, and the dried pattern formation processes modelling.

1. Introduction

There are three basic problems to estimate the colloidal particles self-assembly in evaporating droplets: 1) solute evaporation from the droplet surface to surrounding air (outer problem), 2) hydrodynamic flows in droplet volume (inner problem), 3) particle dynamics into droplet with account of interparticle interactions, particle-surfaces interactions, particle-flow interactions, solvation effects [1,2]. Although self-assembly investigation in evaporating droplet of colloidal solution on smooth surfaces with quite acute contact angles has been widely studied recently for liquids of different properties including binary solute mixtures [3-8], nanoparticles ensemble self-assembly processes in droplet deposited on hydrophobic and superhydrophobic surfaces has not investigated up to date.

A hydrodynamic flow occurs inside the drop as a result of the drop volume change and chemical potential gradients on surface or into volume of the liquid drop. The flow dramatically depends on character of the triple line movement during drop evaporation. Temperature gradient on drop surface can lead to arising of the toroidal vortex flow if Marangoni number is reached a value more than 100 approximately [9]. On the contrary, the flow has the simplest structure without the vortex component when the Marangoni number is small, as it takes place in the case of a little aqueous droplet (with a volume of 10-100 picoliters) at normal conditions [1]. Such a flow, i.e. the Deegan flow, which leads to coffee ring effect [4] we consider for illustration of the 3D generalization approach of 1D

IOP Conf. Series: Journal of Physics: Conf. Series **925** (2017) 012004 doi:10.1088/1742-6596/925/1/012004

approximation of the boundary conditions that are used in the lubrication theory framework to describe not only the droplets with acute contact angles but also in case of obtuse ones. Thus, we neglect the temperature gradient effects and Marangoni convection due to the volume of considered droplet is quite small and Marangoni number is less than 100.

2. Spherical drop evaporation models

The contact angle of aqueous droplet deposited on hydrophobic surface has a value in interval from about 90° to about 180°. In this case, evaporation flow density has no the integrable singularity at contact line, unlike a drop with an acute angle (figure 1). Deegan [4], Hu and Larson [5,6], Popov [7] have elaborated the useable models of evaporation of sessile drop of single liquid based on Maxwell diffusion model.

This model is useful for calculation of the evaporation flow for the contact angle interval from 0° to 90°. In framework of this model, the evaporation flow density on the droplet surface is given by

$$J(r) = J_0(\theta)\left(1-\frac{r^2}{R^2}\right)^{-\lambda(\theta)}, \ \lambda(\theta) = \frac{1}{2}-\frac{\theta}{\pi}. \tag{1}$$

where θ is a contact angle; R is a droplet radius (figure 2).

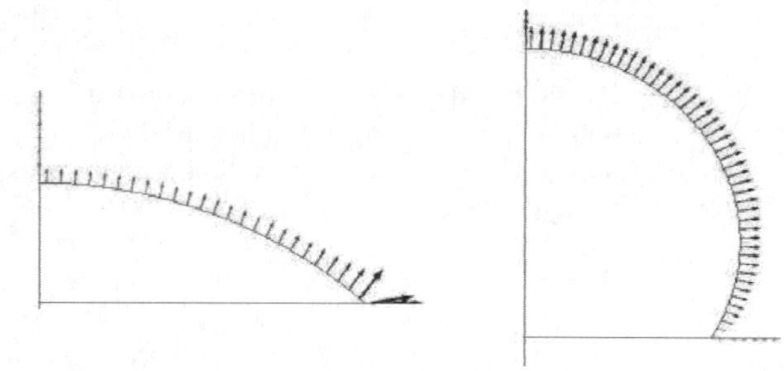

Figure 1. Evaporative flux density (evaporation rate) on a surface of sessile droplet with acute contact angle (left) and with obtuse angle (right); the pictures were estimated by Comsol

To increase the computation speed, computer modeling of self-assembly of an ensemble of nanoparticles in drying droplet requires the using of simple analytic approximation for evaporation flow. A good example of such approximation is the Hu-Larson's [5] solution given by equation 1. But, up to date, such a solution for droplet on solvatophobic substrate is absent.

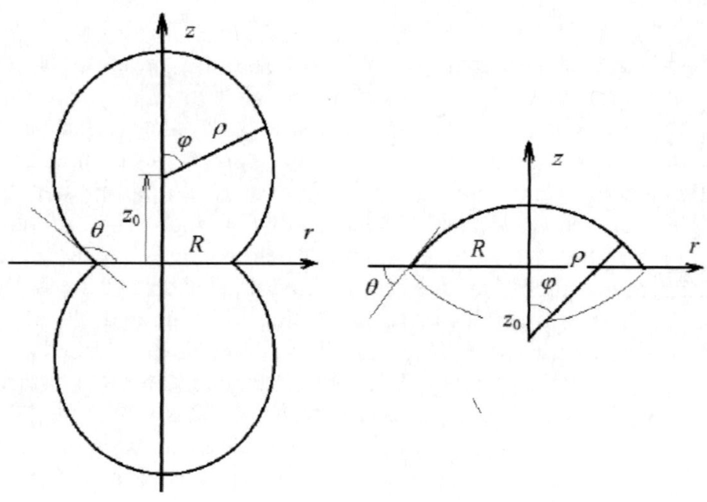

Figure 2. Geometry of liquid capillary-sized droplet on a flat substrate.

IOP Conf. Series: Journal of Physics: Conf. Series **925** (2017) 012004 doi:10.1088/1742-6596/925/1/012004

Popov's model with integral for the evaporation flux [7] can be used for whole open interval of contact angles from about 0° to 180°:

$$W = -\pi RD(n_s - n_\infty)\left[\frac{\sin\theta}{1+\cos\theta} + 4\int_0^\infty \frac{1+\cosh 2\theta\tau}{\sinh 2\pi\tau}\tanh[(\pi-\theta)\tau]d\tau\right], \quad (2)$$

but the numerical approximation of the evaporation flux given by Picknett and Bexon [3] is more applicable for computer modelling.

3. Boundary conditions for hydrodynamic flow

The similar situation takes place for inner problem which is concerned the hydrodynamic flows into droplet. Lubrication approximation analysis given by Oron [10], Deegan et al [4] can be applied only for the flow field produced by droplet evaporation under the additional approximation of a flat droplet (small contact angle). Deegan et al. have used the lubrication approximation to calculate the hydrodynamic flow into a droplet with acute angle and pinned contact line.

Hu and Larson have developed such a semi-analytical approximation in their model of hydrodynamic flows in drying drop. The kinematic boundary condition with a phase change is

$$(\mathbf{n}, \mathbf{V}) = [n_r, n_z][0, \dot{Z}] + \frac{(\mathbf{n}, \mathbf{J})}{\rho}, \quad (3)$$

where (**n, J**) is the evaporation flux along the direction normal to the free surface, where **n** is a unit vector along the normal direction, and n_r and n_z are the r and z components of the unit normal vector **n**, respectively.

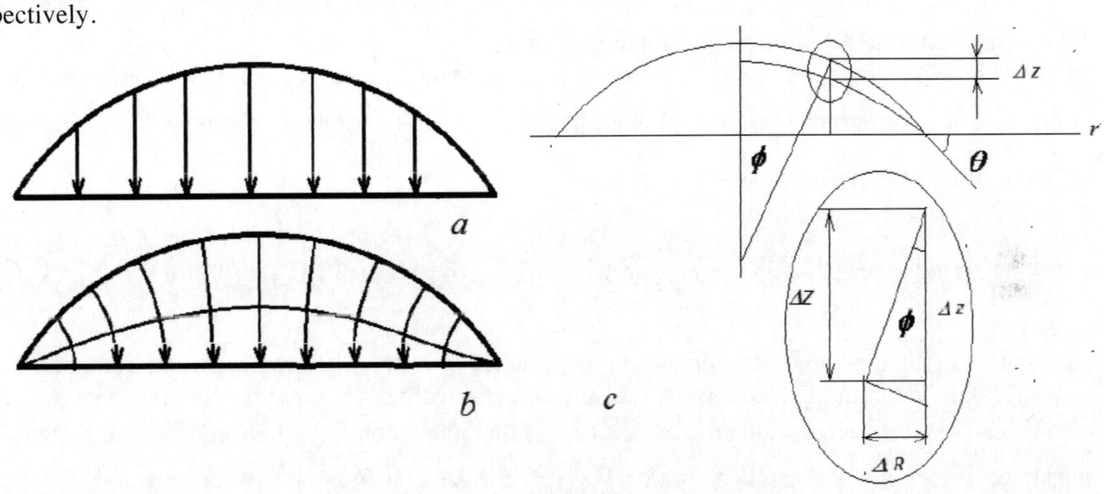

Figure 3. Trajectories of displacement of the points on liquid surface of drying droplets with acute contact angle during evaporation with a pinned contact line: a) lubrication theory (wrong); b) our model [1,2]; c) geometry of displacement of the stationary point on the surface of the evaporating drop [2].

In our work [1], the boundary conditions of a more general form (three-dimensional) for the rate of flow on the top of drop surface were proposed:

$$V_z(\theta,t) = \frac{J(\theta,t)}{\rho_m}\cos\theta - \frac{dZ(\theta,t)}{dt}, \qquad V_r(\theta,t) = \frac{J(\theta,t)}{\rho_m}\sin\theta - \frac{dR(\theta,t)}{dt}, \quad (4)$$

where ρ_m is a mass density of a liquid, $\frac{dZ(\theta,t)}{dt}$ and $\frac{dR(\theta,t)}{dt}$ are z- and r-components of velocity of the geometrical point on a drop surface. Such a shift takes place due to the evolution of the form and size of the evaporated drop with pinned contact line (figure 3).

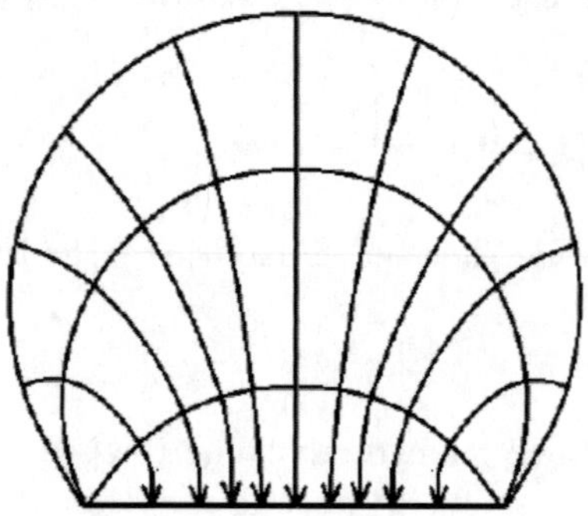

Figure 4. Trajectories of displacement of the stationary mathematical points (not the particles) on surface of drying droplets with obtuse contact angle during evaporation with a pinned contact line according to our new model.

These boundary conditions were successfully used to simulate the self-assembly of nanoparticles in sessile droplets with contact angles having the value smaller than right angle. However, applicability of equations (4) for hydrophobic droplets still has not been investigated. Until now, the boundary conditions for general three-dimensional case which includes the obtuse contact angles are still absent in scientific publications.

We have analyzed the application of existing droplet evaporation models, the boundary conditions for the hydrodynamic flows on the drop surface (figure 4), as well as the nanoparticle dynamics in the hydrophobic droplet, and the dry pattern formation processes modeling.

4. Dissipative particle dynamics into evaporating drop

Dissipative dynamic model [1,2] was used to estimation of the nanoparticles self-assembly in evaporating aqueous droplet on hydrophobic substrate. The i-th particle acceleration $\frac{d\mathbf{v}_i}{dt}$ is calculated by the following Langevin equation:

$$m_i \frac{d\mathbf{v}_i}{dt} = m_i\left(\frac{\partial \mathbf{V}}{\partial t} + (\mathbf{v}_i,\nabla)\mathbf{V}\right) - \nabla\sum_{i\neq j}^{N} U(r_{ij}) + \mathbf{F}_{Si}(r_{Si}) + \mathbf{F}_{Li}(r_{Li}) + \mathbf{F}_{Ri}(r_{Ri}) - 6\pi a\eta(\mathbf{v}_i - \mathbf{V}) + \mathbf{F}_{Bi}, \quad (5)$$

where m_i is an effective mass of i-th particle with radius of a, the first member of the right part is the hydrodynamic force, the second term is the sum of conservative forces with potential $U(r_{ij})$ acting on the i-th particle from other j-th particles, $\mathbf{F}_{Si}(r_{Si})$ is the interaction force between the i-th particle and substrate, $\mathbf{F}_{Li}(r_{Li})$ is the capillary force, $\mathbf{F}_{Ri}(r_{Ri})$ is a friction force acted due to roughness of the substrate surface, $-6\pi a\eta(\mathbf{v}_i - \mathbf{V})$ is a Stokes viscous force, where $\mathbf{V}$ is a field of hydrodynamic flow in the solution, and the Brownian force is $\mathbf{F}_B$.

The calculation result for the pinned drop (included 1000 particles) with initial contact angle of 140^0 is shown in figure 5.

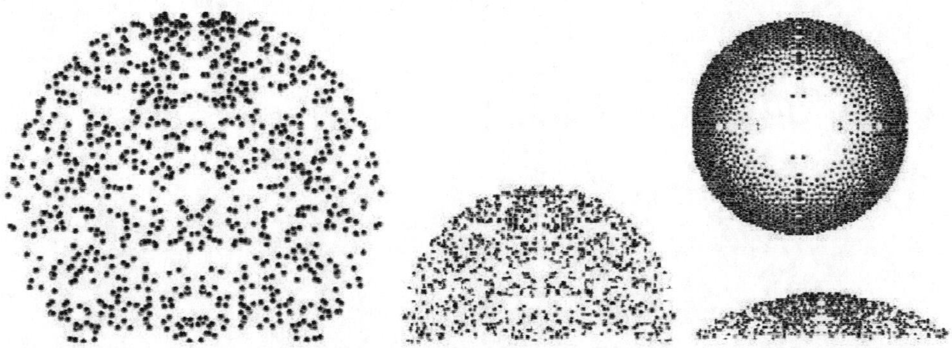

Figure 5. Three stages of evaporating drop with 1000 particles (side view). The final dried pattern is given in upper right corner of the picture (top view).

5. Conclusion

Analysis of application of the existing droplet evaporation models, the boundary conditions for the hydrodynamic flows on the drop surface, as well as the nanoparticle dynamics in the hydrophobic droplet, and the dried pattern formation processes modeling was carried out. Lubrication theory is valid only for small contact angles and absolutely inapplicable to the angles larger than a right angle. We propose the surface boundary conditions for arbitrary contact angles and tested such a model for simply self-assembly calculations which is applicable for aqueous droplet of picoliter volume and small Marangoni number in normal external conditions

Acknowledgements
This work was partially supported by the Russian Foundation for Basic Research, project no. 15-03-08050 and by the Competitiveness Program of National Research Nuclear University "MEPhI".

References

[1] Lebedev-Stepanov P, Vlasov K 2013 *Colloids and Surfaces A : Physicochem. Eng. Aspects.* **432** 132
[2] Lebedev-Stepanov P V 2015 *Introduction to self-organization and self-assembly of ensembles of nanoparticles* (Moscow: National Research Nuclear University MEPhI Press, In Russian) p. 257
[3] Picknett R G, Bexon R 1977 *J. Colloid Interface Sci* **61** 336
[4] Deegan R D, Bakajin O, Dupont T F, Huber G, Nagel S R, Witten T A 2000 *Phys. Rev. E* **62** 756
[5] Hu H, Larson R G 2002 J. *Phys. Chem. B* **106** 1334
[6] Hu H, Larson R G 2005 *Langmuir* **21** 3963
[7] Popov Yu O 2005 *Phys. Rev. E.* **71** 036313.
[8] Talbot E L, Yow H N, Yang L, Berson A, Biggs S R, Bain C D, 2015 *Appl. Mater. Interfaces.* **7** 3782
[9] Duan F, Badam V K, Durst F, Ward C A 2005 *Phys. Rev. E* **72** 056303
[10] Oron A, Davis S H, Bankoff S G. 1997 *Rev.Mod. Phys.* **69** 931.

IOP Conf. Series: Journal of Physics: Conf. Series **925** (2017) 012005 doi:10.1088/1742-6596/925/1/012005

Equilibrium shape of dry spot in isothermal liquid film on a horizontal substrate

L I Maltsev, Y S Podzharov and O A Kabov

Kutateladze Institute of Thermophysics, SB RAS, Novosibirsk

E-mail: maltzev@itp.nsc.ru

Abstract. Thin liquid films formed near dry spots make a big contribution to heat and mass transfer processes due to intensive evaporation. Based on the solution to the problem of determining the shape of the film free boundary in a vicinity of a stationary dry spot formed by gravitational and capillary forces at a given contact angle, a criterion for the development (or shrinkage) of a dry spot in an isothermal liquid film on a horizontal flat substrate was derived without using additional assumptions and postulating the film shape near the spot.

1. Introduction

Liquid films are widely used in industry, in particular for heat and mass transfer processes intensification. The decrease in film thickness allows the method of evaporative cooling to be more efficient. Submicron liquid film, formed on the borders of the dry spots due to the intensive evaporation, can contribute significantly to the total heat and mass transfer [1]. Thus, the authors [2], through the organization of film flow with small-scale wash-off dry spots, could obtain heat flux, one order of magnitude greater than those, taking place in a continuous flow of a film under the same fluid flow. However, when growing, dry spots dramatically reduce the efficiency of heat exchangers and can even lead to their failure [3,4]. Therefore, submicron liquid film formed in the cooling film near the border of dry spots, are of particular interest.

A number of experimental and theoretical works [5-14] is devoted to the analysis of the mechanisms of dry spots formation in the films and to their modeling. In theoretical works on modeling of dry spots rupture and formation of cross section of liquid film free surface in the area adjacent to the dry spot are usually approximated by circular arcs. Based on this approach, the authors of the article proposed a criterion for dry spots formation and development in isothermal liquid films on a horizontal substrate [14].

In this article, a new criterion for dry spot development in the liquid film is given without postulating the shape of the film near the spot. This is done on the basis of the rigorous solution to the problem of the shape of the film free boundary surrounding a stationary dry spot formed under gravity and surface tension forces at a given contact angle.

2. Derivation of the basic equations

Let a stationary film capillary fluid under gravity having a dry spot in the shape of a circle lie on a horizontal flat plate. The spot surroundings cross-sectioned with a plane passing through the center of the spot are illustrated in Figure 1.

IOP Conf. Series: Journal of Physics: Conf. Series **925** (2017) 012005 doi:10.1088/1742-6596/925/1/012005

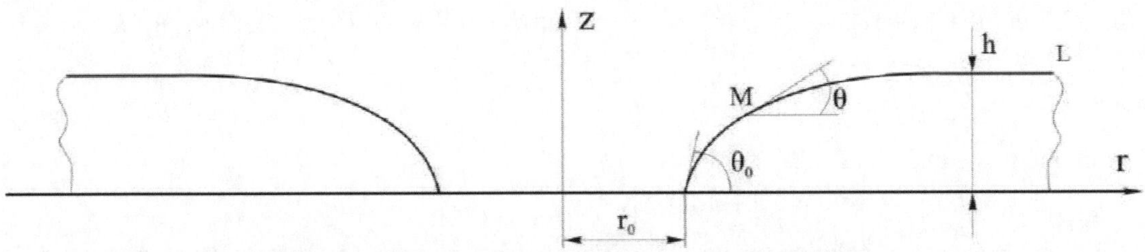

Figure 1. Cross section of the dry spot surroundings.

At each point M on the generating line L a pressure drop occurs due to hydrostatic pressure P in the liquid and to the presence of surface tension forces F. Write expressions for these pressures.

Hydrostatic pressure $P = \rho g(h - z)$

where ρ is the fluid density and g is acceleration of gravity.

The Laplace pressure

$F = \sigma(k_1 + k_2)$ where k_1 и k_2 are the main curvatures of the surface.

The rim will be stable under the condition when the total pressure at the points of the curve L is equal to zero

F + P = 0 or $\sigma(k_1 + k_2) + \rho g(h - z) = 0$.

According to Meusnier theorem [15], the curvature associated with the axial symmetry of the liquid film is $k_1 = \dfrac{\sin\theta}{r}$ where r is the distance from the given point of the generatrix L to the axis of symmetry of a dry spot. The second curvature related to the shape of the contour L is $k_2 = \dfrac{d\theta}{dS}$ where θ is the angle between the substrate and the tangent to the contour L, and S is the arc length along the contour L counted from the point of contact of the contour L with the substrate.

Thus, to find the form of the generating line S = f (θ) of the equilibrium curved surface of the film, we get the equation

$$\frac{d\theta}{dS} = -\frac{\rho g(h - z)}{\sigma} - \frac{\sin\theta}{r}, \text{ or } \frac{d\theta}{d\bar{S}} = Bo(\bar{z} - 1) - \frac{\sin\theta}{\bar{r}} \qquad (1)$$

where $\bar{z} = \dfrac{z}{h}$, $\bar{r} = \dfrac{r}{h}$, – dimensionless coordinates, $Bo = \dfrac{\rho g h^2}{\sigma}$ – Bond number, $\bar{S} = \dfrac{S}{h}$ – dimensionless element of the curve length.

For the solution of differential equations it is necessary to specify boundary conditions. As such conditions we shall take:

$$\theta(0) = \theta_0,\ \bar{r}(0) = \bar{r}_0,\ \bar{z}(0) = 0 \qquad (2)$$

where θ_0 is the contact wetting angle, and $\bar{r}_0$ is the dimensionless radius of the initial dry spot, the parameter unknown in advance.

3. The algorithm for solving the problem

Make sure first that the existence of dry spot for a given combination of θ_0 and Bo is possible. To find out this, it is sufficient to neglect the second term of the right side of equation (1) (assuming $\bar{r}_0 \to \infty$), and solve the equation

$$\frac{d\theta}{d\bar{S}} = Bo(\bar{z} - 1).$$

IOP Conf. Series: Journal of Physics: Conf. Series **925** (2017) 012005 doi:10.1088/1742-6596/925/1/012005

Substituting the expression $d\bar{S} = \dfrac{d\bar{z}}{\sin\theta}$ we get $\sin\theta d\theta = Bo(\bar{z}-1)d\bar{z}$, and having integrated the latter we get $\cos\theta - \cos\theta_0 = Bo(\bar{z} - \dfrac{\bar{z}^2}{2})$.

Assuming in this equation $\theta = 0$, $\bar{z} = 1$, we get:

$$Bo^* = 2(1-\cos\theta_0).$$

Thus, for $Bo < Bo^*$ a solution is possible (i.e., a film with a dry spot exists), otherwise, it is impossible for any $\bar{r}_0$. Conversely, for a given value of Bond number, there is a solution for $\theta_0 > \theta_0^*$, where

$$\theta_o^* = \arccos\left(1 - \frac{Bo}{2}\right). \quad (3)$$

Now we supplement eqn 1 with two obvious equations and get a system

$$\frac{d\theta}{d\bar{S}} = Bo(\bar{z}-1) - \frac{\sin\theta}{\bar{r}}$$
$$\frac{d\bar{z}}{d\bar{S}} = \sin\theta \quad (4)$$
$$\frac{d\bar{r}}{d\bar{S}} = \cos\theta$$

Solving the system (4) together with boundary conditions (2), we obtain the form of the free boundary of the film. This task we shall solve using Runge-Kutta method.

In this case, since the boundary conditions have the unknown parameter, the problem is variational with respect to it. The variation is convenient to be made with the usage of the bisection method, thus, making a solution have a free boundary of the film on a horizontal asymptote at a height equal to the thickness of the liquid film, i.e. fulfilling the condition $\bar{z} = 1$ with $\theta = 0$ to the specified accuracy.

The value of the parameter $\bar{r}_0$, in which the latter condition is met, will give us the critical size of dry spot $\bar{r}_c$, while the solution of the system of equations (4) with this boundary condition will give the opportunity to obtain the desired form of the free boundary of dry spot surroundings.

4. The results of the calculations

In Fig. 2 the equilibrium form of the free boundary of the film adjacent to the spot, built by this method, and the form obtained on the basis of the approximation of the free boundary by an arc of a circle, with θ_0=36° for two values of Bond number are compared. As one can see, the form of the free boundary constructed by this method is markedly different from the arc of a circle: the usage of the approximation increases the critical radius of dry spot; the generating line of the surface of the film adjacent to dry spot achieves the horizontal asymptote faster.

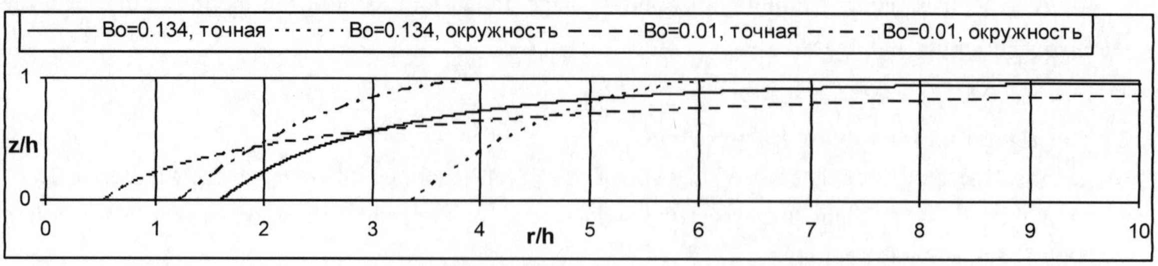

Figure 2. Comparison of the equilibrium shape of the free boundary of the film adjacent to the spot, built by this method, and the form obtained on the basis of the approximation of the free boundary by an arc of a circle, with θ_0=36° for two values of the Bo.

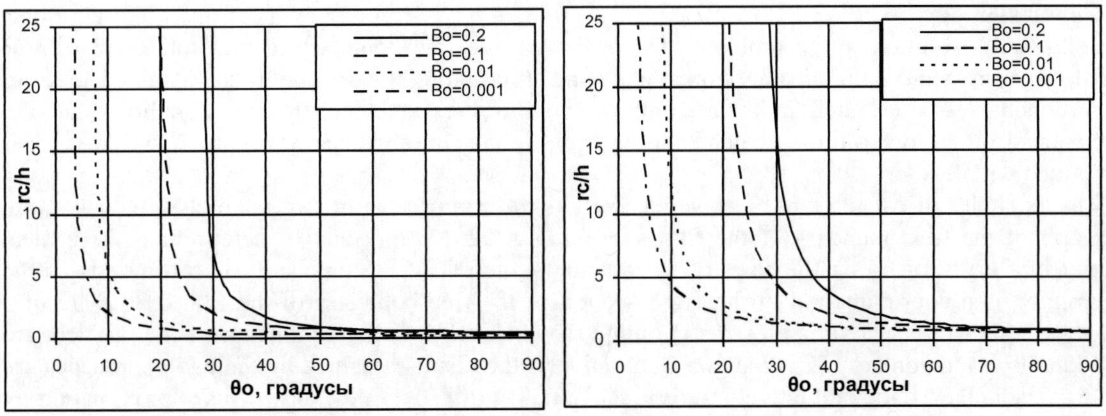

Figure 3. Dependence of r_c on θ_0 for different Bo numbers in precision construction of the free boundary of the film (a) and with its approximation by an arc of a circle (b).

Fig. 3 shows the dependence of the critical radius of the dry spot on Bond number and on the contact wetting angle for the free surface, defined by equations (4) in their exact interpretation (a), and in the case of its approximation by an arc of a circle (b).

With the values of Bo and θ_0 below and to the left of the corresponding curve shown on the chart the dry spots tend to compression, while above and to the right of the curve their tendency is to expand.

As can be seen from the graphs, the critical size of dry spot is small at high angles of wetting, but with the decrease of θ_0 it is growing rapidly, and a function r_c of θ_0 has a vertical asymptote at a particular value of $\theta_0 = \theta_0^*$. Thus, for $\theta_0 < \theta_0^*$ any dry spot tends to compression. Values of θ_0^* are defined by the value of Bond number (3') and are shown in Fig. 4.

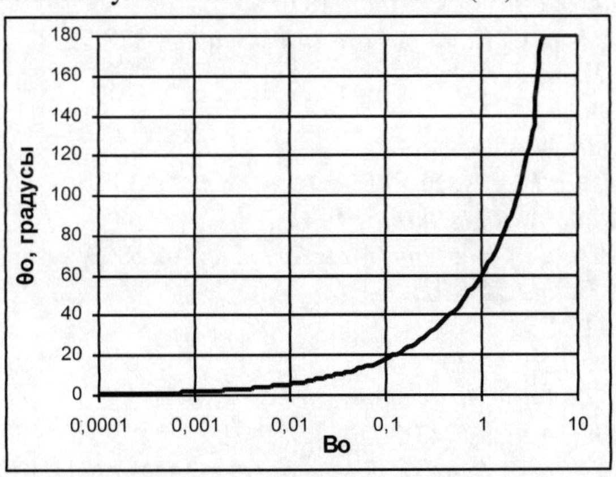

Figure 4. Dependence of a critical angle θ_0^* on the Bond number.

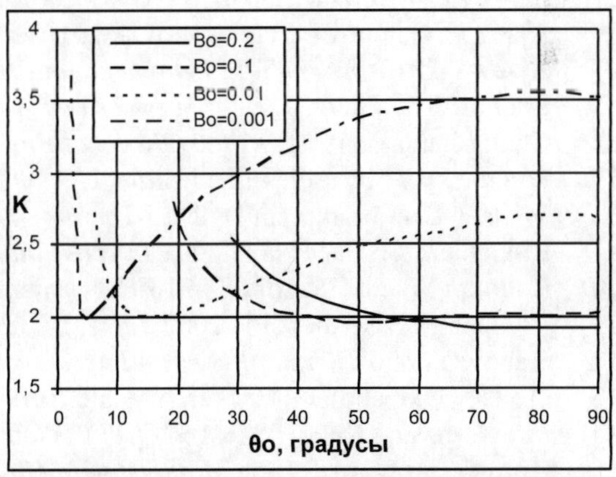

Figure 5. Dependence of the ratio K on the contact wetting angle for different Bond numbers.

In Fig. 5 one can see the correlation of the critical radii $K = \frac{r_c^o}{r_c}$ of dry spot obtained in different ways, where r_c^o is the critical radius obtained with the usage of approximation and $\bar{r}_c$ without it.

IOP Conf. Series: Journal of Physics: Conf. Series **925** (2017) 012005 doi:10.1088/1742-6596/925/1/012005

5. Conclusion

We obtained a solution to the problem of constructing the free boundary of the isothermal film of capillary gravity fluid lying on a horizontal plate and surrounding the dry spot it has within. The result we have achieved contributed, in its turn, to the solution of the problem of the critical radius value of a stationary dry spot, determining which spot, occurred in the film for whatever reasons, will disappear and which will develop.

The examples of calculations are given as well as the comparison of two methods of constructing the form of the free boundary of the film surrounding the dry spot and of determining the critical diameter of dry spot: a) on the basis of the obtained solution of the task and b) on the basis of the commonly used approximation of the cross section of the free boundary of the film by the arc of a circle. It is shown that in variant (a), as in option (b) [14], the defining parameters of the problem are Bond number and contact wetting angle. Qualitatively, the results obtained by the two approaches are close to each other, but quantitatively the critical values of dry spot in the case of its approximation by the arc of a circle are 2-3 times higher than those calculated more rigorously according to this method. We can also postulate a few changes of the form of the free boundary of the film adjacent to the dry spot. The boundary constructed on the basis of approximation lies above the corresponding curve constructed without the use of approximation and goes faster on the horizontal asymptote. This fact should be viewed as important, if to take into account the experiments showing that reduction of the thickness of the liquid film increases the efficiency of evaporative cooling method in heat and mass transfer exchangers and when analyzing the work of such apparatus it is necessary to know the thickness of the film, adjacent to dry spot as precisely as possible.

Acknowledgments

The work was funded by RFBR and Novosibirsk region (research project № 17-48-540124).

References

[1] Ajaev V, Kabov O 2017 *Int. J. of Heat and Mass Transfer* **108** 918
[2] Zaitsev D, Rodionov D and Kabov O 2009 *Technical Physics Letters* **35** 680
[3] Ajaev V, Gatapova E and Kabov A 2016 *Advances in Colloid and Interface Science* **228** 92
[4] Ajaev V 2013 *Interfacial Phenomena and Heat Transfer* **1** 81
[5] Bankoff S 1971 *Int. J. Heat Mass Transf.* **14** 2143
[6] Chung J and Bankoff S 1980 *Chemical Eng. Comm.* **4** 455
[7] Podgorsky T, Flesseles J and Limat L. 1999 *Phys. of Fluids.* **11** 845
[8] Maltsev L and Zavarzin D. 2012 *Thermophysics and Aeromechanics* **19** 607
[9] El-Genk M S , Saber H H 2001 *International Journal of Heat and Mass Transfer* **44** 2809
[10] Kadoura M and Chandra S 2013 *Experiments in Fluids* **54** 1465.
[11] El-Genk M S , Saber H H 2002 *J. Heat Transfer* **124** 39
[12] Kabov O 2000 *Thermophysics and Aeromechanics* **7** 513–520.
[13] Zaitsev D, Kabov O, Cheverda V and Bufetov N 2004 *High Temperature* **42** 450
[14] Maltsev L and Kabov O 2016 *MATEC Web of Conferences* **84** 00024
[15] Korn G and Korn T 1968 *Mathematical Handbook for Scientists and Engineers* (McGrowth Hill Booking Company)

IOP Conf. Series: Journal of Physics: Conf. Series **925** (2017) 012006 doi:10.1088/1742-6596/925/1/012006

Fluid flow inside and outside an evaporating sessile drop

C Bouchenna[1], M Aitsaada[1], S Chikh[1] and L Tadrist[2]

[1] USTHB, Faculty of Mechanical and Process Engineering, LTPMP, Alger 16111, Algeria
[2] Aix-Marseille Université, CNRS, Laboratoire IUSTI, UMR 7343, Marseille 13453, France

E-mail: bouchennachafea@hotmail.fr

Abstract. The sessile drop evaporation is a phenomena which is extensively studied in the literature, but the governing effects are far from being well understood especially those involving movements taking place in both liquid and gas phases. The present work numerically studies the flow within and around an evaporating sessile drop. The flow is induced by the strong mass loss at contact line, the thermo-capillary effect and the buoyancy effect in the surrounding air. The results showed that buoyancy-induced flow in gas phase weakly influences thermo-capillarity-induced flow in the liquid phase. Buoyancy effect can strongly modify the temperature distribution at liquid-gas interface and thus the overall evaporation rate of the drop when the substrate is heated.

1. Introduction

Sessile drop evaporation is a process, which has extensive applications in biochemical assays, thin film coating, spray cooling, microelectronics, nano-devices and others. It is a complex physical problem which involves fluid flow, heat and mass transfer as well as interactions between the solid substrate, the liquid droplet and the surrounding gas throughout moving interfaces and a pinned or receding contact line [1]. Recently Larson [2] summarized analytical, numerical and experimental literature works on drying sessile droplets and deposition of suspended materials. He presented a list of useful dimensionless groups governing mass, momentum, and heat transfer effects in the droplet, the surrounding gas and the substrate.

One of the most important applications of the sessile drop evaporation in thin film technology is the final pattern of the solid particles deposits on substrates. The particle deposition is strongly affected by the flow inside the drop. The competition of the flow induced by the privileged evaporation near the contact line, the thermo-capillarity and the buoyancy effects, is strong during drop evaporation, and the prevalence of one or two effects rules the flow direction inside the drop. Several works were conducted to study the influencing parameters, which define flow patterns in the evaporating sessile drops [3-22]

The present work is a numerical study of the convective internal and external flow of an evaporating water sessile drop. Few previous studies assumed that the surrounding air is in movement [13]. Nevertheless, understanding dynamics of both liquid and gas allows to better analyzing the evaporation kinetic of the drop and handling and controlling the strains of particle deposits after drying on substrates of different natures and in heated or non-heated cases. In this framework, a numerical model is developed by taking into account (1) the flow induced by the strong liquid loss near the contact line, (2) the thermo-capillary flow resulting from the surface tension gradient due to temperature variation at the drop surface, and (3) the flow induced by the buoyancy in the surrounding air. The objective is to analyze the internal and external flow intensity and the flow pattern resulting from the competition of the driving effects as well as its impact on the evaporation rate.

2. Mathematical model

We consider a small water drop on a substrate of high thermal conductivity. The surrounding air is at ambient temperature T_∞= 25°C and relative humidity Ha = 40%. The lower face of the substrate is maintained at a temperature Tw = T_∞. The drop evaporation is assumed to occur with pinned contact line. Buoyancy effect is included in the model by using the Boussinesq approximation. The proposed mathematical model is the same as in our previous works [21, 22], except in gas phase where the momentum equations are added to take into account the flow in the surrounding gas and then the convective term is added to energy and concentration equations. Convective flow in both liquid and gas phases is governed by the conservation equations of mass, momentum and energy coupled with vapor transport equation in the surrounding air and heat conduction equation in the substrate. Mass and momentum conservation and energy balance at the liquid-gas interface are expressed in dimensionless forms as follows:

i) Mass conservation,

$$\left(\vec{\mathrm{W}}_\ell - \vec{\mathrm{W}}_\mathrm{I}\right)\cdot\vec{\mathrm{n}} = R_\rho\left(\vec{\mathrm{W}}_g - \vec{\mathrm{W}}_\mathrm{I}\right)\cdot\vec{\mathrm{n}} = \frac{R_{\alpha_\mathrm{T}}\left(\Delta C/\rho_\ell\right)}{Le\,Ma}\vec{J}^*\cdot\vec{\mathrm{n}} \tag{1}$$

$\vec{\mathrm{n}}$ is a normal unit vector, $\vec{\mathrm{W}}_\mathrm{I}$ is the velocity of the moving interface, $\vec{\mathrm{W}}_{\ell,g}$ is the velocity of a liquid (gas) particle, is evaporation flux, $\vec{J}^*$ is the liquid density and ΔC is equal to ($C_v(T_w)$ - Ha $C_v(T_\infty)$) where Cv(T) is the saturated vapor concentration. Different dimensionless numbers appear in Eq. (1): Marangoni number (Ma), Lewis number (Le), ratio of gas/liquid density (R_ρ), ratio of gas/liquid thermal diffusivity ($R_{\alpha T}$).

ii) Shear-stress balance,

$$\left[\left(\vec{\mathrm{n}}\bar{\bar{\tau}}_\ell\right)-\left(\vec{\mathrm{n}}\bar{\bar{\tau}}_g\right)\right]\cdot\vec{t} = \vec{\nabla}T^*\cdot\vec{t} \tag{2}$$

$\vec{t}$ is a tangential unit vector and $\bar{\bar{\tau}}$ is the dimensionless stress tensor. The temperature gradient ($\vec{\nabla}T^*$) in Eq. (2) represents the term of thermo-capillary effect.

iii) Energy balance,

$$\frac{Ja}{Le}\left(\frac{\Delta C}{\rho_\ell}\right)J^* + \left(R_l^g\nabla T_g^* - \nabla T_\ell^*\right)\cdot\vec{n} = 0 \tag{3}$$

$R^g{}_l = k_g / k_l$ is the ratio of gas/liquid thermal conductivity, $Ja = h_{lg} /(c_{Pl}\,\Delta T)$ is the Jacob number, ΔT is equal to (T_w - T_∞), h_{lg} is the latent heat of vaporization and c_{Pl} is the liquid specific heat. The quasi-study state governing equations with associated boundary and interface conditions are solved numerically. The concentration gradients at the drop surface are then used to evaluate the evaporation rate and deduce the drop lifetime.

3. Numerical procedure

Finite volume method and staggered grid are applied to resolve the mathematical model [23]. Coupling of velocity and pressure fields in fluid phases is addressed by the SIMPLE algorithm. Power law differencing scheme (PLDS) is used to consider the contribution of convection and diffusion in the transport phenomena. The algebraic equations resulting from the finite volume discretization are solved by a combination of the tridiagonal matrix algorithm (TDMA) and the Gauss-Seidel iterative method along with under-relaxation. Solutions of velocity, temperature and concentration fields reach satisfactory convergence during the iterative process once the maximum relative error on the dependent variable (u, v, T, C) is smaller than 0.1%. The maximum allowable absolute residue in the mass conservation equation is less than 10^{-10} and less than 10^{-5} in other conservation equations.

The elaborated computational program is validated with previous works in the literature. Hereafter is presented the comparison of our results with those of the numerical study of Yang et al. [1]. Figure 1 shows flow field in both liquid and gas phases for an evaporating sessile drop at a contact angle of 50°. There is a good agreement between the compared results.

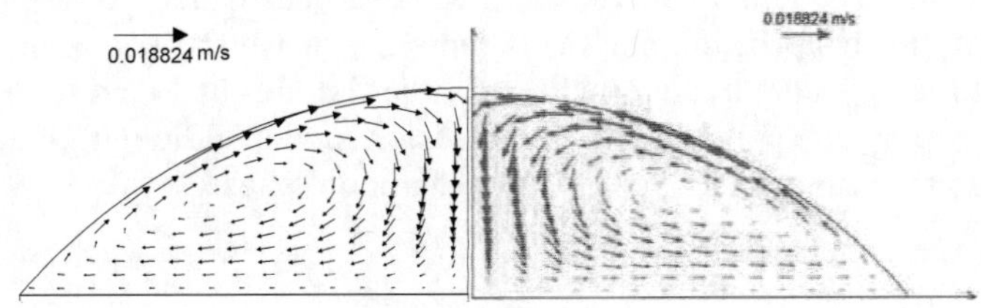

(a) Liquid phase

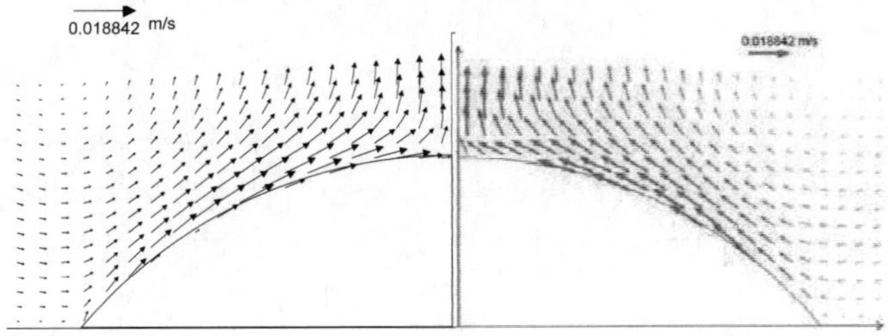

(b) Gas phase

Figure 1. Comparison with numerical results of velocity fields obtained by Yang et al. (water sessile droplet with θ=50° and R=0.95 mm, glass substrate with k_s=096 W/mK and e_w=0.2 mm, T_w=50°C, T_∞=22°C and Ha=50%).

4. Results and discussion

Results are presented for a water drop of 10 mm^3 initial volume deposited on a substrate with high thermal conductivity. Velocity and temperature fields are plotted in Fig. 2 for a contact angle θ = 20° and wall temperature T_w=50°C. A tri-cellular flow appears in the drop due to the thermo- capillary effect in heating conditions and confined geometry of the drop at θ = 20°. The external flow near the drop surface is driven by the internal flow through thermo-capillary effect, whether the buoyancy in the gas is included or not. This is due to the viscosity of the liquid which is much greater than that of the gas. The intensity of fluid flow in gas phase increases with buoyancy effect consideration, and this can affect the temperature distribution also in gas phase. This figure indicates that taking into account the buoyancy effect weakly influences fluid flow and temperature distribution in liquid phase.

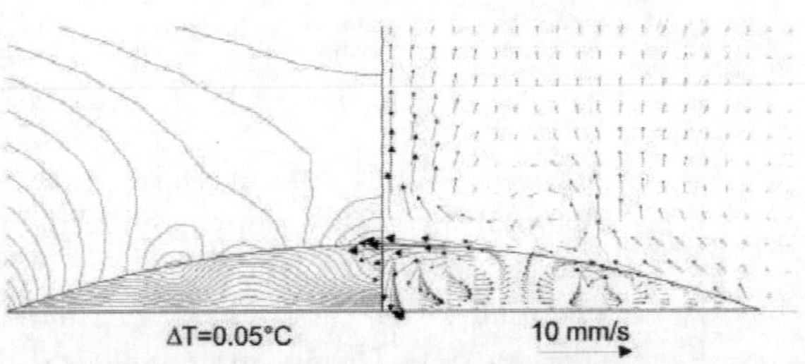

(a) Without buoyancy effect in the gas phase

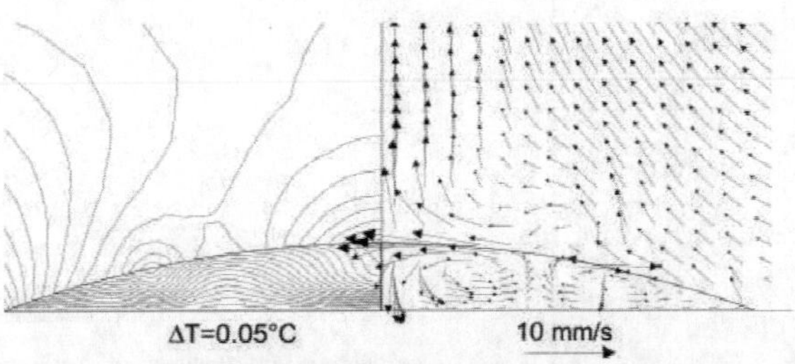

(b) With buoyancy effect in the gas phase

Figure 1. Velocity and temperature fields for a contact angle θ=20° and wall temperature T_w = 50°C.

In Fig. 3, the temperature distribution at drop surface is presented for different models: diffusion model, model taking into account only fluid flow in liquid phase, model taking into account fluid flow in both liquid and gas phases without buoyancy effect, and finally a complete model which takes into account fluid flow in the two phases with buoyancy in surrounding gas. It is clearly shown that the buoyancy in the air can increase the importance of the thermos-capillarity especially under the heating conditions and thus influence the internal flow even if this is weak. The buoyancy effect decreases temperature at drop surface. This is due to the gas movement which takes the water vapour far from the drop and as result the evaporation mass flux increases as well as heat transfer from drop surface to surrounding air. At the opposite, the internal flow rather implies an increase in the surface temperature at the edge of the drop by comparison with the diffusion model, although globally the evaporation rate also increases due to thermal convective effects in the liquid phase.

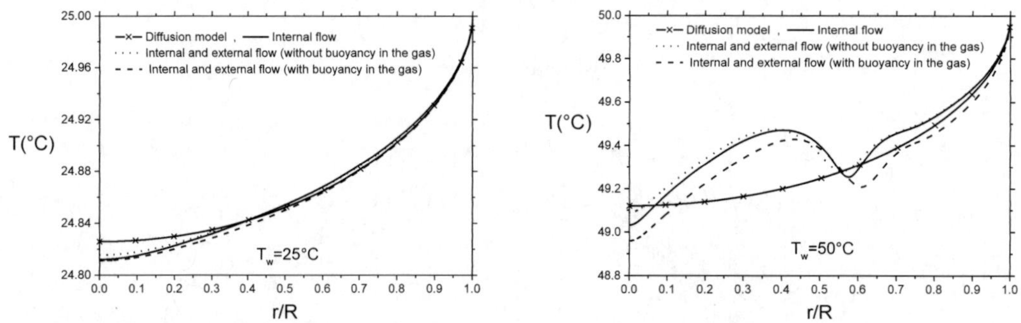

(a) Non-heated substrate at25°C (b) Heated substrate at 50°C

Figure 3. Temperature profile at drop surface for q = 20° and Tw = 25 or 50°C.

Finally, we finish this study by analysing the effect of fluids flow on the evaporation kinetics. The evolution in time of the evaporation rate is of great interest, because of its applications in cooling, painting, coating and others. In several numerical works, the authors were interested on evaporation rate evolution but with models not considering flow in the gas phase. In the figure bellow (Figure 4), we present the evolution in time of the evaporation rate density (evaporation rate per drop surface) in both heated and non-heated cases. The results of two different models are given. In the first model, the flow in liquid and gas phases is taken into account, but without considering the buoyancy effect. In the second model, all effects are considered as well as buoyancy convection in the gas phase. Results show that the buoyancy convection has a significant effect on evaporation rate density especially when the substrate is heated. The effect of drop surface variation on decreasing the evaporation rate is more important than that of liquid or gas flow on its increase.

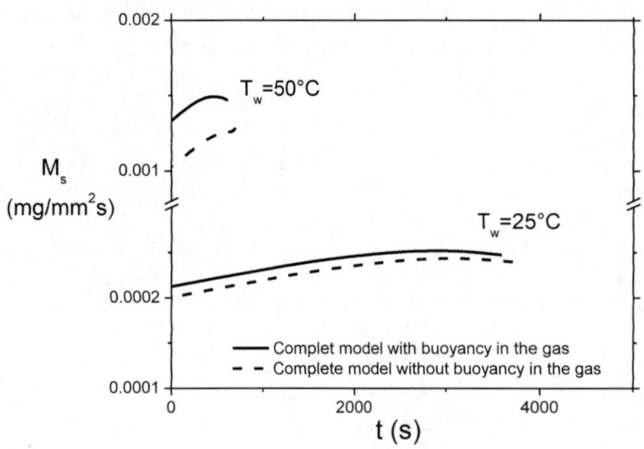

Figure 4. Evolution of the evaporation rate density during sessile drop evaporation on a substrate of high thermal conductivity. The substrate is heated (Tw = 50°C) or non-heated (Tw = 25°C).

5. Conclusions

The results presented in this work showed that, when the fluid flow is considered inside and outside an evaporating sessile drop, the governing effect is the thermo-capillarity; it mainly imposes the direction in both liquid and gas phases. The comparison between results obtained with the model taking into account buoyancy in gas phase, and those obtained with the model neglecting it, showed that the effect of buoyancy convection in the gas on liquid flow intensity and direction is negligible, but its effect on temperature distribution at liquid-gas interface is important. The evaporation rate is underestimated by the model without buoyancy effect in the gas phase, especially in heated substrate cases. From these results we can conclude that the complete and appropriate model to better estimate, fluid flow, temperature distribution, and evaporation rate, is the model taking into account fluid flow inside and outside the evaporating sessile drop.

References

[1] Yang K, Hong F, Cheng P 2014 *Int. J. Heat Mass Transfer* **70** 409
[2] Larson R G 2014 *AIChE Journal* **60**(5) 1538
[3] Lu G, Duan Y, Wang X and Lee D 2011 *Int. J. Heat Mass Transfer* **54** (19) 4437
[4] Picknett R G, Bexon R 1977 *J. Colloid Interface Sci.* **61**(2) 336
[5] Yu Y S , Wang Z, Zhao Y P 2012 *J. Colloid Interface Sci.* **365** (1) 254
[6] Mollaret R, Sefiane K, Christy J R E, Veyret D 2004 *Chem. Eng. Res. Des.* **82**(4) 471
[7] Erbil H Y, McHale G, Rowan S M, Newton M I 1999 *Langmuir* **15**(21) 7378
[8] Erbil H Y, McHale G, Newton M I 2002 *Langmuir* **18**(7) 2636
[9] Birdi K S, Vu D T 1993 *J. Adhes. Sci. Technol.* **7**(6) 485
[10] Yang K, Hong F. Cheng P 2014 *Int. J. Heat Mass Transf.* **70** 409
[11] Barmi M R, Meinhart C D 2014 *J. Phys. Chem. B* **118**(9) 2414
[12] Weon B M, Je J H, Poulard C 2011 *AIP Adv.* **1**(1) 012102
[13] Ait Saada M, Chikh S, Tadrist L 2010 *Phys. Fluids* **22** (11) 112115.
[14] Hu D, Wu H 2015 *Int. J. Therm. Sci.* **96** 149
[15] Sobac B, Brutin D 2012 *Phys. Rev. E* **86** (2) 021602.
[16] Carle F, Sobac B, Brutin D 2013 *Appl. Phys. Lett.* **102** (6) 061603
[17] Carle F, Semenov S, Medale M, Brutin D 2016 *Int. J. Therm. Sci.* **101** 35
[18] Lu G, Duan Y Y, Wang X D, Lee D J 2011 *Int. J. Heat Mass Transf.* **54** (19) 4437
[19] Sodtke C, Ajaev V S, Stephan P 2008 *J. Fluid Mech.* **610** 343
[20] Chen X, Zhu Z Q, Liu Q S, Wang X W 2015 *Micrograv. Sci. Technol.* **27**(5) 353
[21] Bouchenna C, Ait Saada M, Chikh S, Tadrist L 2015 *Interfacial Phenomena and Heat Transfer.* **3**(2) 185
[22] Bouchenna C, Ait Saada M, Chikh S, Tadrist L 2017 Int. J. Heat Mass Transf. 109 482
[23] Patankar S V 1980 *Numerical heat transfer and fluid flow* (McGrawHill, Hemisphere)

IOP Conf. Series: Journal of Physics: Conf. Series **925** (2017) 012007 doi:10.1088/1742-6596/925/1/012007

Cooling of a microchannel with thin evaporating liquid film sheared by dry gas flow

Yu O Kabova[1,2] and V V Kuznetsov[3,4]

[1]Kutateladze Institute of Thermophysics, SB RAS, 630090, 1 Lavrentiev prosp., Novosibirsk, Russia
[2]Novosibirsk State University of Architecture and Civil Engineering, 630008, 113 Leningradskaya Str., Novosibirsk, Russia
[3]Lavrentyev Institute of Hydrodynamics, SB RAS, 630090, 15 Lavrentiev prosp., Novosibirsk, Russia
[4]Novosibirsk State University,630090, 2 Pirogov str, Novosibirsk, Russia

E-mail: kabova@itp.nsc.ru

Abstract. A joint motion of thin liquid film and dry gas in a microchannel is investigated numerically at different values of initial concentration of the liquid vapor in the gas phase, taking into account the evaporation process. Major factors affecting the temperature distribution in the liquid and the gas phases are as follows: transfer of heat by liquid and gas flows, heat loses due to evaporation, diffusion heat exchange. Comparisons of the numerical results for the case of the dry gas and for the case of equilibrium concentration of vapor in the gas have been carried out. It is shown that use of dry gas enhances the heat dissipation from the heater. It is found out that not only intense evaporation occurs near the heating areas, but also in both cases vapor condensation takes place below the heater in streamwise direction.

1. Introduction

Investigation of dynamics and evaporation of locally heated thin liquid films sheared by gas flow becomes one of the most important and most complicated problems in thermal physics, since it is concerned with the problem of cooling of microelectronic equipment [1, 2, 3]. Modern microprocessor represents a flat surface consisting of zones with an inhomogeneous heat release [4]. The shape of these zones is often close to rectangular one [5], and the size can range from a few hundreds of microns to several millimeters. The thermal-power density in hot zones ("hot spots") can reach 1 kW / cm^2 and differs from the average thermal-power density on the chip by 5-10 times. Such "hot spots" as well as the chip itself might be considered as a local heat sources. Spray cooling, boiling in microchannels, thermoelectric modules are just some examples of methods using to solve the problem of heat removal from small size hot zones. One of the promising solution allowing to remove high heat fluxes is technology using processes with phase change, for example evaporation of a thin liquid film moving in a flat microchannel under the action of a gas flow [5, 6], but thin films are subjected to rupture, which drastically reduces the efficiency of the apparatus and can lead to their breakdown [7, 8, 9]. In addition to the thermocapillary effect one of the most important factors affecting the dynamics of the liquid film rupture and the formation of a dry spot in the region of local heating is evaporation [6, 10]. It is worth noting that, as a rule, liquid film is supposed to be sheared by the dry gas flow [7].

2. Mathematical model

In the present work a joint motion of thin film of incompressible viscous liquid and gas in a horizontal microchannel is considered, taking into account the evaporation process at different values of initial concentration of the liquid vapor in the gas phase. The microchannel is assumed to be unbounded in streamwise and spanwise directions. A part of wall at the side of liquid film is heated; temperature at the heater is prescribed. Coupled two-phase problem is simulated using a 3D nonstationary model; the model is based on the previously developed mathematical model [9]. The transport processes in the liquid and in the gas phases are described by the Navier-Stokes, continuity and energy equations and diffusion equation in the gas phase. For the deformable gas-liquid interface the following boundary conditions are posed: the condition of continuity of the temperature (it is equivalent to the fact that we remain in equilibrium thermodynamics) and the tangential components of the liquid and gas velocity vectors, the mass conservation condition, the dynamic condition with the term expressing the mechanical effect of the evaporated matter on the liquid, the thermal boundary condition and condition of local thermodynamic equilibrium. This means that near the liquid surface the partial vapor pressure is equal to the equilibrium pressure corresponding to the local temperature. The upper wall is adiabatic and impermeable. The dynamic viscosity and surface tension are assumed to depend on temperature. The problem is transformed to specially developed new variables, so that continuity equations in the liquid and in the gas retain their form in new variables but kinematic condition at the free interface simplifies and become linear. We employ the lubrication theory to derive reduced set of equations, so we assume that the characteristic film thickness and the characteristic film thickness variation are much smaller than the characteristic length scale of the film in streamwise and spanwise directions, thereby the film aspect ratio is taken asymptotically small $\varepsilon = H_0/l << 1$. Considering inertial terms to be negligible, the system of equations with boundary conditions is rewritten in dimensionless form. Major factors affecting the temperature distribution in the liquid and the gas phases are as follows: transfer of heat by liquid and gas flows, heat loses due to evaporation, diffusion heat exchange. These factors significantly affect each other and, by turn, are heavily dependent on the resulting temperature distribution. Previous investigations [9, 11] have shown that the heat removal from hot spots is mainly determined by evaporation. Therefore, the intensification of the heat dissipation may be achieved by intensifying the evaporation.

An important difference from the previous works is a specification that incoming gas does not contain liquid vapor. Namely, previously [9, 11] it was posed, so as for many problems with the diffusion evaporation, that the vapor concentration must be equal to the saturated concentration on the free interface (vapor concentration corresponding to the pressure of the saturated vapor).

Since a non-stationary problem is solved, for the case of the dry gas, it would be more convenient to calculate the process where, at the beginning of the calculation, the input gas has equilibrium moisture, but for some small time interval the vapor concentration in the input gas flow decreases to zero and remains so (dry gas). Then after a while the heater is turned on, and the heat and mass transfer processes gradually stabilize.

Numerical solution of the problem is implemented by the finite difference method. The alternating directions implicit (ADI) method is used to solve the system of grid equations with boundary conditions. To solve the problem on each fractional grid step the Thomas algorithm is used. For more details refer to [9, 11].

3. Numerical results

Calculations have been performed for the constant channel height equal to 250 mkm. In all calculations the liquid is water and the gas is air. Gas is moving in x-direction and the heater upper edge is located at the origin of coordinate system. The heater size is constant in all calculations and equal to 3 x 6, 4 mm^2 (length x width). Channel has been supposed to be horizontal, so that inclination angle is equal to 0. The gas and the liquid Reynolds numbers have been kept constant in all calculations and equal to Re_g=20 and Re=8, 5 correspondingly (the initial film thickness H_0=92, 2 mkm). The initial temperature is equal to 20ºC. All quantities shown in figures are presented in a dimensionless form, here the time and the temperature are set as $t = l\bar{t}/U$ and $\theta = (T - T_0)/[T]$. Calculations of the total evaporation rate, J_n, have been performed using the following formula

$$J_{total} = \iint_n \frac{k_3(w_g - w)}{1 - k_3} dxdy,$$

here n is the all area of calculations. This follows from mass conservation low at the free boundary.

Figure 1 shows the total heat transfer from the heater for the cases of dry gas (black lines) and for the case when the vapor concentration in the gas phase at the microchannel inlet is equal to the saturated concentration on the free interface (red lines). It can be seen that the use of the dry gas significantly increases the heat transfer from the heater both at high temperature on the heater and at low temperature prescribed on the heater.

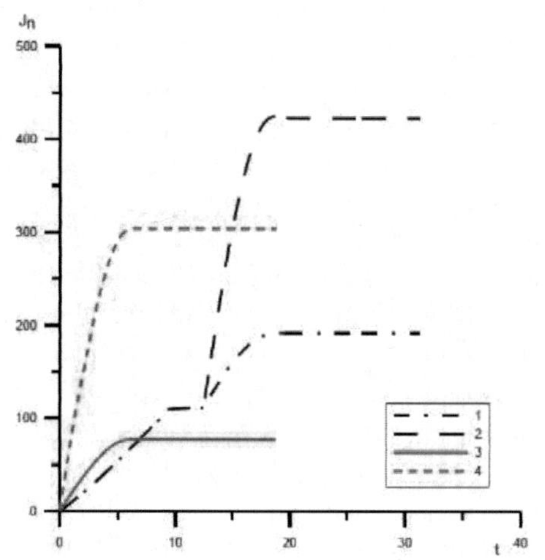

Figure 1. The total heat transfer from the heater vs time. Black lines - dry gas, red lines - equilibrium concentration of vapor in the gas. Lines 1 and 3 correspond to the temperature on the heater equal to 22,5°C, lines 2 and 4 correspond to the temperature on the heater equal to 30°C. All quantities shown on figure are presented in a dimensionless form

Figure 2 shows the distribution of the evaporation intensity over the gas-liquid interface (or vapor condensation). In the case (a) the gas flow entering the channel does not contain liquid vapor, in the case (b) their concentration is equal to equilibrium for the average temperature. It can be seen that the vaporization rate in case (a) is much higher. In addition, it can be seen that in both cases vapor condensation takes place after the heater in streamwise direction and the process of condensation plays an important role in the redistribution of the heat.

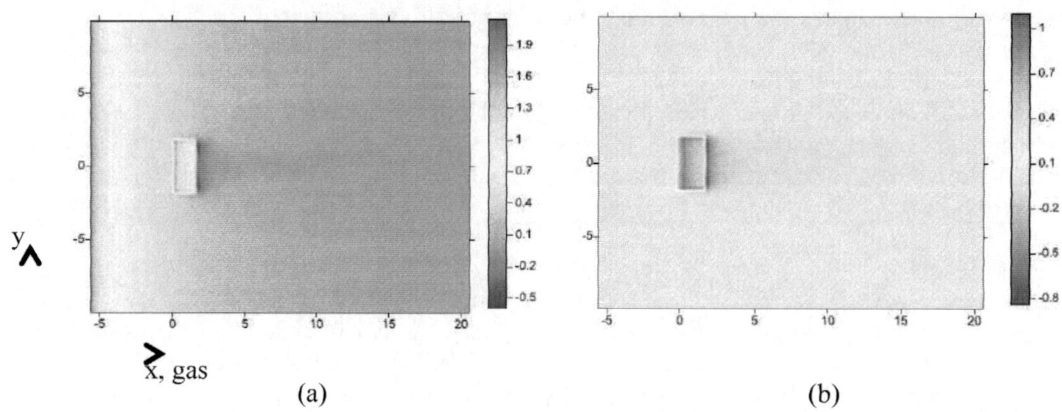

Figure 2. Distribution of the evaporation intensity over the gas-liquid interface (or vapor condensation). (a): dry gas, (b): equilibrium concentration of vapor in the gas. Prescribed temperature on the heater is 30°C.

Figure 3 shows the gas-liquid interface position along the channel at $y = 0$ for the dry gas when the gas flow entering the channel does not contain liquid vapor, and for the case when liquid vapor concentration is equal to equilibrium for the average temperature. Deformation of the film as a bump at the front edge of the heater exists for both cases. The top of the bump is located near the front edge of the heater. The governing factor in this phenomenon is the thermocapillary effect due to the intensive heating and strong temperature gradient along the liquid film near the front edge of the heater. But for dry gas significant film thinning occurs in the area of gas stream entrance to the channel and this overall thinning gradually decreases downstream. This takes place because the incoming dry gas causes evaporation of a liquid that absorbs heat and therefore a longitudinal temperature gradient appears, the value of which decreases downstream. And the resulting Marangoni force causes the film thin. This should be taken into account since significant film thinning may lead to a film rupture.

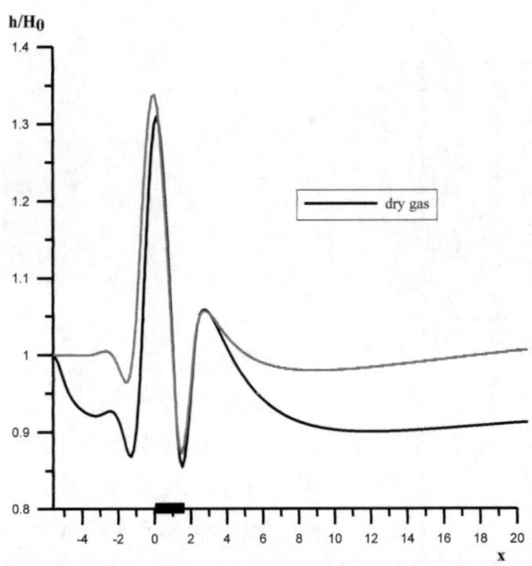

Figure 3. Gas-liquid interface position along the channel at $y = 0$. Black line - dry gas, red line - equilibrium concentration of vapor in the gas. Prescribed temperature on the heater is 30ºC

The minimum film thickness takes place near the heater side downstream. The main reasons are evaporation taking place due to significant increase of temperature on the gas-liquid interface, and thermocapillary effect. Calculations of the minimum values of the film thickness vs time are shown in figure 4. It is seen that the film thinning in the case of dry gas is more significant than in the case of the equilibrium concentration of vapor in the gas. And with increasing the heater temperature, this difference increases.

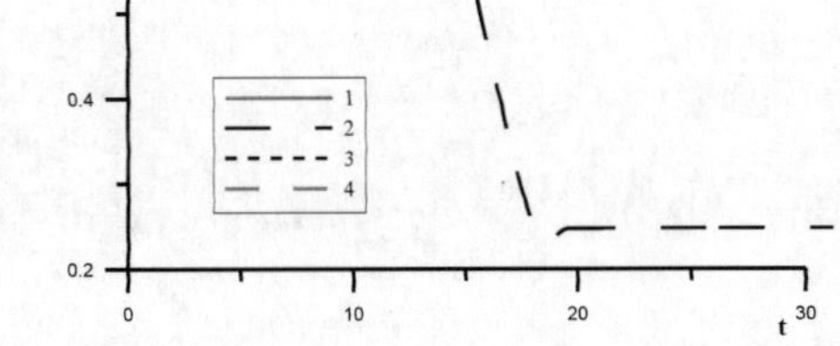

Figure 4. Minimum film thickness vs time. Black lines - dry gas, red lines - equilibrium concentration of vapor in the gas. Lines 1 and 2 correspond to the temperature on the heater equal to 30ºC, lines 3 and 4 correspond to the temperature on the heater equal to 22,5ºC

Figure 5 shows the velocity fields along the channel at y=0 and at the free interface for the case of the dry gas. Deformations of the gas-liquid interface cause distortion of the velocity fields. Calculations show that near the channel entrance, as well as near the heating area, the velocity field differs significantly from the case of co-current flow in a channel with straight streamlines. Therefore the widespread assumption that the dynamic effect of a gas flow on a liquid could be replaced by the action of a certain average shear stress may lead to significant errors while investigating the co-current gas and liquid flow in a channel. In addition, one can see some destabilization of the flow near the heating area and formation of an array of rolls or rivulets descending downstream.

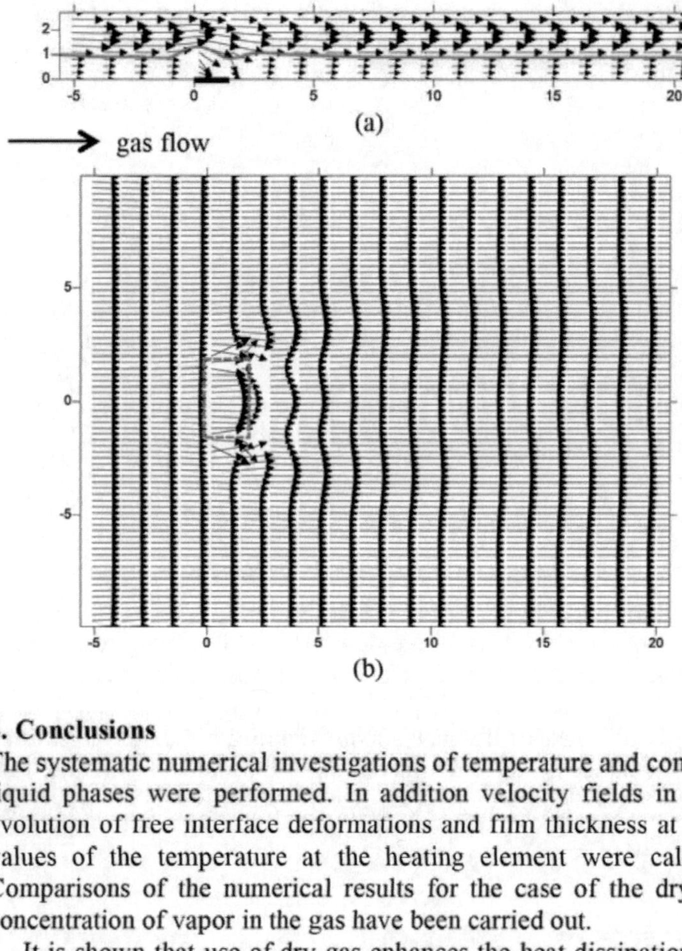

Figure 5. Velocity fields (a): along the channel (y=0) and (b): at the gas-liquid interface.

Prescribed temperature on the heater is 30°C, dry gas.

In the figure (a) the red line shows the position of the gas-liquid interface

4. Conclusions

The systematic numerical investigations of temperature and concentration profiles evolution in gas and liquid phases were performed. In addition velocity fields in the liquid and gas phases as well as evolution of free interface deformations and film thickness at constant liquid flow rate and different values of the temperature at the heating element were calculated for the case of the dry gas. Comparisons of the numerical results for the case of the dry gas and for the case of equilibrium concentration of vapor in the gas have been carried out.

It is shown that use of dry gas enhances the heat dissipation from the heater both at high and low temperatures prescribed on the heater but the film thinning in the case of dry gas is more significant. It is found out that not only intense evaporation occurs near the heating areas, but also condensation of vapor.

Acknowledgments

The work was supported by the Russian Science Foundation, Agreement no. 14_19_01755.

References

[1] Hirokawa T, Murozono M, Kabov O and Ohta H 2014 *Frontiers in Heat and Mass Transfer* **5** 1-8

[2] Nasr M H, Green C E, Kottke P A, Zhang X, Sarvey T E, Joshi Yo K, Bakir M S and Fedorov A G 2017 *Int. Journal of Heat and Mass Transfer* **108** 1702–13

[3] Nasr M H, Green C E, Kottke P A, Zhang X, Sarvey T E, Joshi Yo K, Bakir M S and Fedorov A G 2017 *Journal of Electronic Packaging* **139** 011006

[4] Mahajan R, Chin C and Chrysler G 2006 *Proceedings of the IEEE* **94** 1476-85

[5] Sri-Jayantha S M, McVicker G, Bernstein K and Knickerbocker J U 2008 *IBM Journal, Res. & Dev.* **52** 623-634

[6] Lyulin Yu V, Spesivtsev S E, Marchuk I V and Kabov O A 2015 *Technical Physics Letters* **41** 1034–37

[7] Houshmand F and Peles Y 2013 *Int. Journal of Heat and Mass Transfer* **64** 42-52

[8] Kabova Yu O, Kuznetsov V V, Kabov O A 2008 *Microgravity sci. technol.* **20** 187-192

[9] Kabova Yu, Kuznetsov V, Kabov O, Gambaryan-Roisman T and Stephan P 2014 *Int. J. Heat and Mass Transf.* **68** 527-541

[10] Liu R and Kabov O A 2012 *Physical Review E-Statistical, Nonlinear, and Soft Matter Physics* **85** 066305

[11] Kabova Yu, Kuznetsov V V and Kabov O 2014 *Interfacial Phenomena and Heat Transfer* **2** 185-102

IOP Conf. Series: Journal of Physics: Conf. Series **925** (2017) 012008 doi:10.1088/1742-6596/925/1/012008

Experimental-calculated study of the forced convection magnetic nanofluids

D V Guzei[1,2], A V Minakov[1,2] and M I Pryazhnikov[1,2]

[1] Siberian Federal University, 660041 Krasnoyarsk, Svobodnyi av. 79, Russia
[2] Kutateladze Institute of Thermophysics SB RAS, 630090, Novosibirsk, Lavrentev av. 1, Russia

E-mail: gudimas@yandex.ru

Abstract. Experimental-calculated study of the magnetic nanofluids forced convection in a constant magnetic field was carried out. Dependence study of the magnetic nanofluid heat transfer coefficient on the effect of a constant magnetic field was made. For numerical simulation we have used Euler-Lagrange two-component approach.

1. Introduction

One of the solutions to the problem of heat transfer performance enhancement could be the use of so called nanofluids, which are fluids containing nanoparticles of various composition. The use of electromagnetic fields is a promising way to control the properties of nanofluids. The magnetic field can be influenced by nanofluids prepared from ferromagnetic materials (ferromagnetic nanofluids), for example, from iron nanoparticles, its oxide (Fe_3O_4), nickel etc. Such fluids can be extremely useful in many applications.

Articles [1-3] investigated the nanoparticles transport in the channels with influence of the magnetic field. Sundar et al. [1] studied turbulent forced convection heat transfer and friction factor of Fe_3O_4 magnetic nanofluid in a tube in the absence of magnetic field and developed correlations for the estimation of Nusselt number and friction factor. Their results show that the heat transfer coefficient is enhanced by 30.96 % and friction factor by 10.01 % at 0.6 % volume fraction compared to the base fluid. Recently, Ghofrani et al. [2] investigated the effect of constant and alternating magnetic fields on the forced convection heat transfer in a short tube. They showed that increasing the alternating magnetic field frequency increases the heat transfer up to of 27.6 % in low Reynolds numbers. Lajvardi, et al.[3] experimentally studied the convective heat transfer of water – Fe_3O_4 ferrofluid flowing through a tube in laminar regime in the presence of magnetic field perpendicular to the flow direction. Experiments were carried out on ferromagnetic fluids at different concentrations and in different configurations of the magnetic field. They concluded that by improving the thermophysical properties of the ferrofluids, heat transfer can be attributed in the presence of magnetic field.

Thus, despite the existing experimental data, they are quite contradictory and do not allow comprehending clearly the mechanisms of heat transfer magnetic nanofluids. This circumstance requires additional systematic study of forced convection of magnetic nanofluids.

2. Description of the experimental setup

Experimental study of the forced convection of magnetic nanofluids in a constant magnetic field was carried out. The experimental unit is a loop with a circulation coolant. The experimental unit diagram is shown in Figure 1(a).

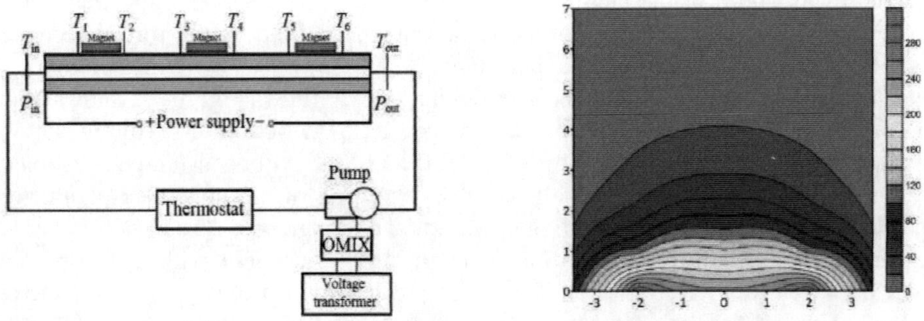

(a) (b)

Figure 1. The experimental unit diagram (a); The magnetic field induction over the magnet (mT) (b).

The working fluid is pumped to the heated test section from the accumulator tank. The gear pump WT30001 was used for pumping a nanofluid. Pump drives provide the ability to pump different fluids at a flow rate from 85.7 to 2571.4 ml/min.

After passing the heated section, the liquid enters the heat exchanger and gives heat to the thermostat. The working fluid flow rate in the circuit is regulated by control valves. Monitoring of the flow rate in the circuit is carried out by flowmeter. The heated section is a stainless steel tube with a diameter 10 mm and a length of 1m. The wall thickness is 1 mm. Electric current is supplied to the wall for heating. This ensures the conditions of constant heat flow on the wall. Heating power is regulated by LATR. 6 copper–constantan thermocouples are fixed on the wall at an equal distance from each other to measure the local temperature of the tube. The temperature is read by the TPM-200 meters. The temperature at the inlet and outlet of the heated section is measured by thermocouples. In this case, the thermocouple for measuring the ambient temperature at the circuit outlet is located at a considerable distance from the heated section end to ensure ambient temperature uniformity at the measurement site. The circuit section from the heater to the ambient temperature measurement site was heat insulated. The drop pressure study at the inlet and outlet of the heated section was also measured. The pressure drop over the length of the working section is measured by a differential pressure gauge OWEN PD200. The experimental unit for measuring heat transfer coefficient was tested using the known empirical data for the heat transfer of pure water.

Three permanent neodymium magnets with dimensions 50x30x10mm were used to create a magnetic field. The magnetic field intensity around the magnet was measured with a milliteslameter TPU-02. The distribution of the magnetic field induction over the magnet is shown in Figure 1(b). The maximum value of the magnetic induction near the magnet is of the order of 0.3T. In experiments, the magnets were located in the immediate vicinity of the heated section at the same distance from each other.

A round glass channel with a diameter of 10 mm and a length of 1 m was used to visualize the flow of a magnetic nanofluid. It was installed in the experimental unit instead of the measuring section.

Iron oxide nanoparticles Fe_3O_4 were used to prepare the nanofluid. The average particle size was 100 nm. Distilled water was used as a base fluid. The volume density of nanoparticles was 0.75%. A standard two-step method was used to prepare the nanofluid. Suspensions were processed in an ultrasonic bath "Sapphire TC-10338". The nanofluids were free of surfactants.

The nanofluid viscosity was measured at a Brookfield DV2T rotary viscometer with a low viscosity ULA adapter (0). The measurement accuracy for the viscosity coefficient was not less than 2.5%. The viscometer was calibrated with distilled water and ethylene glycol before the measurements.

The coefficient of thermal conductivity of nanofluids was measured using a non-stationary hot wire method. The detailed description of the installation and its testing is given in [4].

3. Experimental results

The average and local values of the heat transfer coefficient at the channel walls were measured as a result of experiments. A strong influence of a constant magnetic field on the forced convection of the magnetic nanofluid was shown during the experiments.

The graph of the average heat-transfer coefficient versus time (Figure 2) shows that the heat transfer coefficient increases with the constant magnetic field action. Measurements were made without the magnetic field action in the interval from 0 to 100 s. After that, three permanent magnets were installed near the working section wall, and the recording of the measured parameters was not interrupted. The average heat transfer coefficient was shown to increase.

The magnetic field action on the nanofluid heat transfer coefficient with a nanoparticle volume density of 0.75% was investigated. The influence of the magnetic field intensifies the average heat transfer coefficient by 55% relative to the base fluid at a fixed flow rate, while the nanofluid heat transfer coefficient without the magnetic field influence is 10% below pure water. The dependence of the heat transfer coefficient on the flow rate is shown in Figure 3.

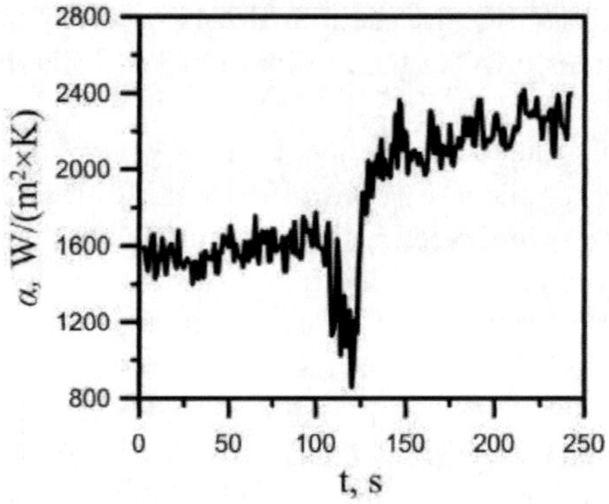

Figure 2. Dependence of the average heat transfer coefficient on time (Fe_3O_4 100nm, 0.25%, Re=1400).

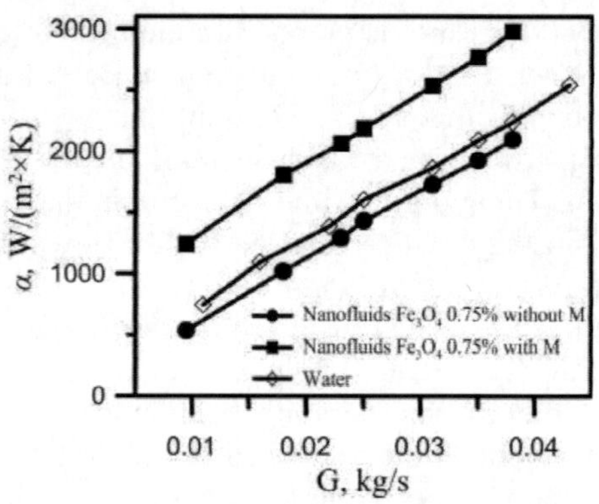

Figure 3. Dependence of the average heat transfer coefficient on flow rate with and without the magnetic field action (Fe_3O_4 100nm, 0.75%).

The flow structure was visualized under the action of a magnetic field. The least particle concentration of 0.01% was used for better visibility. Figure 4 shows photographs of the deposits formation on the channel wall from the time at the Reynolds number of 1750. Analysis of the deposits formation dynamic shows that the deposits shape continues to change for about 100 s. After this a dynamic balance is established between deposition and particles entrainment, and the deposits shape stabilizes. Figure 5 presents experimental photographs of nanoparticle deposits on the channel wall a minute after the deposits beginning. The volume concentration of the nanoparticles was 0.125%, the Reynolds number varied from 2000 to 5000.

Visualization shows that the nanoparticle deposits significantly distort the flow pattern in the channel. The fluid must flow around deposits, as a result, there is a local velocity increase in the gap between the channel wall and deposits. In addition, flow recirculation zones behind deposits from nanoparticles are well visible. For large values of the Reynolds numbers, the flow in the flow around the deposits becomes nonstationary and accompanied by an intense vortex separation and significant flow fluctuations.

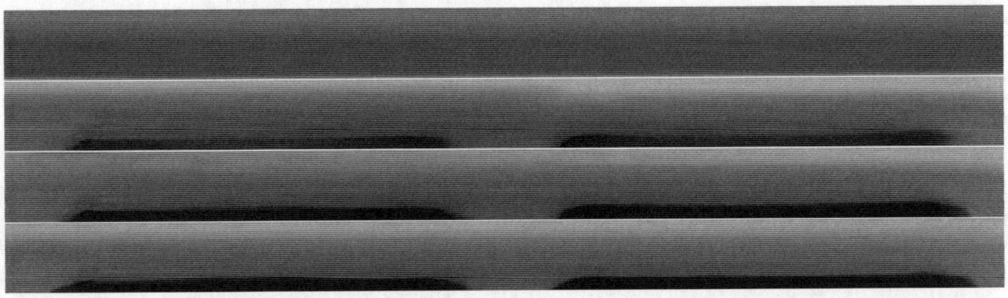

Figure 4. The deposits form on the tube wall at different times for Re = 1750. Top-down: 0c, 5c, 20s, 60s. The nanoparticles concentration in water is 0.01%, the magnetic field induction is 325 mT.

Figure 5. The deposits form on the tube wall a minute after the start of the deposition process. Top-down: Re = 2000, Re = 3000, Re = 4000, Re = 5000. The nanoparticles concentration in water is 0.125%, the magnetic field induction is 325 mT.

4. Numerical simulation

To simulate the transport of nanoparticles in the magnetic field we used a Euler-Lagrange combined two-component approach. As part of the Lagrange approach for the base fluid the equations of continuity, momentum and energy were being solved, and the movement of the particles was modeled on the basis of Newton's second law by solving an ODE. Euler two-component model describes nanoparticles in a liquid as a binary mixture, one of which components is nanoparticles. Statement of the problem completely repeated the experimental study. The computational domain is a circular channel with an inner diameter of 10 mm and a length of 1 m. A computational grid consisting of 850 thousand nodes, was used for the simulation.

Figures 6-7 show the distribution of the magnetic induction components in the channel and the distribution of the magnetic force in the channel, when the magnets are as close as possible to the channel wall.

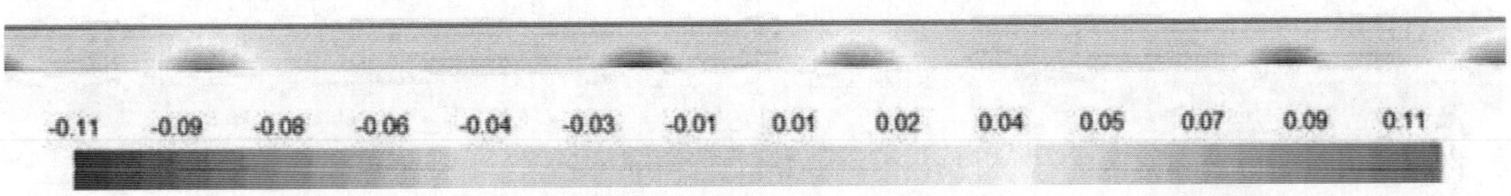

Figure 6. Bx component of the magnetic induction in the plane y = 0.

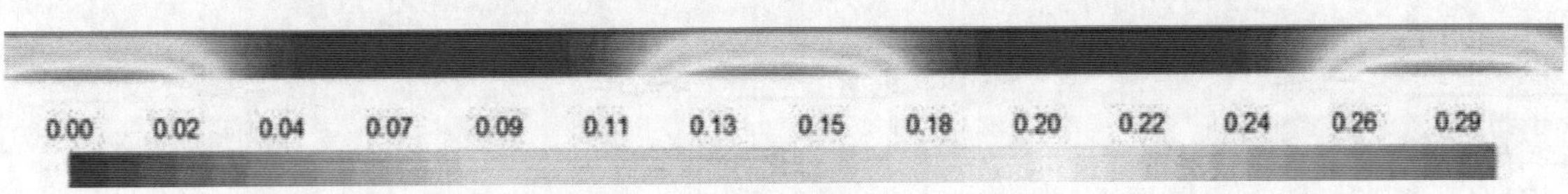

Figure 7. Bz component of the magnetic induction in the plane y = 0.

IOP Conf. Series: Journal of Physics: Conf. Series **925** (2017) 012008 doi:10.1088/1742-6596/925/1/012008

Figures 8,9 show the vector velocity field in the channel and the temperature field without a magnetic field. There is no the nanoparticles deposition on the channel surface.

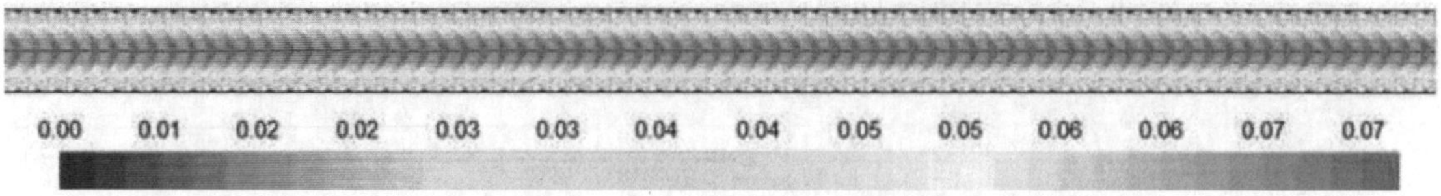

Figure 8. The vector velocity field in the channel for the Reynolds number of 1500 without a magnetic field.

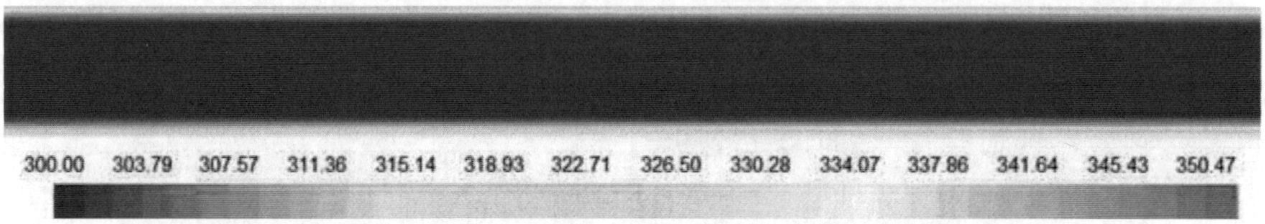

Figure 9. The temperature field in the channel for the Reynolds number of 1500 without a magnetic field.

A series of calculations of heat transfer in a magnetic field was carried out. Figures 10-12 show the results of calculations for the Reynolds number 1500. Powerful deposits of particles are formed on the channel wall. The nanoparticles path for this variant are shown in Figure 11. In addition, the interaction of the magnetic field with the particle flow leads to the formation of recirculation zones. It was found that the nanoparticle deposits formed in the channel significantly distort the velocity profile of the flow. The fluid must flow around deposits, as a result, there is a local velocity increase in the gap between the channel wall and deposits.

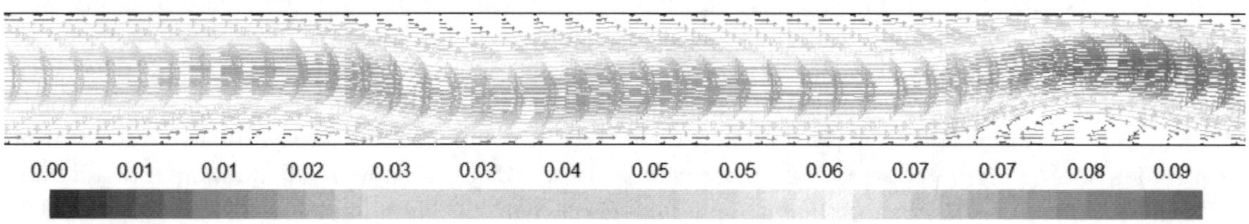

Figure 10. The vector velocity field in the channel for the Reynolds number of 1500 with a magnetic field.

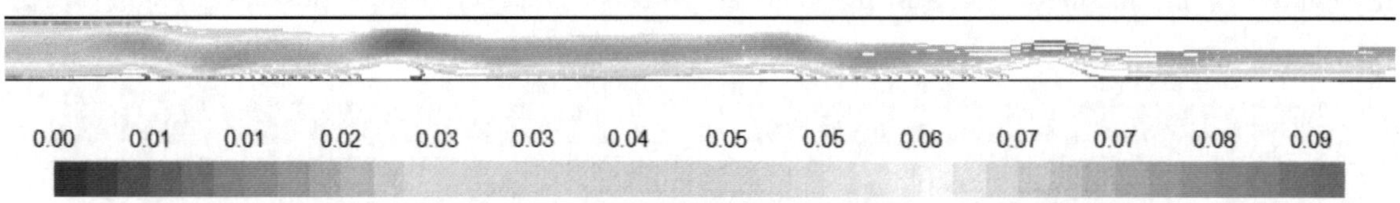

Figure 11. The nanoparticles path for the Reynolds number of 1500 with a magnetic field.

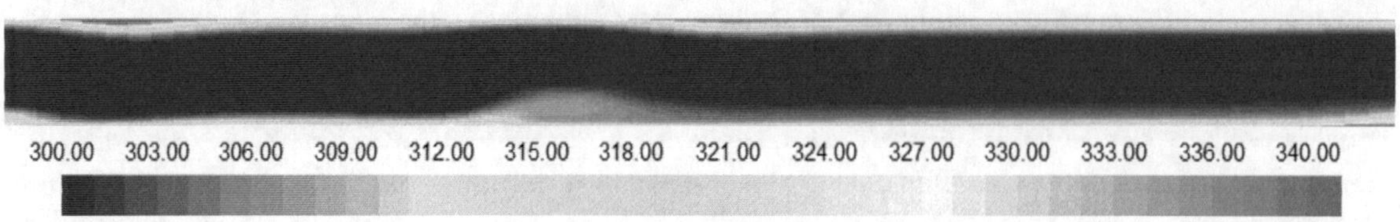

Figure 12. The temperature field in the channel for the Reynolds number of 1500 with a magnetic field.

Figure 13 shows the calculated distribution of the increment of the local heat transfer coefficient at the tube wall for Reynolds number of 1500 with a magnetic field. The recirculation zones of the flow and local inhomogeneities of the velocity field in the scope of the magnetic field lead to a significant increase in the local heat transfer coefficient.

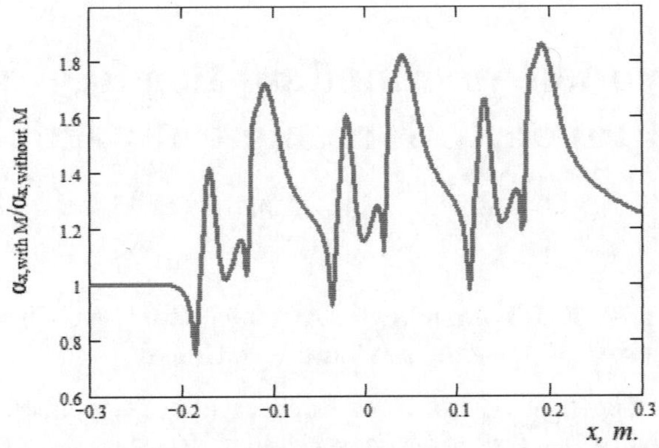

Figure 13. Exceeding the local heat transfer coefficient under the action of the magnetic field for Re = 1500.

5. Conclusions

As a result of the experiments, the mean and local values of the heat transfer coefficient at the channel walls were measured. The dependences of the heat transfer coefficient and the flow rate of the nanofluid are obtained. The experiment established that the local heat transfer coefficient of the magnetic nanofluid increases with influence of a constant magnetic field. At the locations of the magnets, the local heat transfer coefficient increases to 1.8 times relative to the nanofluid without the influence of the magnetic field. It is shown that the average heat transfer coefficient of magnetic nanofluids with iron oxide nanoparticles Fe_3O_4 is increased by up to 1.55 times in comparison with the base fluid by means of a stationary magnetic field., It was shown that the interaction of the magnetic field with the particle flow leads to the deposits formation of nanoparticles on the tube walls at the locations of the magnets by numerical simulation. These deposits lead to a significant distortion of the local flow structure in the channel and to the formation of flow separation. It is the main reason for increasing the heat transfer coefficient.

Acknowledgments

The reported study was funded by Russian Foundation for Basic Research, Government of Krasnoyarsk Territory, Krasnoyarsk Region Science and Technology Support Fund, the research project №16-48-243061.

References

[1] Sundar L S, Naik M T, Sharma K V, Singh M K, Reddy T Ch 2012 *Exp. Therm. Fluid Sci.* **37** 65-71
[2] Ghofrani A, Dibaei M H, Hakim Sima A, Shafii M B 2013 *Exp. Therm. Fluid Sci.* **49** 193-200
[3] Lajvardi M, Moghimi-Rad J, Hadi I, Gavili A, Isfahani T D, Zabihi F, Sabbaghzadeh J 2010 *J. Magn. Magn. Mat.* **322** 3508-3513
[4] Minakov A V, Rudyak V Ya, Guzei D V, Pryazhnikov M I, Lobasov A S 2015 *J. Eng. Phys. Thermophys* **88** 149-162

IOP Conf. Series: Journal of Physics: Conf. Series **925** (2017) 012009 doi:10.1088/1742-6596/925/1/012009

Damage to formation surrounding flooding wells: modelling of suspension filtration with account of particle trapping and mobilization

S A Boronin[1], K I Tolmacheva[1], A A Osiptsov[1], A N Sitnikov[2], A A Yakovlev[2], B V Belozerov[2], E V Belonogov[2] and R R Galeev[2]

[1] Skolkovo Center of Hydrocarbon Recovery, Skolkovo Institute of Science and Technology, Nobelya st. 3, Moscow, 143026, Russia
[2] Gazpromneft-NTC, r. Moika emb. 75-79, St. Petersburg 190000

E-mail: s.boronin@skoltech.ru

Abstract. We consider the filtration of raw water in a formation surrounding injection wells in oilfields of Western Siberia. The mathematical model for suspension filtration developed earlier on the basis of tree-continua approach allows to describe the permeability damage and recovery due to trapping and mobilization of externally-introduced fines. As compared to classical deep-bed filtration models, the proposed model takes into account the filtration of the carrier fluid through the pack of trapped fines and uses only two free parameters to describe the particle trapping and mobilization rates. It has gone through a thorough validation campaign against experimental data with contamination of porous samples by external fines and mobilization of pre-seeded particles in sand packs. Simulations of permeability dynamics in the zone surrounding injection wells are carried out using the values of free parameters obtained by tuning the model against available lab experiments. Both continuous and periodic water flooding/cleanup is modelled. It is found that there are periodic regimes of water injection, in which the permeability of the rock is not damaged. The study will be continued after the calibration of the model against thorough laboratory tests with natural cores and field tests of injection rate dynamics in flooding wells.

1. Introduction

Filtration of suspensions and particle transport in porous media are widely encountered in various natural and technological processes. In particular, there are several oilfield technological processes, in which the transport and trapping of fine particles in porous medium plays an important role, namely, dynamics of permeability of a formation during well drilling and cleanup as well as degradation of hydraulic fracture conductivity during cleanup and consequent production. Another important oilfield process is the operation of flooding (injection) wells, which are used to increase the oil recovery in a nearby production wells by injecting a low-viscosity fluid into formation to maintain high reservoir pressure. The flooding is usually accompanied by a contamination of the formation by external fines contained in raw water injected into formation and by internal (mineral) fines detached from pore matrix: fine particles can deposit on pore walls and plug the pore throats, which leads to a reduction in the rock permeability [1]. The study of particle capture and entrainment into a flow allows one to minimize the reduction in the well injection rate due to the contamination of a surrounding porous medium and optimize the operation of injection wells.

A family of continuum models usually referred to as Deep-Bed filtration (DBF) models are most widely-used in open literature to describe particle transport in porous medium. The DBF models are based on three-continuum description of a particle-laden filtration [2-4]. The flow is represented as three continua, namely, carrier fluid (usually Newtonian incompressible liquid), suspended particles (transported through a porous medium) and trapped particles (deposited in a porous medium). The rate of particle trapping, which is the volume of particles being trapped in a porous medium per time unit, is assumed to be proportional to the volumetric flux of suspended particles. The coefficient of proportionality depends on the concentration of trapped particles, which describes the filter clogging effect – change in particle trapping rate due to increase in the concentration of trapped fines. Overview of different expressions for the filter clogging functions and reduced permeability due to trapped fines can be found in [5]. The number of tuning parameters of DBF models describing both particle trapping and mobilization is typically in the range from two to five. The original concept of DBF model was generalized in [6]. The key novel element is a parallel-pathway approach, so that the porous medium is represented as the collection of "plugging" and "non-plugging" channels. In "plugging" pathways (small pores), the particles can only be trapped, while in the "non-plugging" pathways (large pores), the particles both deposit on pore walls and are mobilized into the flow. It was obtained experimentally that the particles are mobilized only when the filtration velocity exceeds a certain value (usually referred to as critical velocity). Above critical velocity, the rate of particle mobilization is found to be proportional to the product of concentration of trapped fines and the difference between the filtration velocity and the critical velocity. The model was successfully used to describe experiments on contamination of sand packs and sandstones, as well as mobilization of pre-seeded fines in a porous medium. Overall, the model contains four tuning parameters describing particle trapping and mobilization rates as well as four tuning parameters describing the reduced permeability of "plugging" and "non-plugging" pathways due to trapped fines. We believe that the number of tuning parameters is excessively large, which makes it difficult to apply the model [6] for prediction of the permeability dynamics of a porous medium during the transport of fines in a wide range of flow conditions typical for oilfield applications.

One of the key free parameters of DBF models is the trapping coefficient λ describing the particle trapping rate. It depends on the combination of properties of the porous medium and the particles including characteristic size of the pores and the particles. In [7], the analytical expression for λ is obtained by modelling the particle transport in porous medium at pore-scale level. The porous medium was represented by a collection of isolated Happel cells (spherical collectors), on which particles are deposited. By comparing with experimental data, it is found that the result is valid only for particle-to-grain diameter ratio not larger than 0.18. Another limitation for application of the model [7] for describing the particle trapping at macroscale level is that it ignores bridging and plugging of small porous channels. To the best of our knowledge, this is a unique analytical expression for trapping coefficient in terms of the flow parameters derived at microscale level, which is available in the open literature. One of the most reliable methods to obtain the trapping coefficient is through the tuning of the model against experiments on core flooding followed by micro-CT scanning and image processing to obtain distribution of trapped fines inside the porous medium (e.g., [4]).

The aim of the present study is to apply the three-continua suspension filtration model developed earlier to describe the permeability dynamics in the vicinity of fractured injection wells. The free parameters describing particle trapping and mobilization rates are obtained by tuning the model against the experiments with filtration of colloidal suspension through the sand columns.

2. Problem formulation

The one-dimensional multiphase filtration of the particle-laden fluids in a porous medium is considered [8]. The new element of our approach as compared to traditional DBF models [2-4] is that the trapped fines inside a porous medium form a packed bed through which the carrier fluid is filtered (figure 1). A porous medium is represented by the two classes of channels: large pore channels and small pore channels. The large channels are formed by the porous matrix with exclusion of trapped particles and a void space between them. The small channels are formed by the gaps between the trapped particles. Fluid can flow both in large and small pore channels, while the suspended particles can be transported in large pore channels only. In order to describe this effect, we introduce two porosities: ϕ_c describing the volume fraction of large pore channels, and φ_t describing the total volume fraction of large and small pore channels:

$$\varphi_c = \varphi_0 - \frac{\sigma}{C_{\max}}, \quad \varphi_t = \varphi_0 - \sigma \tag{1}$$

Here, φ_0 is the initial porosity, $C_{\max}$ is the maximum concentration of random close packing and σ is the volume concentration of trapped particles.

The multiphase filtration of suspension is described using the three-continua approach with different continua being a carrier fluid with the density ρ_γ^f and the filtration velocity u_γ^f, suspended particles with the volume concentration C (the ratio of a volume occupied by suspended particles to the pore volume) and filtration velocity u_γ^p and trapped particles with the volume concentration σ (the ratio of a volume occupied by trapped particles to the total volume of the elementary porous sample). Mass balance equations for the continua are as follows:

$$\frac{\partial}{\partial t}\left[\rho_\gamma^f s_\gamma(\varphi_t - C\varphi_c)\right] + \frac{\partial}{\partial x}\left(\rho_\gamma^f u_\gamma^f\right) = 0, \ \gamma = 1, 2 \tag{2}$$

$$\frac{\partial}{\partial t}\left[\varphi_c s_\gamma C\right] + \frac{\partial}{\partial x} u_\gamma^p = -q_\gamma, \ \gamma = 1, 2 \tag{3}$$

$$\frac{\partial \sigma}{\partial t} = q_\gamma, \ \gamma = 1, 2 \tag{4}$$

Here, s_γ is the phase saturation and q_γ is the rate of particle trapping and mobilization (the volume of particles trapped per time unit in a certain elementary volume of the porous medium), and $\gamma = 1, 2$ is the fluid index. Filtration velocities u_γ^f and u_γ^p of the carrier fluid and particles are expressed in terms of the suspension filtration velocity in large pore channels u_γ and carrier fluid filtration velocity in small pore channels u_s as follows:

$$u_\gamma = -\frac{kk_\gamma}{\mu_\gamma}\frac{\partial p}{\partial x}, \ u_s = -\frac{k_s k_\gamma}{\mu_{\gamma,0}}\frac{\partial p}{\partial x}, \ u_\gamma^p = Cu_\gamma, \ u_\gamma^f = (1-C)u_\gamma + u_s \tag{5}$$

Here, k is the permeability of the porous medium; k_γ is the relative permeability; μ_γ is the viscosity of the suspension formed by the fluid γ and suspended particles; $\mu_{\gamma,0}$ is the carrier fluid viscosity; p is the fluid pressure. Equations (1) – (5) require closure relations for the particle trapping and mobilization rates, suspension viscosity and permeabilities of the large and small pore channels, which are described below.

The permeability $k_{s,0}$ of a pack formed by trapped particles of diameter d_p is defined after Kozeny-Carman equation [9], while the permeabilities of large and small pore channels are defined as functions of trapped particle concentration after [3, 4]:

$$k_{s,0} = \frac{(1-C_{\max})d_p^2}{180C_{\max}^2}, \ k = k_0\left(1 - \frac{\sigma}{\varphi_0 C_{\max}}\right)^3, \ k_s = k_{s,0}\left(\frac{\sigma}{\varphi_0 C_{\max}}\right)^3 \tag{6}$$

During injection of a raw water into formation surrounding flooding wells, the volume fraction of particles is typically small, so that the suspension viscosity can be defined according to Einstein formula:

$$\mu_\gamma = \mu_{\gamma,0}(1 + 2.5C) \tag{7}$$

A key element of continuum models describing particle transport in porous media is the closure relation for the source term q_γ describing the particle trapping $q_{t,\gamma}$ and mobilization $q_{m,\gamma}$ rates

($q_\gamma = q_{t,\gamma} + q_{m,\gamma}$). We consider trapping of fines only due to plugging (and bridging) of the pore channels [2-5], while the particle mobilization rate $q_{m,\gamma}$ is described after [6]:

$$q_{t,\gamma} = \lambda_\gamma u_\gamma C,\ q_{m,\gamma} = -\alpha_\gamma \sigma (u_\gamma - u_{c,\gamma}) \theta (u_\gamma - u_{c,\gamma}) \tag{8}$$

Here, λ_γ and α_γ are the particle trapping and mobilization coefficients, respectively (determined experimentally), $u_{c,\gamma}$ is the critical velocity, and θ is the Heaviside function (takes zero value if its argument is negative and unity otherwise). The critical velocity $u_{c,\gamma}$ can be obtained analytically by considering the balance of forces acting on a single sphere attached to the wall of a round channel [10].

3. Simulations of fines transport in the formation surrounding an injection well

The wells are often turned to flooding regime after their oil production rate falls below a certain level so that they are hydraulically fractured. Therefore water flooding occurs not in the vicinity of the well, but it takes place along the surface of the attached hydraulic fracture. As the study is preliminary, we assume that the flooding runs in the linear reservoir regime, so that the streamlines of suspension in the rock surrounding the hydraulic fracture are straight lines perpendicular to the fracture surface. The flow configuration in this case is similar to linear reservoir regime during early stage of production [11], while the direction of the flow is opposite. As we consider cyclic flooding-cleanup regime of a well, it is reasonable to assume that the pressure drop along the hydraulic fracture is much smaller as compared to the difference between the bottom-hole pressure and far-field pressure in the formation. This is justified by extremely low permeability of the formation (1 – 100 mD) as compared to that of open fracture during flooding and proppant-filled fracture during cleanup. This allows us to apply 1D suspension filtration model formulated above for estimation of the impact of particle-laden water injection on the permeability dynamics of the rock surrounding hydraulic fracture in flooding wells and, consequently, to estimate the corresponding dynamics of water flooding rate.

In the framework of the model (1) – (8), we carried out numerical simulations of suspension filtration in a porous medium representing the rock formation surrounding an injection well. The length of the porous medium is assumed to be 1000 m, the permeability is $k = 10$ mD, the porosity is $\varphi_0 = 0.1$. At this stage we neglect the compressibility of the fluid and rock matrix to study the effect of permeability dynamics only, the flow is single-phase. Free parameters of the model (particle trapping λ and mobilization α coefficients) are varied in the range, which corresponds to that obtained by tuning the model against the experimental data on suspension filtration through sand columns [12].

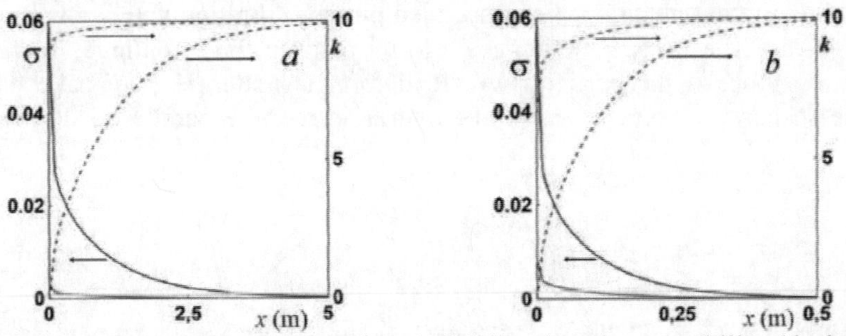

Figure 1. Trapped particle concentration (solid lines) and permeability (dashed lines) as a function of coordinate. Stage 1 is flooding with $\lambda = 1\,m^{-1}$, $t = 60\,s$, stage 2 is clean up with $\alpha = 0.1\,m^{-1}$, $t = 30\,s$ (a); stage 1 is flooding with $\lambda = 10\,m^{-1}$, $t = 60\,s$, stage 2 is clean up with $\alpha = 1\,m^{-1}$, $t = 20\,s$ (b).

The results of simulations of cyclic flooding and cleanup of the porous medium are shown in figure 1a. The first stage is flooding (particle-laden suspension is injected, only particle trapping is considered), while the second stage is cleanup (flow is reversed, only particle mobilization is considered). In figure 1b we show the simulation of a similar sequence, but with the smaller filtration velocity as well as the larger trapping and mobilization coefficients. The profile of trapped particle concentration for these two cases are similar, while the penetration distance in the latter case is an order of magnitude smaller as compared to the former one. It is found that in both filtration conditions the permeability is almost completely restored, while the period of cleanup is significantly smaller than that of the injection.

Simulations of injection and cleanup process of the near-wellbore zone taking into account both particle trapping and mobilization are carried out (figure 2). During the suspension filtration, trapping and mobilization rates eventually reach an equilibrium, which corresponds to a certain concentration of trapped particles σ_{eq}. This value can be calculated analytically by considering a steady-state distribution of trapped particle concentration in (4) assuming $u_c = 0$:

$$0 = \lambda u C - \alpha \sigma_{eq} u, \ \sigma_{eq} = \frac{\lambda C}{\alpha} \tag{9}$$

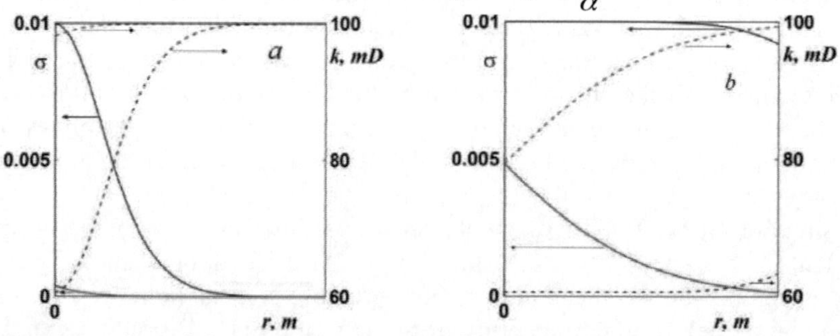

Figure 2. The trapped particle concentration (solid lines) and permeability (dashed lines) as a function of coordinate. Stage 1 (blue lines) $U = 10^{-4}\ m/s$, $C_0 = 10^{-3}$, $U = 10^{-4}\ m/s$, $\lambda = 10\ m^{-1}$, $\alpha = 1\ m^{-1}$, $t = 20.8\ h$ (*a*), $t = 5.8\ d$ (*b*), stage 2 (red lines) $U = 10^{-2}\ m/s$, $t = 33$ min (*a*), $U = 1\ m/s$, $t = 50\ s$ (*b*).

We carried out numerical simulations of the suspension filtration up to the moment when at the inlet boundary the concentration reached the value σ_{eq}. During further filtration, the concentration of trapped fines will reach the similar value along the whole porous sample (figure 2b). The results presented in figures 1,2 shows that the concept of periodic injection seems to be realistic and requires further investigation.

In the framework of the model (1) – (8), the total pore plugging due to trapped particles in the vicinity of the inlet is modelled ($\sigma(x=0) = \varphi_0 C_{\max}$, see figure 3). According to Eqs. (5), at the time instant corresponding to total plugging, the suspended particle filtration velocity turns to zero, and the particles stop to invade the porous medium, while the carrier fluid continues to flow. During the consequent filtration process, the particles will tend form an external filter cake provided that the component of the suspension velocity inside the hydraulic fracture parallel to the fracture surface is small enough.

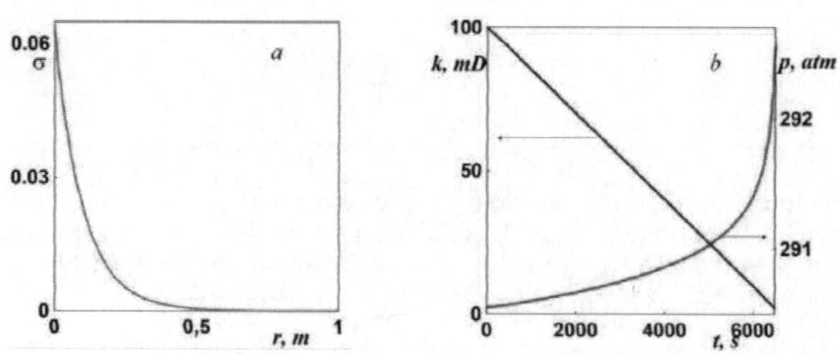

Figure 3. Particle concentration along the porous medium (*a*) as well as inlet pressure (red curve) and permeability (blue curve) against time (*b*) during simulations of total pore plugging. $U = 10^{-4}\ m/s$, $C_0 = 10^{-2}$, $\lambda = 10\ m^{-1}$, $\alpha = 0$, $t = 1h\ 48m$ (*a*).

The model (1) – (8) was developed with the aim to describe permeability dynamics of porous medium during transport of externally-introduced fines. The injection of particle-free fluids into natural cores can also cause a significant permeability reduction due to migration of mineral fines detached from pore matrix. The formation damage can be severe depending on ionic conditions and rock composition, so that it can exceed significantly the damage due to injection of external fines [1, 6, 10]. Therefore, the model (1) – (8) needs to be generalized to take into account both permeability reduction due to injection of external fines, as well as generation and transport of internal mineral fines. The current study will be continued after the generalization of the suspension filtration model to describe the permeability dynamics both by external and internal fines, and after finding the values of free parameters by tuning against experimental data produced from natural cores.

4. Conclusions

The study presents a methodology for optimization of hydraulically fractured water flooding wells based on mathematical modeling of suspension filtration in the formation surrounding the attached hydraulic fracture. Cycling regimes of injection and production are investigated. When particle trapping during injection and particle mobilization during cleanup are in effect, then cleanup during production is feasible for half the time of injection. When both trapping and mobilization are taken into account during injection and cleanup, then at a fixed flow rate the cleanup stage takes the time by two orders of magnitude longer. As a recommendation, the production needs to be done with much higher rate than injection in order to cleanup the formation effectively. With a increase in the trapping coefficient, the radius of the damaged zone near the hydraulic fracture decreases (by the same factor). The study will be continued after generalization of the model to describe formation damage due to migration of internal fines and tuning against proper experiments with natural cores.

Acknowledgements

The authors are grateful to the management of Gazpromneft NTC for the support of this work. Startup funds of Skolkovo Institute of Science and Technology are gratefully acknowledged.

References

[1] Civan F 2007 Reservoir Formation Damage (Oxford: Gulf Professional Publishing)
[2] Iwasaki T 1937 *J. Am. Water Work Ass.* **29** 1591–602.
[3] Herzig J P, Leclerc D M and Le Goff P L 1970 *Industrial and Engineering Chemistry* **625** 8–35
[4] Boek E S, Hall C and Tardy P.M.J. 2012 *Transport in Porous Media* **91** 479–508
[5] Zamani A and Maini B 2009 *Journal of Petroleum Science and Engineering* **691** 71–88
[6] Gruesbeck C and Collins R E 1982 *SPE Journal* **22** 847–56
[7] Rajagopalan R and Tien C 1976 *AIChE Journal* **22** 523–33
[8] Boronin S A, Osiptsov A A and Tolmacheva K I 2015 *Fluid Dynamics* **50** 759–68
[9] Carman P C 1937 *Transactions-Institution of Chemical Engineers* **15** 150–66
[10] Boronin S A, Tolmacheva K I, Osiptsov A A, Orlov D M, Koroteev D A, Sitnikov A N, Yakovlev A A, Belozerov B V, Belonogov E V and Galeev R R 2017 *SPE Russian Petroleum Technology Conference* (187806-EN).
[11] Cinco-Ley H and Samaniego-V F 1981 wells *J. Pet. Tech.* **33** 1749–66
[12] Bradford S A and Mehdi B 2006 *Journal of Contaminant Hydrology* **82** 99–117

IOP Conf. Series: Journal of Physics: Conf. Series **925** (2017) 012010 doi:10.1088/1742-6596/925/1/012010

Varieties of the gas driven water rivulet flow regimes in the minichannels

O V Svetlichnaya[1,2], V V Cheverda[1,2] and E. O. Kirichenko[1,2]

[1] Kutateladze Institute of Thermophysics SB RAS, 1, Ac. Lavrentyieva ave., 630090 Novosibirsk, Russia
[2] Novosibirsk State University, Pirogova st., 2, 630090, Novosibirsk, Russia

E-mail: sovatomsk.93@gmail.com

Abstract. This paper reports on an experimental study of the gas (nitrogen) driven water rivulet flow regimes in the minichannel. The width of the liquid nozzle is 10 mm, the height and width of minichannel are 1.61 ±0.02 mm and 30 mm correspondently. The new flow regime has been found by authors. Its occurrence is connected with influence of relatively large static wetting angle of water on the flow and comparably small height of the channel. It was named "rivulet bridge". The paper presents a map of the various rivulet flow regimes, characterized by different geometry and different kinds of surface deformations at temperature 40°C.

1. Introduction

Solving of complex computing tasks requires the use of powerful electronics. Due to the large number of transistors on a limited surface, such electronic devices have a high heat dissipation, that should be compensated by cooling systems of high efficiency. Thus, it is possible to prevent the transistors from burning out and, consequently, it will help to protect electronic devices from breakdown.

Two-phase systems have a great potential for use in cooling systems. The first studies, dedicated to the development of a compact, advanced cooling technology based on two-phase flows can be found here [1]. The macroscopic interface shape is shown to be sensitive for slip length comparable with the initial film thickness. The slip is found to increase evaporative mass flux with the maximum velocity value in the micro-region. The review on the regimes of two-phase flows in the channels of different geometries under different conditions is presented in [2]. The experimental investigations of two-phase flows in rectangular channels are given in [3, 4]. In particular, the rivulet flow, a kind of film flow, has a number of advantages useful for cooling. The main advantage of rivulet flow is the presence of two contact lines, in the area of which the heat transfer coefficient is several times greater than the average [5]. Cheverda et al in [6] also showed, that the maximum heat flux density takes place in the area of the contact line and exceeds the average heat flux density from the entire. The lower energy consumption for pumping the gas-liquid mixture through the channel is also an important advantage of the rivulet flow in comparison with the film flow.

A liquid film flow refers to a separate flow in which the fluid flows over the entire surface of the substrate. At a rivulet flow only a portion of the substrate is wetted by liquid, while the other part remains dry. Due to abnormally high heat transfer coefficient in the area of three phase contact line [7, 8], a rivulet flow is perspective solution for cooling purposes. The fluid viscosity is significantly

IOP Conf. Series: Journal of Physics: Conf. Series **925** (2017) 012010 doi:10.1088/1742-6596/925/1/012010

higher than the gas viscosity. Because of it rivulet flow has lower energy consumption comparing with film flow. Despite the increased interest to this kind of problems, a number of available experimental [9, 10] and theoretical works [11] is quite small and extremely insufficient for comprehension of the physics of the processes associated with the heat transfer in gas driven rivulet flow of water in the minichannel.

Studies related to rivulet flows can also be useful for studying the film flow. For example, a flowing down film can split into several rivulets due to the influence of temperature, external disturbances, periodic artificial perturbations, etc [12, 13].

To create an effective cooling system, a research of water rivulet flow regimes in minichannels at different temperatures is required. It is also important to obtain the geometric parameters of the rivulet flow and to measure the evaporation of water under different conditions.

2. Experimental setup

Experimental setup is presented in Fig.1. The following are the main elements of the working setup: gas supply system; fluid delivery system; the system for maintaining a predetermined pressure in the test section and a regenerative system; the power supply system (not shown); the water thermostabilization system; the data collecting system; the optical system; the replaceable test section with the temperature-controlled substrate of the minichannel (from 10 to 100°C).

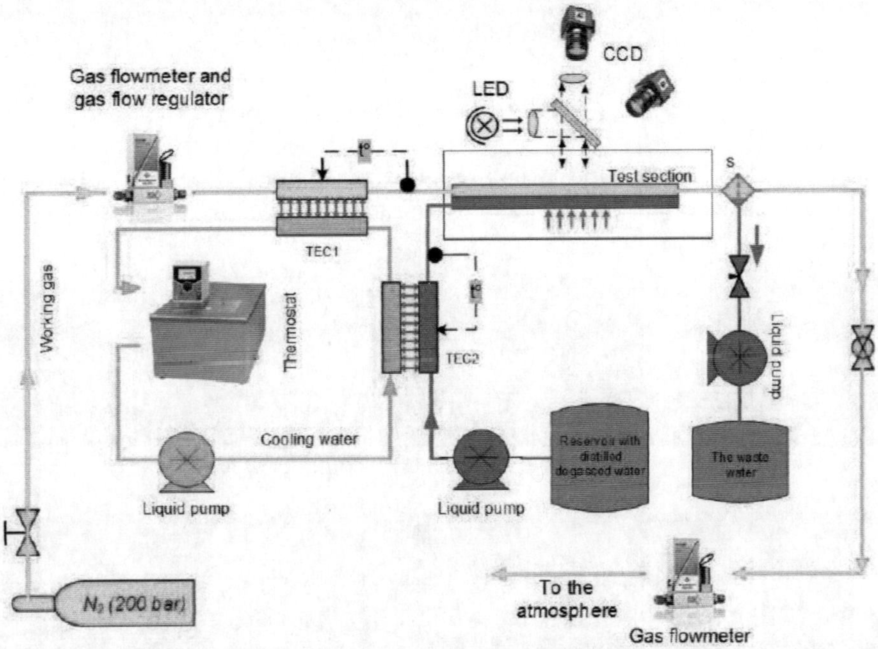

Figure 1. The sketch of experimental setup for studying the nitrogen driven isothermal rivulet flow regimes of the working fluid (water) in a horizontal minichannel

The basis of the test section (Figure 2) is the frame (5) and cover (4) made of textolite. With the help of previously calibrated thermistors, initial temperature of the liquid (10) and gas (11) at the inlet of minichannel, as well as temperature of the substrate (3) are maintained at the value of 40°C with an accuracy of 0.2°C using Peltier elements (7) which are cooled by water heat exchanger (8). The inlet temperatures of the gas and liquid as well as outlet temperature of the mixture, the temperature of the substrate (2) and the ambient temperature are measured by thermal sensors.

The top wall of the channel should be transparent for visualization of the process and because of it test cell is covered by optical glass (1). Geometrical parameters of the channel are the following: the

IOP Conf. Series: Journal of Physics: Conf. Series **925** (2017) 012010 doi:10.1088/1742-6596/925/1/012010

width is 30 mm and a height is 1.61±0.02 mm. Liquid from the inlet gets to the liquid nozzle (9) and after that gets into the minichannel. The nozzle width for rivulet flow is equal to 10 mm. Gas (nitrogen) is fed into the test section under pressure through the nozzle (11) and drives the fluid rivulet along a rough substrate made of stainless steel (3). The surface was processed with abrasive M-40. Further, mixture of water and nitrogen is released through the orifice (6).

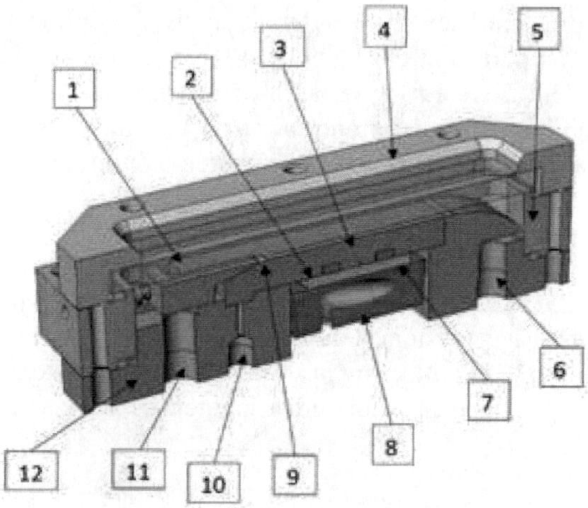

Figure 2. Sketch of the test section

Employing DSA-100 system and using sessile drop method (see Fig. 3) we measured the advancing wetting contact angle 140° (Figure 3) and receding wetting contact angle 10° (Figure 4). Hysteresis amounted to 130°.

Figure 3. Advancing contact angle

Figure 4. Receding contact angle

Also, we measured surface roughness for a better understanding of the interaction between the substrate and the liquid. The arithmetic average of the absolute values of the profile deviations within the basic length (Ra) was 0.458 μm.

3. Results and Discussion

We obtained regime map for various gas driven water rivulet flow regimes in the minichannel. For convenience geometrical and surface deformations are presented on different Figures (Figure 5 and Figure 6). Maps were obtained at a temperature of 40°C. We observed a new form of geometric deformations, which has not previously been detected for the FC-72 fluid flow. We called it "rivulet

IOP Conf. Series: Journal of Physics: Conf. Series **925** (2017) 012010 doi:10.1088/1742-6596/925/1/012010

bridge" (Figure 6(1),6(2),6(3)). In this regime, the rivulet does not have a contact with the side walls of the channel, however, it touches the upper and lower walls of the channel.

These kinds of geometric deformations were revealed: divergent (Figures 6(2), 6(5)), in which the flow width increases by more than 10% compared to the initial width; straight (Figures 6(3), 6(6)), the width of which remains within 10% compared to the initial one; pulsating rivulet bridge (Figure 6(1)), in which the appearance of the drop completely filling the channel alternates with the drainage of the channel; and unstable flow regime (Figure 6(4)), which represents two or more narrow rivulets.

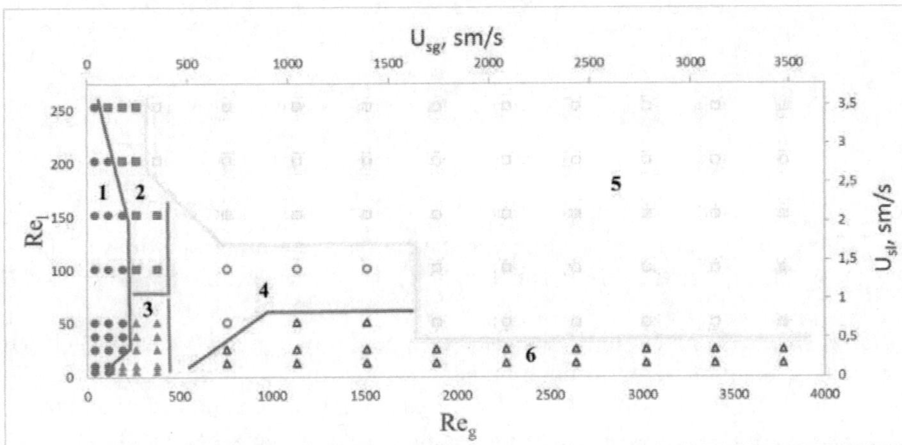

Figure 5. Flow regime map representing varieties of the geometry of the nitrogen driven rivulet flow of water in the minichannel at 40°C. 1 – divergent rivulet bridge, 2 – straight rivulet bridge, 3 – pulsating rivulet bridge, 4 – divergent flow regime, 5 – straight flow regime, 6 – unstable flow regime

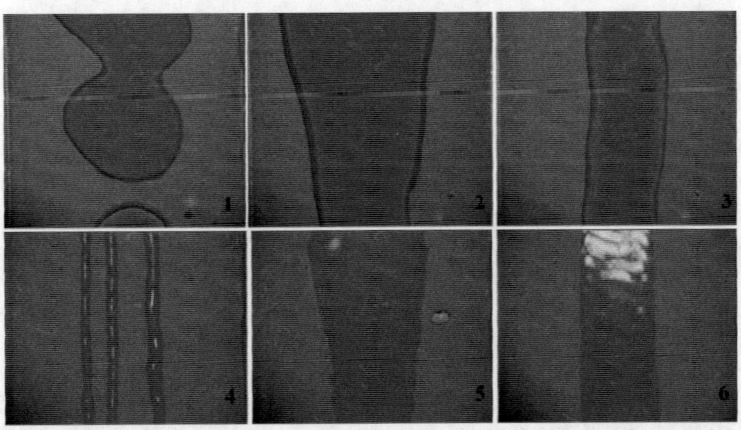

Figure 6. Varieties of the geometry of the nitrogen driven rivulet flow of water in the minichannel. 1) - pulsating rivulet bridge, 2) - divergent rivulet bridge, 3) - straight rivulet bridge, 4) - unstable flow regime, 5) - divergent flow regime, 6) - straight flow regime

For better visualization we have also constructed a map representing various surface deformation regimes (Figure 7). We observed both the regimes already observed earlier for the flow of a shear-driven rivulet flow of FC-72 in minichannel (2D waves and 3D waves), and a new type of surface deformation braiding regime [14] that is caused by a larger surface tension coefficient.

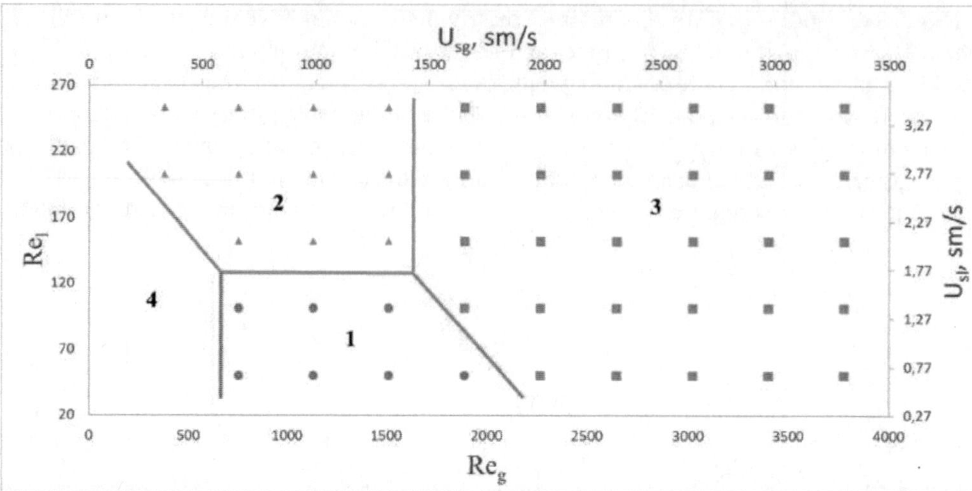

Figure 7. Flow regime map representing varieties of the surface deformations of the nitrogen driven rivulet flow of water in the minichannel at 40°C. 1 – 2D waves, 2 – braiding, 3 – 3D waves

In conclusion it should be noted, the emergence of a new flow regime we associated with the fact that the static wetting angle of water is much higher than that in the fluorinert fluid FC-72. It means that the height of the water rivulet when driven by the gas in the minichannel is much greater than that of FC-72 fluid. Hence, we can conclude that the static wetting angle significantly affects the rivulet flow regimes.

Acknowledgments

The work was financially supported by the Ministry of Education and Science of the Russian Federation (Agreement 14.613.21.0067, project identifier RFMEFI61317X0067).

References

[1] Gatapova E., Kabov O. 2007 *Microgravity science and technology* **19** 132–134
[2] Chinnov E.A., Ron'shin F.V., Kabov O.A. 2015 *Thermophysics and Aeromechanics* **22** 265-284
[3] Chinnov E.A., Ron'shin F.V., Guzanov V.V., Markovich D.M., Kabov O.A. 2014 *High Temperature*, **52** 681-687
[4] Chinnov E.A., Ron'shin F.V., Kabov O.A. 2014 *Thermophysics and Aeromechanics* **21** 759-762
[5] Marchuk I., Karchevsky A., Surtaev A., Kabov O. 2015 *Int. J. Aerospace Eng*. **2015**
[6] Cheverda V., Marchuk I., Karchevsky A., Orlik E., Kabov O. 2016 *Thermophys. Aeromech*. **23**
[7] Stephan P., Brandt C. 2004 *Heat Transfer Engng*. **25** 78–85.
[8] Potash M., Wayner P.C. 1972 *Int. J. Heat Mass Transfer* **15** 1851–1863.
[9] Kuznetsov V.V., Vitovskii O.V., Krasovskii V.A. 2007 *High Temp*. **45** 68.
[10] Cheverda V. V., Glushchuk A., Queeckers P., Chikov S. B., Kabov O.A. 2013 *Microgravity science and technology* **25** 73-81.
[11] Bartashevich M.V., Kabov O.A., Kuznetsov V.V., 2010 *Fluid dynamics*, **45** 924-929.
[12] Chinnov E. A., Shatskii E. N. 2014 *High Temperature*, **52** 461-464
[13] Chinnov E. A., Shatskii E. N. 2014 *Technical Physics Letters*, **40** 817-819
[14] Mertens K., Putkaradze V., Vorobieff P. 2005 *Journal of Fluid Mechanics*, **531** 49-58.

Experimental study and empirical prediction of fuel flow parameters under air evolution conditions

E E Kitanina[1], E L Kitanin[1], D A Bondarenko[2], P A Kravtsov[3], M M Peganova[4], S G Stepanov[4] and V L Zherebzov[4]

[1]Peter the Great St.Petersburg Polytechnic University, Polytechnicheskaya 29, Saint-Petersburg, 195251 Russia
[2]Moscow Aviation Institute, 4 Volokolamskoe shosse, Moscow, 125993 Russia
[3]National Research Center "Kurchatov Institute" Petersburg Nuclear Physics Institute, Gatchina, 188300 Russia
[4]RSC "Applied Chemistry", Krylenko street 26A, Saint-Petersburg, 193232 Russia

E-mail: kkitanina@mail.ru

Abstract. Air evolution in kerosene under the effect of gravity flow with various hydraulic resistances in the pipeline was studied experimentally. The study was conducted at pressure ranging from 0.2 to 1.0 bar and temperature varying between -20°C and +20°C. Through these experiments, the oversaturation limit beyond which dissolved air starts evolving intensively from the fuel was established and the correlations for the calculation of pressure losses and air evolution on local loss elements were obtained. A method of calculating two-phase flow behaviour in a titled pipeline segment with very low mass flow quality and fairly high volume flow quality was developed. The complete set of empirical correlations obtained by experimental analysis was implemented in the engineering code. The software simulation results were repeatedly verified against our experimental findings and Airbus test data to show that the two-phase flow simulation agrees quite well with the experimental results obtained in the complex branched pipelines.

1. Introduction

Air evolution in kerosene flow may cause issues in aircraft fuel systems. The fuel supplied to the aircraft during fueling is usually saturated with air under atmospheric pressure and ground temperature conditions. In cruise flight the ambient pressure drops, leading to the fuel oversaturation with air. Moreover the oversaturation of the fuel with air is large due to the temperature decrease down to -30°C. In the case of pump deselection, pressure in the pipeline with pure gravity flow is contingent on the ambient pressure and fuel level in the tank only. Air can evolve from fuel, resulting in a flow rate decrease. The air flow rate decline can be due to both the flow conditions (pressure, temperature) and the pipeline design. Substantial air evolution may occur as oversaturated fuel passes through the flow limiter (diaphragm) or other local resistance.

The gas phase effect on pressure losses related to friction and the pressure leveling component can be quite reliably handled within the Chisholm model based on the Lockhart and Martinelli approach [1]. The much less studied issues are the calculation of the pressure losses at the local resistances

IOP Conf. Series: Journal of Physics: Conf. Series **925** (2017) 012011 doi:10.1088/1742-6596/925/1/012011

under air release conditions and establishing the gas fraction in the flow downstream of the local resistances [2]. Moreover the transition from stratified to foamy flow can occur in a tilted pipeline segment. However the standard calculation methods mentioned above are devised for the foamy flow mode and do not allow the transition point coordinate determination. Numerical simulation of two-phase flow in such systems still remains an intractable task. As an alternative to this approach, we developed a flow simulation tool that can take into account the experimental data on air evolution in pipeline elements and pressure losses in individual segments.

2. Results and discussion

2.1. Experimental technique and pipeline designs

More than 200 experiments were performed to study the dissolved air evolution process and pressure losses in individual pipeline elements. The experiments were conducted using TS-1 jet fuel (substitute for Jet A1 fuel) in several test pipelines. The study was performed at pressure ranging from 0.2 to 1.0 bar and temperature varying between -20°C and +20°C. All test rigs included two fuel tanks and a working section of pipeline, isolated by a valve. In all cases the test rig was prepared for the test following the same steps. First, fuel was saturated by dry air at atmospheric pressure and temperature level in the supply tank. Then air was evacuated from the entire system. As a result the fuel in the supply tank became oversaturated with air. Finally, the valve was opened to let the fuel flow into the experimental pipeline. Fuel temperature and pressure along the pipeline working section and fuel flow rates and pressures at the pipeline inlet were measured for every experiment. A special technique based on the measurement of the dissolved air concentration was applied for air evolution calculation. The mass flow quality of evolved air was equated with the difference between dissolved air concentration values at the pipeline inlet and at the section under consideration. The dissolved air concentration was measured by chromatographs within the range of $5 \cdot 10^{-5} \div 2.5 \cdot 10^{-4}$ kg/kg. A fuel sample was supplied to chromatographs by a specifically designed sampling system. In addition to instrumental measurements, the two-phase flow modes were observed through clear pipeline segments.

Initially, the fuel flow was studied in an unbranched pipeline [3]. The experimental pipeline included vertical and horizontal segments connected by a smooth bend (see figure 1). A diaphragm was installed in the upper part of the working pipeline section. Two diaphragms with holes of different diameters were used. Temperature, pressure and dissolved air concentration were measured at several sections located after the diaphragm.

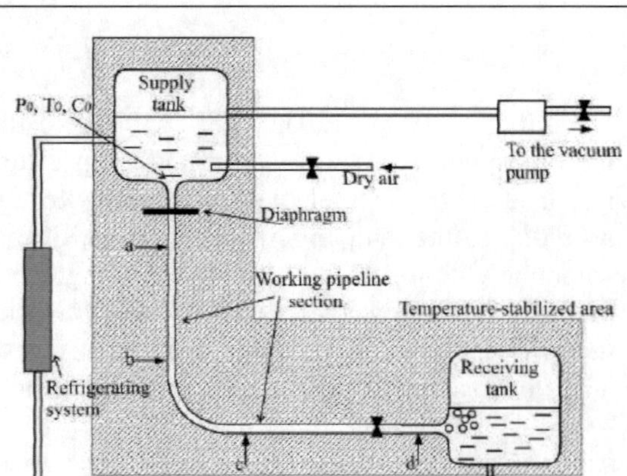

Figure 1. Test pipeline with various local loss segments (view from above)

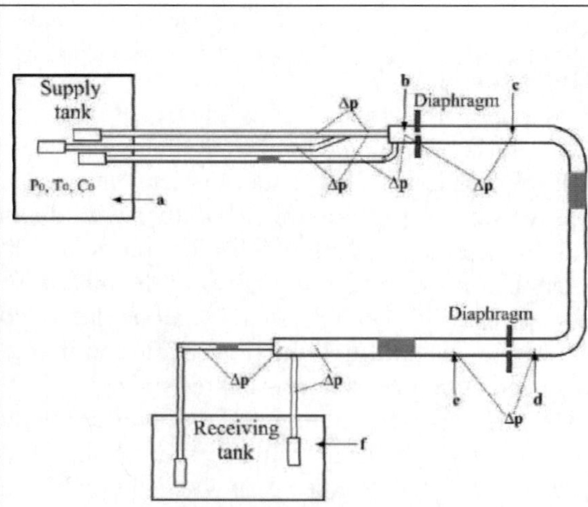

Figure 2. Branched pipeline simulating the refuelling system (view from above)

The second test rig [4] was built specifically for the study of air release at local resistances and assessment of their resistance value, therefore measurement of the flow mass quality downstream of each resistance appeared as a big challenge in the testing. A diaphragm installed downstream of the inlet diffuser was followed by a clear section of the same inner diameter, a set of local resistances, including a ball valve, a pipe expansion, a contraction, 90^0 bends and a ball valve. Specific tests were performed to study the effect of the tilted segment on the two-phase flow rate. The pipelines had clear sections for viewing and recording the two-phase flow behaviour downstream of the local resistances.

In further experiments, we built a branched pipeline to simulate a real refueling system [5]. A schematic view of the pipeline is shown in figure 2. The working pipe section contains two diaphragms, expansions, contractions and bends. The pipeline has branchings at the upper tank outlet and the lower tank inlet. Pressure, temperature and mass flow quality were measured at the sections a-f. In addition, differential pressure sensors were placed along the pipeline. A pipeline segment located between sections c and d was tilted by 10 and 20 degrees to verify the equations obtained by previous experiments.

2.2. Key experimental outcomes

Our measurements indicate that air evolution in oversaturated fuel starts as pressure drops abruptly to a certain boundary value. Figure 3 illustrates a fuel velocity dependence on environment pressure under gravity flow conditions. As can be seen from the figure, the fuel velocity decreases slightly due to pressure change from 100 kPa to 40 kPa. When the pressure value reaches the level of 20 kPa a dramatic velocity drop takes place. The flow drop is accompanied by an intensive air evolution process. The driving force is decreased due to effective flow density reduction. It turned out that the flow kinetic energy and the fuel temperature have an additive influence on this value. The method elaborated for boundary pressure evaluation is discussed in detail in [3].

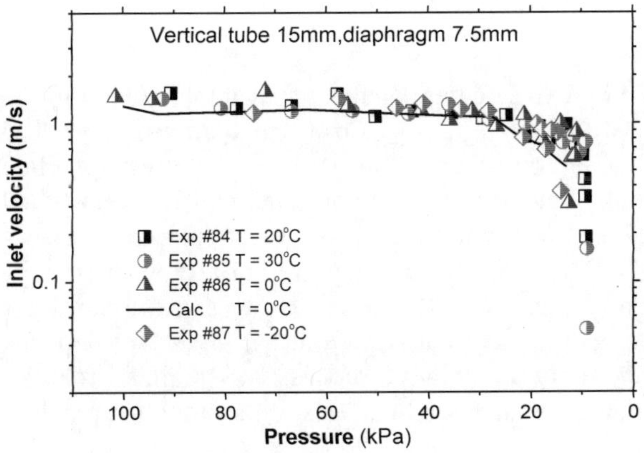

Figure 3. Fuel velocity depending on environment pressure value by gravity flow in a pipeline with a diaphragm

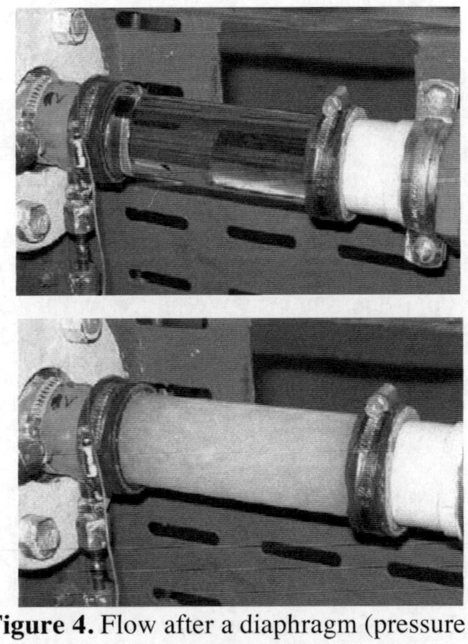

Figure 4. Flow after a diaphragm (pressure in the supply tank equal to 60kPa (above) and 20kPa (below)

We notice that almost no air evolves downstream of the segments generating a minor pressure drop, e.g. bends, contractions, expansions and branching, whereas the diaphragm provokes intensive air evolution in fuel flow (see figure 4). A large set of experimental data was analyzed to produce empirical correlations for the pressure drop and air evolution after the diaphragm. It was discovered

IOP Conf. Series: Journal of Physics: Conf. Series **925** (2017) 012011 doi:10.1088/1742-6596/925/1/012011

that the diaphragm pressure loss coefficient depends on the volume flow quality downstream of the diaphragm. This correlation provides adequate results in a wide range of diaphragm to pipeline diameters ratios [4].

As a result of experimental data analyze a special method of calculating the two-phase flow pressure in a titled pipeline segment was developed [5]. Conventional methods prove to be inefficient under the conditions at hand (very low mass flow quality and fairly high volume flow quality) due to co-existence of two flow modes in different parts of the titled segment. In many experiments we observed stratified flow in the upper part and foamy (bubble) flow in the lower part (figure 5). Though it is easy enough to calculate the pressure in each individual part, the flow mode transition point was a priori unknown. We built a complete calculation chain including pressure drop equations for both flow modes, an equation taking into account the change in kinetic energy during transition and an empirical equation for calculating the pressure drop in the tilted pipeline segment. This helped to estimate the stratified to foamy flow transition point coordinate.

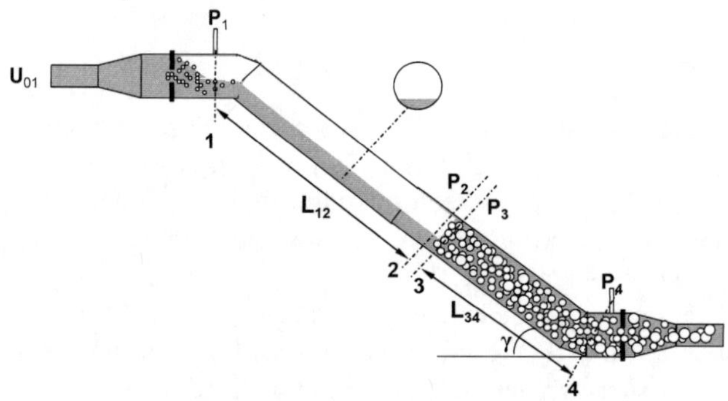

Figure 5. Flow modes in the tilted segment

3. Comparison of calculations and experimental data

The complete set of empirical correlations obtained by experimental analysis was implemented in the engineering code specifically developed for two-phase gravity flow calculations [6]. The computational algorithm is based on the balance (Bernoulli) equations written for every pipeline segment and for the whole pipeline system. Hence the velocity at the supply tank exit is proportional to the difference in fuel levels in the tanks and inversely proportional to the total pressure loss including two-phase flow acceleration loss, local loss and friction loss that are initially evaluated for each pipeline segment. We notice that the flow parameters (e.g. density, velocity and pressure loss coefficients) strongly depend on the evolved air content in the fuel. Air evolution in its turn depends on pressure, temperature and fuel velocity. Consequently an iterative process was realized for fuel velocity calculations. A postponed recursion technique provided computations of the flow in a pipeline with branching and junctions.

3.1. Flow parameters after a diaphragm

Our experiments on different test rigs demonstrated that in most cases a pressure drop after the diaphragm provokes intensive air evolution. The two-phase flow modes downstream of the diaphragm can lead to considerable changes in fuel flow rate in a pipeline.

Figures 6 and 7 show the distributions of pressure and evolved air mass flow quality along the pipeline (second test rig). The calculation results are compared with the experimental data. The charts reveal tendencies typical for the entire experimental series. A dramatic pressure drop is clearly seen after the diaphragm (figure 6). This corresponds to an increase in the amount of the evolved air (figure 7). As can be seen from the figure, the other local loss segments located downstream of the diaphragm have much weaker influence on the flow parameters as compared to the diaphragm.

Generally, we observe strong consistency between the calculated and experimental results for the pressure value downstream of the diaphragm (difference less than 8%) and fuel velocity (relative error not exceeding 10%). The flow quality is calculated with higher errors of 10-35%. This can stem from both the calculation model constraints and experimental errors.

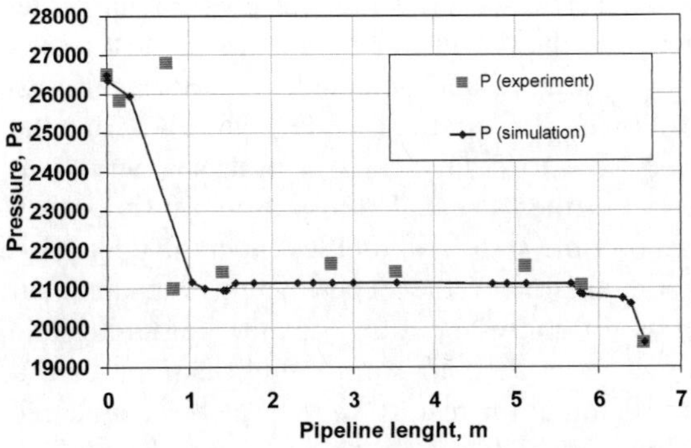

Figure 6. Distribution of pressure along the pipeline

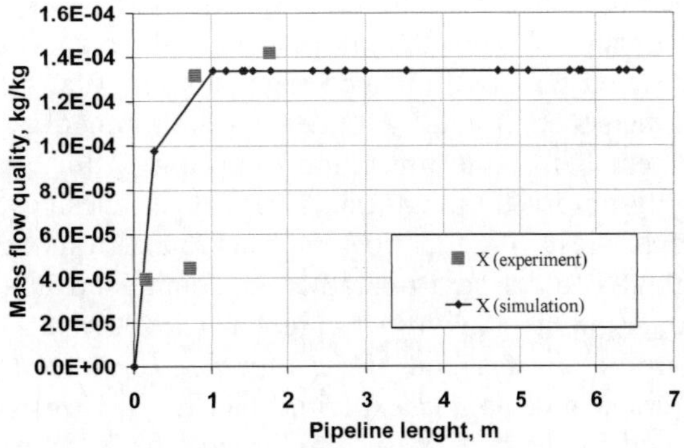

Figure 7. Distribution of mass flow quality along the pipeline

3.2. Flow in pipelines simulating refuelling system

This paragraph features the flow simulation results obtained for the pipelines that were specially designed to simulate refueling systems. All the tests addressed the gravity flow in the pipelines comprising two restrictors (diaphragms), expansions, contractions, bends, tilted parts and branches near inlet and outlet tanks.

Figures 8 and 9 illustrate the pressure and mass flow quality distribution along the main branch of the tilted pipeline corresponding to the third test rig used in our experiments. An abrupt pressure drop is observed downstream of the diaphragms. The two-phase flow regime changes between the first and second diaphragms, with the pressure remaining constant in the laminated flow segment and growing higher in the rest of the tilted segment where the foamy flow develops.

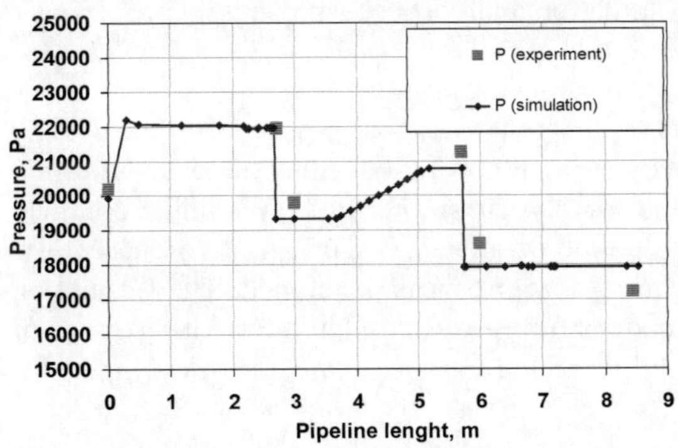

Figure 8. Distribution of pressure along the main branch of the horizontal pipeline along the main branch of the tilted pipeline

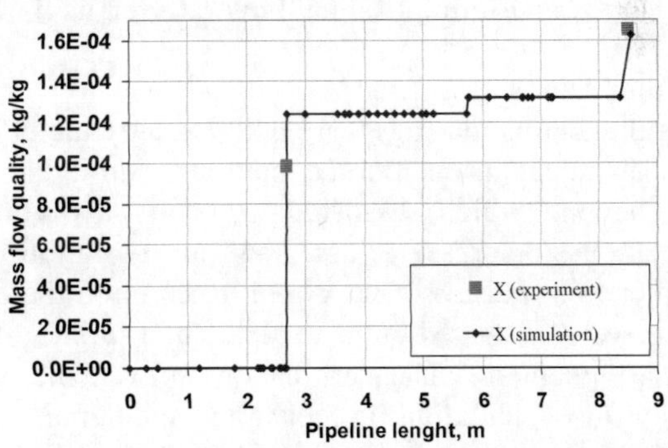

Figure 9. Distribution of mass flow quality along the main branch along the main branch of the tilted pipeline

The comparison of the calculated and experimental data for the tilted pipeline reveals a 4 to 23% error in the mean fuel flow velocity calculation. The error is largely due to the inaccuracy in calculating the flow transition coordinate, which proved to be quite a challenge for the test pipeline in question. It is evident that the uncertainties in the simulation of the two-phase flow mode transition in a tilted pipeline can result in inaccurate fuel flow rate values both for the entire pipeline and in its individual branches. On the other hand, these experiments revealed high flow pulsations that affected the measurement accuracy of the flow meters.

Our empirical equations and calculation results were further verified against the independent experimental results provided by Airbus. The test rig represents the left wing and centre tank refuel sub-system. Besides the full scale size, the main difference of the Airbus pipeline from our test rigs is connected with two inclined segments, the first tilted by 5 degrees and bounded by two diaphragms and the second tilted by 39 degrees and immediately following the first and ending with a horizontal segment before the branching. A change from stratified flow to foamy flow in both segments was clearly observed though transparent pipes. Besides, the experiments revealed high flow pulsations. In Airbus's tests, the fuel flow rate was found from the data on time steps and fuel levels in the tank. This method could lead to a rather substantial errors. A comparison of the experimental data and our calculations is shown in figure 10. The flow rate results differ by about 25% at the beginning and about 20% at the end of the process. The uncertainty may result both from simplifications of our empirical models and experimental errors. However the simulation results show that the developed method can be applied to complicated pipelines similar to real refueling systems.

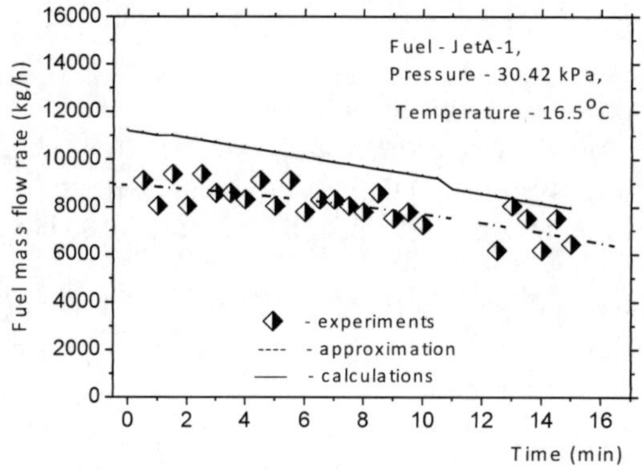

Figure 10. Fuel flow rate versus time. Simulation and Airbus test results

4. Conclusions

Several experimental rigs were built to study the effects of different local loss segments on the gravity fuel flow under air evolution conditions. Through these experiments, we established the oversaturation limit beyond which dissolved air starts evolving intensively from the fuel. We also obtained correlations for pressure losses and air evolution on pipeline segments. To predict two-phase flow parameters a special software tool based upon the empirical formulas was developed. The simulation results were compared to the experimental studies of steady and unsteady flow in different pipelines of complex geometry. The verifications proved the reliability of the empirical methods developed for gravity flow calculations in fuel systems under air evolution conditions.

References

[1] Chisholm D 1983 *Two-phase Flow in Pipelines and Heat Exchangers* (Godwin G)

[2] *Multiphase Handbook* 2005 Edited by Clayton T Crowe (Taylor and Francis)

[3] Kitanin É L, Kitanina E É, Zherebtsov V A, Merkulov O A and Peganova M M 2013 *J. Eng. Phys. Thermophys.* **86** 1426–33

[4] Kitanin É L, Kitanina E É, Zherebtsov V A, Peganova M M, Stepanov S G, Bondarenko D A and Morisson D 2016 *J. Eng. Phys. Thermophys.* **89** 1212–21

[5] Kitanin É L, Kitanina E É, Kravtsov P A, Bondarenko D A and Morrison D 2014 *J. Eng. Phys. Thermophys.* **87** 588–97

[6] Kitanina E E, Bondarenko D A and Morrison D 2014 *Proc. Int. Council Aeronaut. Sci.* **2014**_0254

IOP Conf. Series: Journal of Physics: Conf. Series **925** (2017) 012012 doi:10.1088/1742-6596/925/1/012012

Thermocapillary structures in a heated liquid film

S P Aktershev[1], E A Chinnov[1,2]

[1] Kutateladze Institute of Thermophysics, Lavrentiev Ave. 1, Novosibirsk 630090, Russia
[2] Novosibirsk State University, Pirogova St. 2, Novosibirsk 630090, Russia

E-mail: sergey-aktershev@mail.ru

Abstract. The system of the equations which in long-wave approach describes a non-stationary 3D non-isothermal film flow in the presence of thermocapillarity is derived. The used model, unlike lubrication theory, is applicable not only for small, but also for a moderate Reynolds number in a wide range of hydrodynamic and thermal parameters of a flow and does not assume a priori set a temperature profile in a film. Based on the derived equations the numerical simulation of stationary 3D thermocapillary rivulets in a liquid film flowing over a "semi-infinite" heater is executed.

1. Introduction

The study of non-isothermal liquid film flow is an important problem of hydrodynamics and heat transfer not only from a theoretical point of view, but also from practical standpoint, since the falling films are used in many industrial devices. In contrast to the isothermal falling liquid films, the heated film flows are poorly understood both in theoretically and experimentally. Dependence of the surface tension on temperature (thermocapillary Marangoni effect) leads to occurrence of the stress shear which influences velocity of a fluid. Thus, the hydrodynamics and a heat transfer appear the interconnected processes, it in essence complicates the theoretical analysis of a problem. Experimental investigations [1-4] of thin films falling down locally heated plates reveal the rivulet structures, horizontal platen and lateral waves. Thermocapillary instabilities in a heated liquid films induce a film thinning process causing the film rupture in regions of high surface temperature. This paper presents the results of numerical simulation of stationary 3D thermocapillary rivulets in a heated liquid film flowing over a "semi-infinite" heater.

2. Theoretical model

Let us consider a vertical laminar liquid film flowing over a "semi-infinite" heater and use Cartesian coordinate system with Ox axis directed downward, Oy axis directed normal to the plate, and Ox axis directed in spanwise. Let us assume the following basic simplifications, acceptable for a wide range of practically important flow conditions. 1) Temperature of the heater's surface $T_w = const$, the liquid surface is in contact with stagnant gas with temperature $T_g = const$. 2) The perturbation in the film is considered to be long-wave. 3) Density ρ, dynamic viscosity μ, liquid conductivity λ are assumed constant, but surface tension depends on temperature linearly $\sigma = \sigma_0 - \gamma(T - T_0)$. Based on the IBL model [5], taking into account thermocapillarity, we derive a system of equations which describes the unsteady 3D film flow

$$\frac{\partial q}{\partial t}+\frac{\partial J_1}{\partial x}+\frac{\partial J_{1,2}}{\partial z}=\frac{3}{Re_m}\left(h\left(\sin\theta-\cos\theta\frac{\partial h}{\partial x}\right)-\frac{Ma}{2}\frac{\partial T_s}{\partial x}-\frac{q}{h^2}\right)+Weh\left(\frac{\partial^3 h}{\partial x^3}+\frac{\partial^3 h}{\partial x\partial z^2}\right),$$

$$\frac{\partial m}{\partial t}+\frac{\partial J_2}{\partial z}+\frac{\partial J_{1,2}}{\partial x}=-\frac{3}{Re_m}\left(h\cos\theta\frac{\partial h}{\partial x}+\frac{Ma}{2}\frac{\partial T_s}{\partial z}+\frac{m}{h^2}\right)+Weh\left(\frac{\partial^3 h}{\partial z^3}+\frac{\partial^3 h}{\partial z\partial x^2}\right), \quad (1)$$

$$\frac{\partial h}{\partial t}+\frac{\partial q}{\partial x}+\frac{\partial m}{\partial z}=0,$$

$$\frac{\partial T}{\partial t}+u\frac{\partial T}{\partial x}+w\frac{\partial T}{\partial z}+\frac{V}{h}\frac{\partial T}{\partial \eta}=\frac{1}{h^2\,\mathrm{Re}_m\,\mathrm{Pr}}\frac{\partial^2 T}{\partial \eta^2}, \quad (2)$$

$$\left.\left(\frac{\partial T}{\partial \eta}+BihT\right)\right|_{\eta=1}=0,\quad T\big|_{\eta=0}=1. \quad (3)$$

Here $h(t,x,z)$ is the film thickness, $T(t,x,\eta,z)$ is the liquid temperature, T_s is the interface temperature, $q(x,z,t)=\int_0^h udy$ and $m(x,z,t)=\int_0^h wdy$ are the liquid flow rates along the Ox and Oz axes, $u=\frac{3q}{h}\left(\eta-\frac{\eta^2}{2}\right)-\frac{Ma}{4}(3\eta^2-2\eta)h\frac{\partial T_s}{\partial x}$ and $w=\frac{3m}{h}\left(\eta-\frac{\eta^2}{2}\right)-\frac{Ma}{4}(3\eta^2-2\eta)h\frac{\partial T_s}{\partial z}$ are velocity components along Ox and Oz axes, $\eta=y/h$,

$$J_{1,2}=\frac{6mq}{5h}-\frac{Ma}{40}h\left(q\frac{\partial T_s}{\partial z}+m\frac{\partial T_s}{\partial x}\right)+\frac{h^3Ma^2}{120}\frac{\partial T_s}{\partial x}\frac{\partial T_s}{\partial z},\quad J_2=\frac{6m^2}{5h}-Ma\frac{mh}{20}\frac{\partial T_s}{\partial z}+\frac{h^3}{120}\left(Ma\frac{\partial T_s}{\partial z}\right)^2,$$

$$J_1=\frac{6q^2}{5h}-Ma\frac{qh}{20}\frac{\partial T_s}{\partial x}+\frac{h^3}{120}\left(Ma\frac{\partial T_s}{\partial x}\right)^2,$$

$$V=\left(\frac{\partial q}{\partial x}+\frac{\partial m}{\partial z}\right)\left(\eta-\frac{3\eta^2}{2}+\frac{\eta^3}{2}\right)-\frac{Ma}{4}(\eta^2-\eta^3)\left(\frac{\partial}{\partial x}\left(h^2\frac{\partial T_s}{\partial x}\right)+\frac{\partial}{\partial z}\left(h^2\frac{\partial T_s}{\partial z}\right)\right),\quad \Delta h=\frac{\partial^2 h}{\partial x^2}+\frac{\partial^2 h}{\partial z^2}.$$

Here, unperturbed film thickness h_m is used as the distance scale. We also introduced scales for velocity $u_m=gh_0^2/3\nu$, time $t_m=h_m/u_m$, flow rate $q_m=u_m h_m$, and temperature $T_m=T_w-T_g$. Relation (3) gives boundary conditions for the equation (2).

Non-isothermal film flow is determined by the following dimensionless criteria: Reynolds number $\mathrm{Re}_m=gh_m^3/3\nu^2$; Weber number $We=(3Fi/\mathrm{Re}_m^5)^{1/3}$; Kapitza number $Fi=\sigma^3/\rho^3 g\nu^4$; Marangoni number $Ma=\gamma T_m/\mu u_m$; Bio number $Bi=\alpha h_m/\lambda$; and Prandtl number Pr. Since h_m~$\mathrm{Re}_m^{1/3}$, u_m~$\mathrm{Re}_m^{2/3}$, than $Bi=Bi^*\,\mathrm{Re}_m^{1/3}$ and $Ma=Ma^*/\mathrm{Re}_m^{2/3}$, where the dimensionless complexes of Bi^* and Ma^* are determined only by the liquid properties and heating conditions $Bi^*=\alpha(3\nu^2/g)^{1/3}/\lambda$, $Ma^*=\gamma(3/g\nu)^{2/3}(T_W-T_g)/\mu$.

3. Simulation results

Thermocapillary rivulets in heated vertical film are obtained by solving of the equations (1) and (2) with applying the finite difference method. The numerical algorithm is the same used in [5, 6] for 2D film flow. The formation of rivulets was simulated through small perturbation periodic along Oz axes, which was imposed on unperturbed two-dimensional flow by two ways: 1) by periodic roughness of edge of the heater, i.e. $x_{heater}=x_0+A\cos(2\pi\cdot z/L_z)$; 2) by periodic variation of the flow rate in entry

IOP Conf. Series: Journal of Physics: Conf. Series **925** (2017) 012012 doi:10.1088/1742-6596/925/1/012012

of a calculation area, i.e. $q = q_0(1 + A\cos(2\pi \cdot z / L_z))$. The equations (1), (2) were solved by the relaxation method; unperturbed 2D heated film flow on "semi-infinite" heater was used as the initial condition. The calculation was ended upon establishing 3D rivulet structure, not changing in time. Periodic rivulet structure was calculated at various values of the parameters Re_m, Bi^*, Ma^* and the period L_z. The calculations were performed for a film section with a width equal to a single period, i.e., the computational domain represented a rectangle $0 \le x \le X_{end}$, $-L_z/2 \le z \le L_z/2$. On the x-axis where the rivulet crest is arranged, the symmetry condition $\partial/\partial z = 0$ was provided, while the lateral boundaries of the computational domain (at $z = \pm L_z/2$) were provided by periodicity condition along z: $m = 0$, $\partial q/\partial z = 0$, $\partial h/\partial z = 0$, $\partial T/\partial z = 0$. Unperturbed thickness, flow rate, and temperature were set on the inlet (at $x = 0$).

In Fig. 1 the rivulet structure is visible, that branch off from the horizontal platen available in the edge of a heater (two periods along z are shown). In a gap between rivulets a film thickness is much less than unperturbed value hm. At so high value Ma* = 120 the structure immediately behind the platen is already developed, and amplitude of rivulet almost does not change downstream.

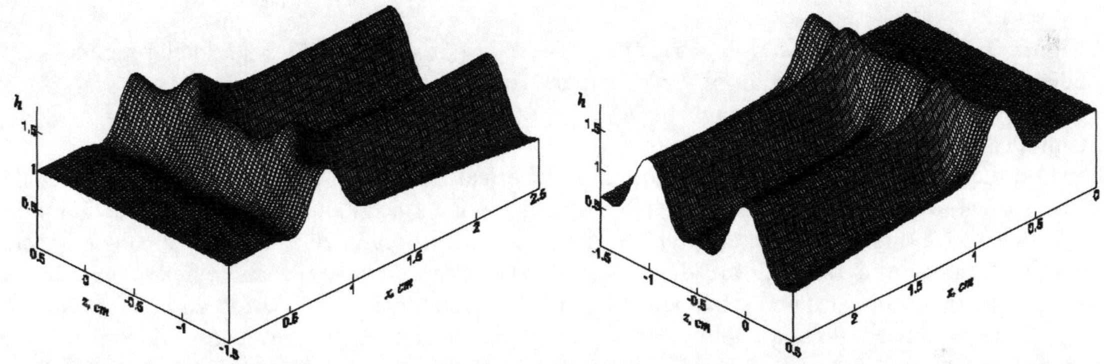

Figure 1. Rivulet structure at $Re_m = 1$, $Ma^* = 120$, $L_z = 1$ cm, $Bi^* = 0.2$ (periodic roughness of edge of the heater is set). The front view (left) and rear view (right)

In Fig. 2 the rivulet structure is shown at same value Ma *, as on Fig. 1, but at essential large value Rem. From figure it follows, that growth of Rem leads to essential increasing of the distance of the development of perturbation, and amplitude of rivulet increase downstream. This effect is quite clear, as parameter Rem characterizes inertial properties of a film.

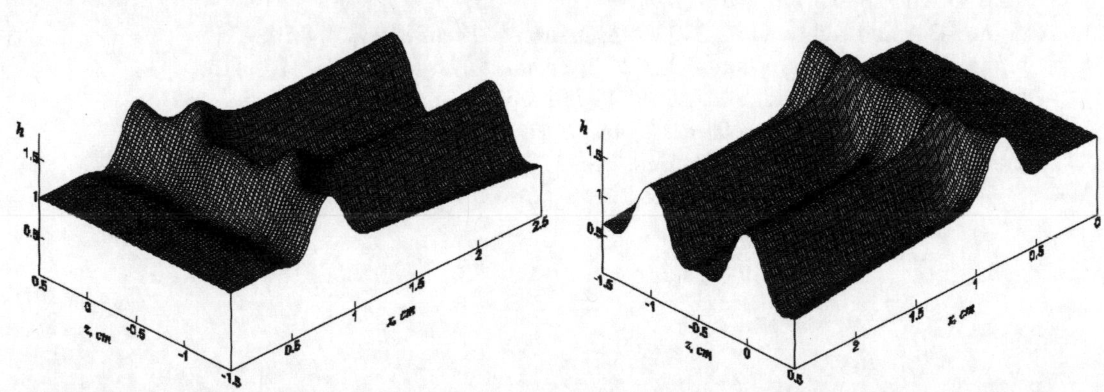

Figure 2. Rivulet structure at $Re_m = 5$, $Ma^* = 120$, $L_z = 1.2$ cm, $Bi^* = 0.2$ (periodic variation of the flow rate in entry of a calculation area is set). The front view (left) and rear view (right)

Development of rivulet structures happens only if period L_z belongs to range of instability with respect to spanwise perturbations. In Fig. 3 the case, when flow parameters do not belong the range of instability (for the spanwise perturbations) is shown. Provided that the spanwise perturbation, set on an edge of a heater, does not develop therefore the rivulets are damped downwstream.

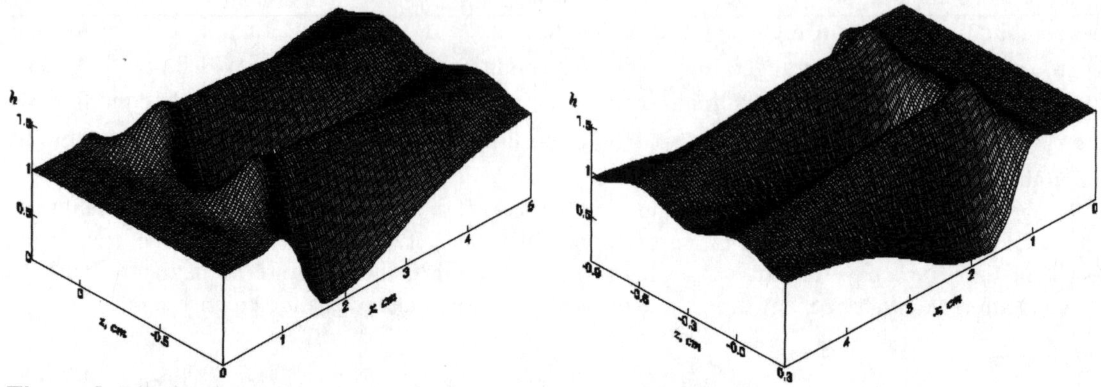

Figure 3. Rivulet structure at $Re_m = 51$, $Ma^* = 80$, $L_z = -.26$ cm, $Bi^* = 0.2$ (periodic roughness of edge of the heater is set). The front view (left) and rear view (right).

4. Conclusions

Based on the developed model, we carried out a numerical simulation of thermocapillary 3D rivulets on the surface of a liquid film flowing over a heater with a constant temperature. The numerical simulation of 3D rivulet structure which develops downstream due to the small transversal perturbation is carried out. At preset values of Marangoni number and Reynolds number the development of rivulet structures happens only in the event that space period L_z belongs to some range. If the parameters do not belong to this range, then the rivulet structure damps downwards on a stream.

Acknowledgments
The authors acknowledge the support by the Russian Foundation for Basic Research (projects No. 15-01-04320 and No. 15-01-06702).

References

[1] Kabov O A 1998 *Thermophys. Aeromech.* **5** 547–551
[2] Chinnov E A and Kabov O A. 2003 *J. Appl. Mech. Techn. Phys.* **44** 708–715
[3] Chinnov E A and Zhukovskaya O V 2006 *Techn. Phys. Letters* **32** 410–413
[4] Demekhin E A and Shkadov V Ya 1984 *Fluid Dynamics.* **19** 689–695
[5] Aktershev S P 2013 *Thermophys. Aeromech.* **20** 1–16
[6] Aktershev S P 2012 *J. Engin. Thermophys.* **21** 36

IOP Conf. Series: Journal of Physics: Conf. Series **925** (2017) 012013 doi:10.1088/1742-6596/925/1/012013

Evaporation and condensation of rarefied gas between two parallel condensed phases

A Polikarpov[1] and I Graur[2]

[1] Ural Federal University, 620000, str. Lenina 51, Ekaterinbourg, Russia
[2] Aix-Marseille Université, CNRS, IUSTI UMR 7343, 13453 Marseille, France

E-mail: Alexey.Polikarpov@urfu.ru

Abstract. Evaporation and condensation processes between two parallel condensed phases are simulated using the kinetic models (S-model and BGK model). The compressible Navier-Stokes equations subjected to the temperature and pressure jump conditions on the liquid-vapor interphase are also solved numerically. The results obtained by the different models are compared. It was found that the Navier-Stokes model subjected to the jumps conditions provide the results close to the kinetic simulations results.

1. Introduction

The evaporation and condensation phenomena are interesting not only from the fundamental point of view [1], these phenomena are also largely present in various industrial applications. The thin layer, Knudsen layer is formed on the liquid-vapour interface. In this layer a gas is in non-equilibruim state even at atmospheric conditions. This why the correct and detailed description of physical processes of evaporation and condensation requires the application of the kinetic theory based approaches [2], [3], [4]. However, the implementation of the kinetic methodology for the practical problems is not so easy. This is why it is interesting to use the classical Navier-Stokes equations with the temperature and pressure jumps boundary conditions [5], [6] to simulate the evaporation-condensation problems.

The main objective of this study is the development of the mathematical model to simulate the behavior of the vapor phase at the liquid-gas interface. First the S-model [7] and BGK kinetic equations are used to simulate the evaporation-condensation phenomena appearing between two parallel condensed phases maintained at different temperatures. Then, these kinetic results are compared with that obtained from the Navier-Stokes equations subjected to the special temperature and pressure jumps boundary conditions [5]. These conditions were derived to take into account the non-equilibrium behaviors of a gas during the evaporation and condensation on the vapor-liquid interface. The comparison between two solutions allows us to determine the domain of the applicability of the Navier-Stokes equations with the jump conditions.

2. The problem statement

We consider two parallel plane gas - liquid interfaces at rest, maintained at temperature T_1 and T_2, ($T_2 > T_1$) at the bottom ($y = 0$) and top ($y = H$) interfaces, respectively, see Fig. 1. We investigate the behavior of the steady gas flows caused by evaporation and condensation on the condensed phases

IOP Conf. Series: Journal of Physics: Conf. Series **925** (2017) 012013 doi:10.1088/1742-6596/925/1/012013

on the basis of kinetic theory. Let p_1 be the saturation gas pressure at temperature T_1 and p_2 corresponds to temperature T_2. The problem is characterized by the parameters p_2/p_1 and T_2/T_1 and by the rarefaction parameter δ. The saturation gas pressure is a function of the saturation gas temperature given by the Clausius-Clapeyron equation [8].

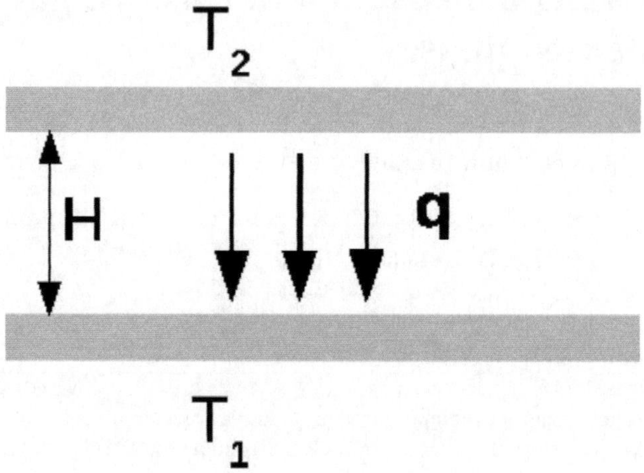

Figure 1. Sketch of 1D geometry

The S-model kinetic equation is be written as

$$\frac{\partial f}{\partial t} + v\frac{\partial f}{\partial r} = Q(f, f), \tag{1}$$

where $f(t,r,v)$ is the velocity distribution function of gas molecules, v is the molecular velocity vector, r is the position vector, t is a time. In the frame of S-model kinetic equation the collision term $Q(f, f)$ has the form

$$Q(f, f) = \nu(f^S - f), \tag{2}$$

where ν is the collision frequency, f^S is an equilibrium distribution function defined by

$$f^S(r,v,t) = f^M\left[1 + \frac{2m(v-u)q}{15n(r)(kT(r))^2}\left(\frac{m(v-u)^2}{2kT(r)} - \frac{5}{2}\right)\right], \tag{3}$$

here $T(r)$ is a gas temperature, $n(r)$ is a gas number density, u is a bulk velocity, q is a heat flux, m is a mass of gas molecule, k is the Boltzmann constant.

The boundary conditions for the plane vapor-liquid interface can be written as

$$f^+ = (\alpha n_i + (1-\alpha)n_{wi})f_w^i, \quad i = 1,2, \tag{4}$$

$$f_w^i(v) = \left(\frac{m}{2\pi kT_i}\right)^{3/2} \exp\left(-mv^2/(2kT_i)\right), \tag{5}$$

n_i is the number density of the saturated vapor near the plane interface, it can be calculated from the equation of state using the values of the saturated pressure at the saturated temperature as $n_i = p_i/(kT_i)$. α is the evaporation and condensation coefficient, here we supposed that both coefficients are equal. The number density n_{wi} ($i = 1,2$) can be found from the relation:

$$n_{wi}\sqrt{\frac{kT_i}{2\pi m}} = \int_{nv<0} v_y f^- \mathrm{d}\mathrm{v}. \quad i = 1,2. \tag{6}$$

If the function f is known the macroscopic parameters can be calculated:

$$n(r,t) = \int f(r,v,t)\mathrm{d}v, \tag{7}$$

$$T(r,t) = \frac{m}{3kn}\int f(r,v,t)V^2\mathrm{d}v, \tag{8}$$

$$q(r,t) = \frac{m}{2}\int f(r,v,t)V^2 V\mathrm{d}v, \tag{9}$$

here $V = v - u$. The mass flow rate can be calculated as

$$J_N = \int v_y f(r,v,t)\mathrm{d}v. \tag{10}$$

The distance H between the two interfaces is taken as the characteristic dimension of the problem. The Knudsen number and the rarefaction parameter are defined as following:

$$Kn = \frac{\ell}{H}, \quad \delta = \frac{1}{Kn}, \quad \ell = \frac{\mu_1\sqrt{2RT_1}}{p_1}, \tag{11}$$

here ℓ is the equivalent mean free path, μ_1 is the dynamic viscosity of the vapor phase and $\sqrt{2RT_1}$ is the most probable molecular speed, both at the temperature T_1, R is the specific gas constant.

3. Results

The evaporation-condensation between two condensed planes is simulated by two kinetic models (BGK and S-model) and by using the compressible Navier-Stokes equation with the temperature and pressures jumps boundary conditions [2], [4] for several sets of the rarefaction parameter (Knudsen number), temperature and pressure ratios.

As an example the evaporation and condensation of Argon is considered between two parallel condensed phases. The temperature and pressure ratios are equal to $T_2/T_1 = 1.036$ and $p_2/p_1 = 1.459$, respectivelly. The temparture and pressure are interrelated through the Clausius-Clayperon relation [8]. The comparison between the results obtained with the kinetic models (S-model and BGK model) and the Navier-Stokes equations with the temperature and pressure jumps boundary conditions are provided on Figure 2. The rarefaction parameter δ is equal to 100 which corresponds to the slip flow regime. Relatively good agreement is found between all three approaches.

Under several temperature and ptessure ratios the inverse temperature gradient phenomenon [1] can be observed: the vapor temperature near the colder (condensed phase) is hotter than that near the hotter phase.

Further simulations are needed to find the conditions of the applicability of the Navier-Stokes equations with the jump boundary conditions.

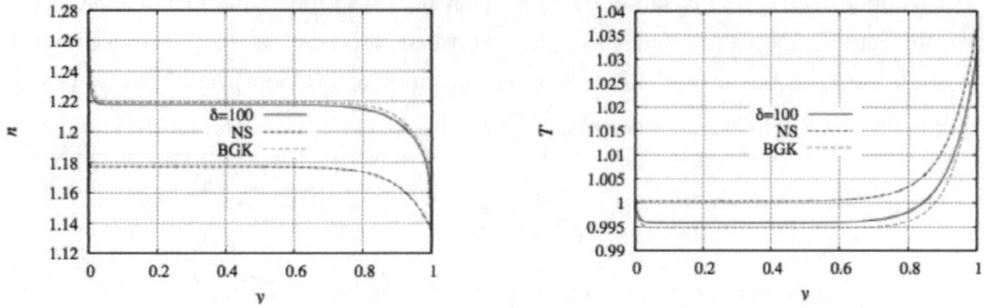

Figure 2. The number density (left) and temperature (right) profiles

4. Conclusion

The evaporation-condensation problem between two parallel condensed phases are simulated using the kinetic approach. The profiles of the macroscopic parameters such as temperature, pressure and number density of the vapor are obtained. The system of compressible Navier-Stokes equations completed by the energy conservation equation is solved also numerically with the pressure and temperature jumps boundary conditions. It was shown that the continuum modeling results agree relatively good with the kinetic solution when the rarefaction parameter is large enough.

Acknowledgement

The research was supported by the Ministry of Education and Science of the Russian Federation through the base part of the State task for high educational institutions (the Research Project No. 3.6064.2017/8.9) and Agreement No. 02.A03.21.0006.

References

[1] Sone Y 2007 *Molecular gas dynamics. Theory, techniques and applications* Birkhäuser, Boston
[2] Aoki K and Masukawa N 1994 *Phys. Fluids* **6** 1379–1395
[3] Sone Y 2002 *Kinetic Theory and Fluid Mechanics* Birkhäuser, Boston
[4] Onishi Y and Sone 1979 *J. Phys. Soc. Japan* **47** 1676–1685
[5] Gatapova E, Graur I, Sharipov F and Kabov O 2015 *Int. J. Heat Mass Transfer* **83** 235–243
[6] Gatapova E, Filipenko R, Lyulin Y, Graur I, Marchuk I, Kabov O 2015 *Thermophysics and Aeromechanics* **22** 701–706
[7] Shakhov E 1968 *Fluid Dyn.* **3** 95–96
[8] Landau L and Lifshitz E *Course of theoretical physics* 1969 **5** Pergamon Press

IOP Conf. Series: Journal of Physics: Conf. Series **925** (2017) 012014 doi:10.1088/1742-6596/925/1/012014

Thermography of flame during diesel fuel combustion with steam gasification

I S Anufriev[1], S S Arsentyev[1,2], M V Agafontsev[3], E P Kopyev[1], E L Loboda[3], E Yu Shadrin[1] and O V Sharypov[1]

[1] Institute of Thermophysics. S.S. Kutateladze SB RAS, Novosibirsk, 630090, Russia
[2] Novosibirsk State University, Novosibirsk, 630090, Russia
[3] Tomsk State University, Tomsk, 634050, Russia

E-mail: anufriev@itp.nsc.ru

Abstract. The paper represents a study concerning the combustion of liquid hydrocarbon fuel in a perspective burner device with the controlled forced supply of overheated steam into the combustion zone, using diesel fuel. The thermal imaging measurements are conducted for the outer flame of the burner device in the wide range of regime parameters (flow rate and temperature of steam). A thermal imaging camera (FLIR, JADE J530SB) is used in the experiments. The effective emissivity coefficient of flame is obtained versus the flow rate of steam supplied. The steam parameters are found to influence on the temperature in the outer flame of the burner device.

1. Introduction

Fuel gasification is the conversion of solid or liquid fuel into combustible gases by using the partial oxidation during the high temperature heating [1, 2]. The gasification of organic fuels is used in some cases for processing of low grade fuels to produce a synthesis gas [3]. The synthesis gas obtained is used as a final product for manufacturing purposes, for example as gaseous fuels for thermal electric power stations or as a hydrogen source (for example, the production of ammonia), carbon black, etc. [4-8]. In addition, this gaseous fuel releases a smaller amount of toxic gases during combustion compared to original fuel. Steam is often used as an oxidizer for fuel gasification and the synthesis gas produced is called a water gas.

Also, steam is used as a sprayer in fuel nozzles [9]. This improves the solubilizing of black oil and intensifies its vaporization and combustion. There are the other examples of using steam during the fuel combustion, for example, the supply of steam into the combustion chamber of a gas turbine installation allows the content of toxic nitrogen oxides to be reduced in combustion products due to decreasing the flame temperature [10, 11].

In addition, the injection of water or steam into the combustion chambers of thermal electric power stations (using liquid or solid fuel) is also used to reduce NO_x emissions. There are the examples of the water injection into the internal-combustion engines in automobiles, in order to suppress detonation, decrease carbonization and reduce NO_x and CO concentrations [12, 13]. In all these applications, water (steam) acts as a cooling component used to dilute a fuel mixture, which allows the temperature and the formation of thermal NO_x to be reduced in the combustion zone. However, this method leads

IOP Conf. Series: Journal of Physics: Conf. Series **925** (2017) 012014 doi:10.1088/1742-6596/925/1/012014

to the decrease in the power and efficiency of equipment, and its use is limited since CO emissions and other products of incomplete fuel combustion start increasing along with the reduction of NO_x. Therefore, the use of steam during the fuel combustion in a certain device requires a comprehensive scientific evidence to select the optimal (according to environmental and thermotechnical specifications) modes and structural parameters.

At the Institute of Thermophysics SB RAS a new method is proposed for fuel combustion with the use of steam, when there is the gasification of carbon-containing particles during the incomplete combustion of liquid hydrocarbons [14]. The supply of overheated steam to the liquid hydrocarbon combustion zone was shown to sharply intensify combustion [15, 16]. This provides stable ignition, high completeness of fuel combustion and a low content of toxic components in combustion products. This method of combustion is promising for utilization of low-grade liquid hydrocarbon fuels and combustible industrial wastes to produce heat energy. In the previous papers [15-17] the authors used an autonomous burner device with constant steam parameters (flow rate and temperature) [18, 19]. However, these characteristics should be adjusted to control the combustion process. In this work a new burner device equipped with an electric steam generator is used to obtain a flow of overheated steam for the wide range of flow rates and temperatures. To study the effect of steam parameters on the main characteristics of the liquid hydrocarbons combustion process (the composition of combustion products and the specific heat release capacity), it is necessary to obtain the data for different combustion modes.

This paper represents the thermal imaging study concerning the flame of a perspective burner device during the combustion of diesel fuel along with the supply of overheated steam to the combustion zone for the wide range of steam parameters.

2. Experimental setup and measurement procedure

The studies were conducted using a combustion stand (figure 1) equipped with a new burner device (10 kW), an electric steam generator (average power of 1.5 kW), a plunger metering pump, an automatic steam generator control system, a fuel supply system, an electronic balance to control the flow of water and fuel, as well as with necessary control and measurement instruments.

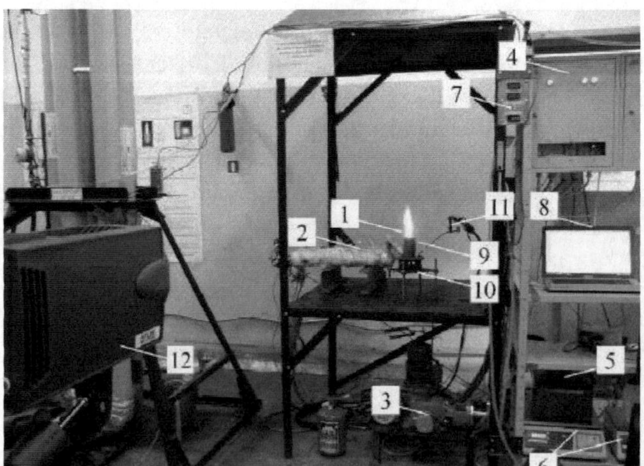

Figure 1. Combustion stand for the study of liquid hydrocarbon combustion with steam gasification: 1 is the burner device, 2 is the electric steam generator, 3 is the plunger metering pump, 4 is the automatic steam generator control system, 5 is the fuel supply system, 6 is the electronic balance, 7 is the indicators of control and measurement instruments, 8 is the computer, 9 is the gas-generation chamber of the furnace unit, 10 is the combustion chamber of the furnace unit, 11 is the fuel pipe, 12 is the thermal imager

IOP Conf. Series: Journal of Physics: Conf. Series **925** (2017) 012014 doi:10.1088/1742-6596/925/1/012014

The main elements of the new burner device are: a cylindrical body, a combustion chamber, a steam nozzle and a gas-generation chamber. The combustion chamber has holes for the air inflow from the atmosphere. Fuel is supplied into the combustion chamber through the fuel pipe. The stable rate of fuel is provided by the fuel nozzle and the pump and controlled by the electronic balance. A steam nozzle is mounted coaxially above the combustion chamber near the base of the gas-generation chamber and is oriented vertically (outlet diameter is 0.5 mm). The steam nozzle is connected with an external steam generator. The test burner is a vaporizing burner according to the operation principle. The combustion process is similar to the operation of an autonomous burner [17].

The steam generator consists of three heating units connected in sequence. Each unit is a metal tube (38 mm in external diameter) with a length of 0.5 m and a wall thickness of 8 mm. An electric tubular U-shaped heater (maximum power of 700 W) was mounted inside the pipe. The space between the heater and the pipe wall was filled with periclase DM-1 to increase heat transfer. In the wall of each pipe there are the closed rectangular cross-section channels (4×2 mm and 6 m in length for each block) in the form of spiral. The pipes are insulated outside. The monitoring temperature sensors are installed on the surface of each block. The mass flow of water (steam) is provided by the plunger metering pump and controlled by the electronic balance. The steam temperature is regulated by changing the power of the heaters. The pressure in the steam generator is recorded by the digital pressure sensor. The developed laboratory electric steam generator produces the overheated steam in the range of the temperature and flow rate of 100÷600 °C and 0.25÷1.5 kg/h (pressure up to 10 atm).

The thermal imaging camera (FLIR, JADE J530SB) was used to measure the temperature in the high temperature flame of the burner device. This device has a high temporal resolution: the frame rate up to 177 Hz with a maximum resolution of 320x240 pixels and up to 18 kHz with a resolution of 320x4 pixels. The minimum time of frame exposure is 6 μs. The operating range of the thermal imager is in the middle infrared range of 2.5-5.0 microns. The same spectral range contains the powerful flame radiation lines caused by the radiation of hot combustion products (including water vapors, CO_2, and CO). Based on the results of previous studies [20], a narrow spectral band dispersion optical filter F0616 with bandwidth of 2.5-2.7 microns was selected for operation.

The filter was selected due to the powerful emission lines of water vapors and CO_2 in this spectral range. The temperature measurement range of the thermal imager is determined by calibration and reaches 583-1773 K for the filter selected. The data collection and initial processing of thermograms were carried out using a specialized program, Altair. The factory calibrations were used for the F0616 filter with an exposure time of 9, 64, and 350 μs for the operation of the thermal imager. The previous spectral analysis of the flame temperature and pressure [20] demonstrated the absence of peak periodic oscillations in the test flame. To obtain an average temperature value, the measurements were conducted with a frequency of 50 Hz for 1 min.

To determine the effective emissivity coefficient simultaneously with the thermal imager measurements at the control points (on the axis of the burner device r = 0, at the height x = 30 mm from the edge of the burner), the temperature was recorded using a platinum-platinum rhodium type B thermocouple (600…1600 °C) with the thermoelectrodes 300 microns in diameter. The thermal inertia of the thermal converter was not more than 5 seconds. The permitted deviation limits of thermoelectromotive forces were ±0.005 compared with the standard static characteristics of the converter for the value of the measured temperature. Using the thermogram, the average temperature was determined in the area of hot junction. The effective emissivity coefficient, the values of which were ranged from 0.3 to 0.65 depending on the steam flow, was selected using the Altair software. The value of the effective emissivity coefficient obtained was used for the entire area.

3. Measurement results and analysis

Measurements were carried out under different operating conditions of the burner device. For the constant rate of fuel $F_f = 0.82$ kg/h, the value $\gamma = F_v/(F_v+F_f)$ was varied in the range of 0.24÷0.56, where F_v is the average mass flow rate of steam. The steam temperature vas varied in the range of T_v = 160÷560 °C for each value of F_v.

IOP Conf. Series: Journal of Physics: Conf. Series **925** (2017) 012014 doi:10.1088/1742-6596/925/1/012014

Figure 2 shows the time-average temperature profiles in the outer flame of the burner device for different flow rates of steam at T_v = const. It should be noted that the flame is a translucent three-dimensional optical medium, so the thermal imager detects the integral radiation of all inner medium layers. This integral radiation is required to determine the temperature in the observation plane, using the calibrations and the effective emissivity coefficient. The thermograms shown in figure 2 demonstrate a significant effect of the supplied steam flow both on the temperature distribution in the outer flame and the flame dimensions (photographs in figure 2).

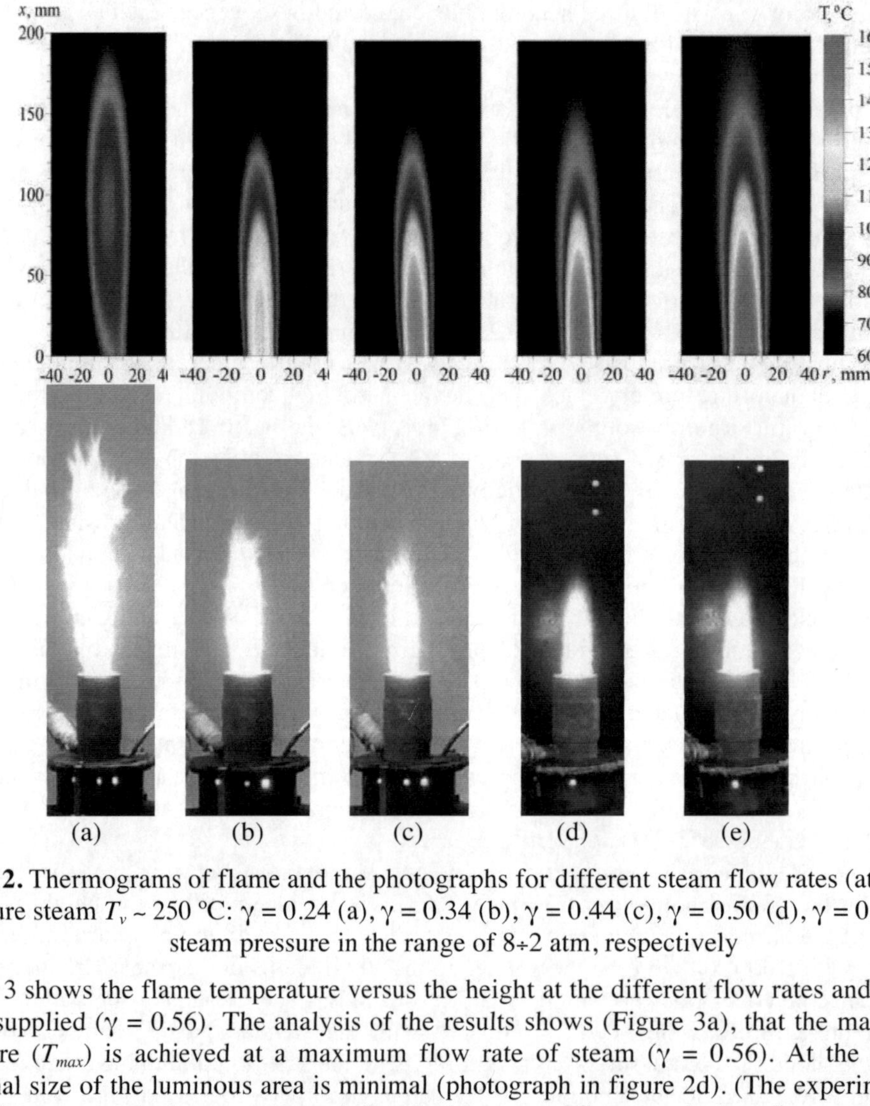

Figure 2. Thermograms of flame and the photographs for different steam flow rates (at a constant temperature steam $T_v \sim 250$ °C: $\gamma = 0.24$ (a), $\gamma = 0.34$ (b), $\gamma = 0.44$ (c), $\gamma = 0.50$ (d), $\gamma = 0.56$ (e), and a steam pressure in the range of 8÷2 atm, respectively

Figure 3 shows the flame temperature versus the height at the different flow rates and temperatures of steam supplied ($\gamma = 0.56$). The analysis of the results shows (Figure 3a), that the maximum flame temperature (T_{max}) is achieved at a maximum flow rate of steam ($\gamma = 0.56$). At the same time, a longitudinal size of the luminous area is minimal (photograph in figure 2d). (The experiments showed that a further increase in the γ parameter led to the blowout of flame, and a low flow rate of steam ($\gamma <$ 0.3) led to the incomplete combustion of fuel). The maximum flame temperature increases with the increase in the steam temperature (figure 3b).

For the modes with a high value of the parameter γ the height dependence of the flame temperature is monotonous, and the maximum is reached at the outlet of the burner. When $\gamma \leq 0.5$ the maximum temperature is observed at some distance from the outlet of the burner, which indicates the incomplete mixture combustion before the release into the atmosphere.

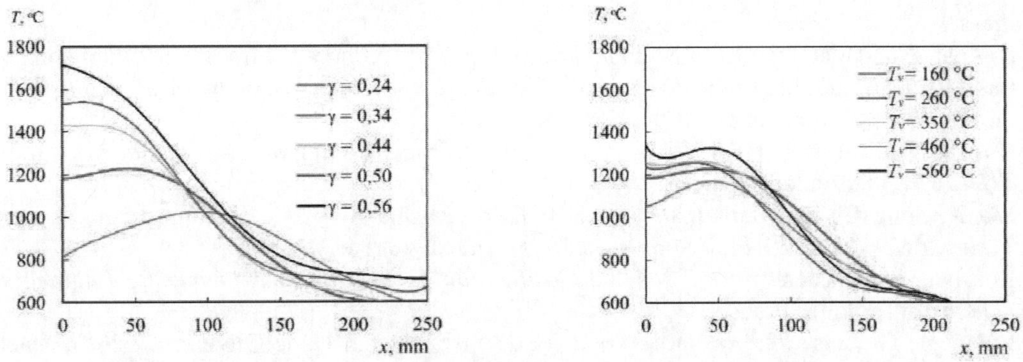

(a) (b)

Figure 3. Distribution of the time-average temperature in the outer flame along the vertical center line of the flame for different mode parameters: (a) the steam flow rate is varied, $T_v = 250$ °C; (b) the steam temperature is varied, $\gamma = 0.34$, the steam pressure is 4 atm

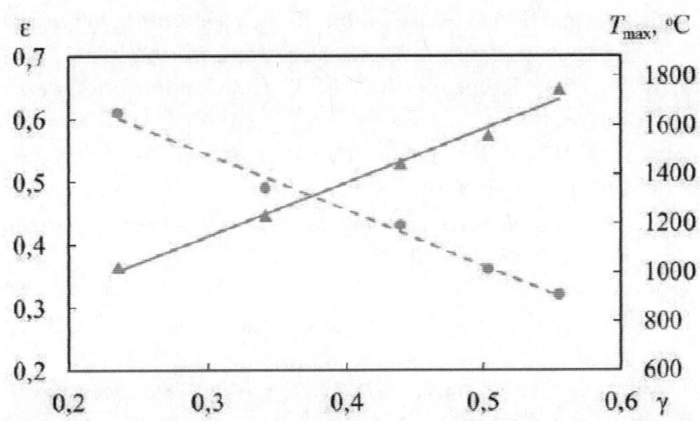

Figure 4. Effective emissivity coefficient of flame (●) and maximum temperature (▲) in the outer flame versus the γ parameter, T_v=250 °C

The emissivity coefficient of flame ε was determined versus the flow rate of the steam supplied (figure 4): the value ε decreases with increasing the γ parameter. The γ parameter dependence of the maximum flame temperature is also linear, but increases with increasing γ (figure 4).

4. Conclusion

With the use of thermography, thermal characteristics in the flame of the prospective burner device were studied when diesel fuel was burned with forced regulated feeding of superheated steam to the combustion zone. Thermal imaging measurements were carried out over a wide range of the operating parameters (flow rate and temperature of the supplied water vapor). It has been established that the consumption of superheated water vapor has a significant influence both on the temperature distribution in the flame and on its dimensions. The maximum temperature was reached at the maximum steam flow rate and was 1700 °C. The dependence of the temperature of the flame on the temperature of the vapor turned out to be weaker. The dependence of the effective coefficient of flame emission on the flow rate of the supplied steam was also established.

The experimental data obtained can be used for the measurement analysis of the gas composition in combustion products and the heat release, in order to provide a scientific evidence for the selection of the energy-efficient and environmentally friendly utilization ways for off-grade liquid hydrocarbons to produce thermal energy.

References

[1] Higman C and Van Der Burgt M 2008 *Gasification. (2nd ed.)* (Gulf Professional Publishing)
[2] Lavrov N V and Shurygin A P 1962 *Introduction to the theory of fuel combustion and gasification* (Acad. Sciences of the USSR, Moscow)
[3] Rostrup-Nielsen J and Christiansen L J 2011 *Concepts in Syngas Manufacture* (Imperial College Press, London)
[4] Liu K, Song Ch and Subramani V 2010 *Hydrogen and Syngas Production and Purification Technologies* (John Wiley & Sons, Inc., Hoboken, New Jersey)
[5] Rezaiyan J and Cheremisinoff N 2005 *Gasification Technologies: A Primer for Engineers and Scientists* (Taylor & Francis)
[6] Emanuele Graciosa Pereira, Jadir Nogueira da Silva, Jofran L. de Oliveira, Cssio S. Machado 2012 *Renewable and Sustainable Energy Reviews* **16** 4753
[7] Ryabtsev I I and Volkov A E 1968 *Gas production from liquid fuels for synthesis of ammonia and alcohols* (Khimiya, Moscow)
[8] Andreev F A, Kargin S I, Kozlov L I and Pristavko V F 1974 *Combined nitrogen technology* (Khimiya, Moscow)
[9] Khzmlyan D M and Kagan J A 1976 *Combustion theory and furnace arrangements* (Energia, Moscow)
[10] Tsanev S V, Burov V D and Remezov A N 2002 *Gas turbine and steam-gas plants of thermal power stations* (MEI, Moscow)
[11] Ivanov A A, Ermakov A N and Shlyakhov R A 2010 *Izv. RAN. Energetika* **3** 119
[12] Lerner M O 1981 *Khimia i zhizn'* **5** 16
[13] Efremov P K 1977 *Proc. of the All-Union Scientific Conference «Protection of air pool from pollution by toxic emissions vehicles»* **1** 93
[14] Vigriyanov M S, Salomatov V V and Alekseenko S V 2003 *RF Patent* **2219435**M.S.
[15] Alekseenko S V, Pashchenko S E and Salomatov V V 2010 J. *Engng. Physics and Thermophysics* **83** 729
[16] Alekseenko S V, Anufriev I S, Vigriyanov M S, Dulin V M, Kopyev E P and Sharypov O V 2014 *Thermophys. Aeromech* **21** 393
[17] Alekseenko S V, Anufriev I S, Vigriyanov M S, Kopyev E P and Sharypov O V 2016 *Combust. Expl. Shock Waves* **52** 286
[18] Vigriyanov M S, Alekseenko S V, Anufriev I S and Sharypov O V 2014 *RF Patent* **2523591**
[19] Vigriyanov M S, Alekseenko S V, Anufriev I S and Sharypov O V 2016 *RF Patent* **2579298**
[20] Anufriev I S, Kopyev E P and Loboda E L 2014 *Proc. SPIE 9292 «20th International Symposium on Atmospheric and Ocean Optics: Atmospheric Physics»* **929226**

IOP Conf. Series: Journal of Physics: Conf. Series **925** (2017) 012015 doi:10.1088/1742-6596/925/1/012015

Numerical simulation of heat transfer at unsteady heat generation in falling wavy liquid films

A N Chernyavskiy and A N Pavlenko

Kutateladze Institute of Thermophysics SB RAS, Acad. Lavrentiev ave. 1, Novosibirsk, 630090, Russia

E-mail: cherny@ngs.ru

Abstract. The mathematical model which allows the calculation of the wave surface profile as well as velocity and temperature fields has been presented. The numerical simulation of heat transfer in falling wavy films of liquid nitrogen has been performed. The dependencies of boiling expectation time and total local evaporation time on heat flux density for different inlet Reynolds numbers have been calculated. The regime map which describes the different mechanisms of film decay was obtained by summing up the simulation results. The results of numerical simulation are in satisfactory agreement with the experimental data.

1. Introduction

The liquid flow over a vertical surface is inevitably accompanied by development of wave perturbations on the free film surface which bring considerable contribution to the heat transfer processes and development of crisis phenomena [1–3].

The area of highly efficient heat transfer at the film flow of liquid is limited by the heat flux and depends significantly on the heat release law. When the system reaches the critical heat flux, this leads to decay of the film flow, formation of large-scale dry zones, reduction of heat transfer intensity and as a consequence to a drastic increase in temperature of the heat-releasing surface. All these can cause destruction of the heat-releasing element. The ability to calculate the critical heat fluxes and maximal times of their impact for a given system is required for the design of stably working heat exchangers.

The time of crisis beginning and scenario of film flow decay are characterized by such parameters as time of boiling expectation and time of total local evaporation of the falling liquid film. These values determine maximal times of heat flux influence on the film with certain parameters.

The mathematical model which allows the calculation of the wave surface profile as well as velocity and temperature fields has been presented in the current work. This model allows calculations of the wave characteristics and parameters of liquid film decay at different regime parameters. The results of numerical simulation performed in the framework of the suggested model have been compared with the experimental data obtained for the liquid nitrogen in [2].

2. Mathematical model

Simulation has been carried out within a 2D calculation area. The abscissa axis is directed along a vertical wall, downstream the liquid film flow. The ordinate axis is orthogonal to the abscissa axis and directed from the wall to the free film surface. The detailed description of the calculation area and the mathematical model has been presented in [4].

1.1. Hydrodynamic model

To describe the processes of liquid flow and evolution of the wave surface profile, in the framework of the current study the 2D hydrodynamic model of Kapitza-Shkadov has been used:

$$\frac{\partial Q}{\partial t}+\frac{6}{5}\frac{\partial}{\partial x}\left(\frac{Q^2}{h}\right)=-\frac{3\nu Q}{h^2}+gh+\frac{\sigma}{\rho}h\frac{\partial^3 h}{\partial x^3} \quad (1)$$

$$\frac{\partial h}{\partial t}+\frac{\partial Q}{\partial x}=0 \quad (2)$$

$$u=\frac{3Q}{h}\left[\frac{y}{h}-\frac{1}{2}\left(\frac{y}{h}\right)^2\right]=0 \quad (3)$$

$$v=-\int_0^y\frac{\partial u}{\partial x}dy \quad (4)$$

The method of liquid flow description by means of flow rate characteristic Q allows excluding the dependence on the ordinate and reducing the problem to the pseudo-single-dimensional one. Velocity components in the whole calculation area can be obtained from h and Q by means of expressions (3) and (4). The system of equations (1–4) supplemented with the initial and boundary conditions allows calculation of flow dynamics and evolution of surface profile within the 2D calculation area.

Perturbations at the inlet of the calculation area were set through the addition to the flow rate of the undisturbed film under the condition of keeping the thickness undisturbed:

$$h|_{x=0}=h_0=\left(\frac{3}{4}\frac{\nu^2}{g}\mathrm{Re}_{\mathrm{in}}\right)^{1/3} \quad (5)$$

$$Q_0=\frac{gh_0^3}{3\nu}=\frac{\nu\,\mathrm{Re}_{\mathrm{in}}}{4} \quad (6)$$

$$Q|_{x=0}=Q_0(1+RA),\ A(t)\in[-1;1] \quad (7)$$

Values h_0 and Q_0 set by expressions (5) and (6) are also the initial conditions. In other words, at zero time moment the film is assumed to be waveless and its thickness and flow rate are constant and determined by the conditions of the problem. The additive member in (7) consists of two multipliers: A is a random real number from −1 to 1 and R is the relative perturbations amplitude.

1.2. Heat transfer model

In natural variables the boundaries of the calculation area are not constant because of the unsteady character of the free surface. The following coordinate transformation converting the calculation area into the rectangular one has been used to solve this problem:

$$\eta(t,x,y)=y/h(t,x) \quad (8)$$

After the coordinates transformation the unsteady heat conductivity equation in the liquid film looks like the following:

$$\frac{\partial T}{\partial t}+u\frac{\partial T}{\partial x}+\frac{1}{h}\left[v-\eta\frac{\partial h}{\partial t}-u\eta\frac{\partial h}{\partial x}-2a\frac{\eta}{h}\left(\frac{\partial h}{\partial x}\right)^2+a\eta\frac{\partial^2 h}{\partial x^2}\right]\frac{\partial T}{\partial \eta}=$$
$$=a\frac{\partial^2 T}{\partial x^2}-2a\frac{\eta}{h}\frac{\partial h}{\partial x}\frac{\partial^2 T}{\partial x\partial\eta}+\frac{a}{h^2}\left[1+\eta^2\left(\frac{\partial h}{\partial x}\right)^2\right]\frac{\partial^2 T}{\partial\eta^2} \quad (9)$$

At the inlet of the calculation area and at the free film surface, the temperature was assumed equal to the temperature of liquid saturation. The following boundary condition was used on the heating surface:

IOP Conf. Series: Journal of Physics: Conf. Series **925** (2017) 012015 doi:10.1088/1742-6596/925/1/012015

$$\frac{\partial T}{\partial \eta}\Big|_{x=0} = -h\frac{q(t,x)}{\lambda} + h\frac{\delta_h C_h \rho_h}{\lambda}\frac{\partial T}{\partial t}\Big|_{\eta=0} \tag{10}$$

To take into account evaporation from the free surface, equation (2) has been modified as follows:

$$\frac{\partial h}{\partial t} + \frac{\partial Q}{\partial x} = \frac{\lambda}{L\rho}\left(\frac{1}{h}\frac{\partial T}{\partial \eta}\Big|_{\eta=1} - \frac{\partial h}{\partial x}\frac{\partial T}{\partial x}\Big|_{\eta=1}\right) \tag{11}$$

3. Results and conclusions

Boiling expectation times have been calculated as the functions of heat flux density in the falling films of liquid nitrogen upon various input Reynolds numbers (Figure 1).

The boiling expectation time was determined as the minimum time of reaching the given overheat by the heating surface. In the framework of the present work three threshold values of overheating for the liquid nitrogen under conditions of atmospheric pressure were used: two theoretically calculated values, $\Delta T_{b.e} = 26$ K, corresponding to 95% and $\Delta T_{b.e} = 32$ K, corresponding to 100% of probability of boiling under the conditions of homogeneous nucleation [5], and one approximation for experimentally retrieved dependence of overheat on heat flux density obtained for boiling-up under the condition of heterogeneous nucleation:

$$\Delta T_{b.e}(q) = 32 - 21.1585 \cdot 0.93^{q}, q[W/cm^2] \tag{12}$$

For high heat fluxes, the calculated dependences are almost identical for all considered Reynolds numbers. In the region of low heat fluxes, discrepancy between the calculated curves is observed, which can be explained by a considerable contribution of evaporation to heat transfer upon the given parameters of the problem.

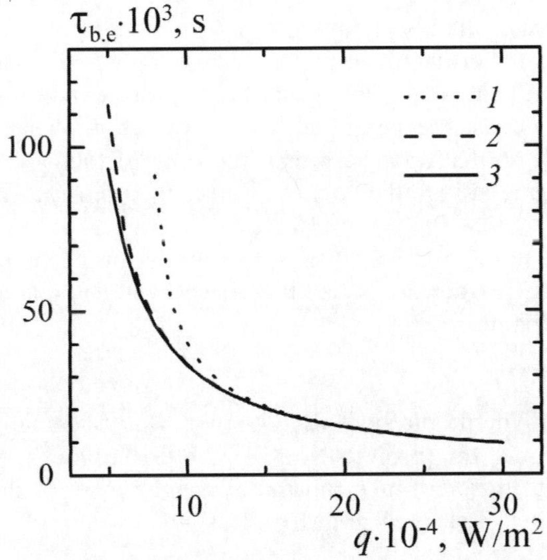

Figure 1. Calculated boiling expectation times, $\Delta T_{b.e} = 26$ K: *1* — $Re_{in} = 38$; *2* — $Re_{in} = 230$; *3* — $Re_{in} = 630$.

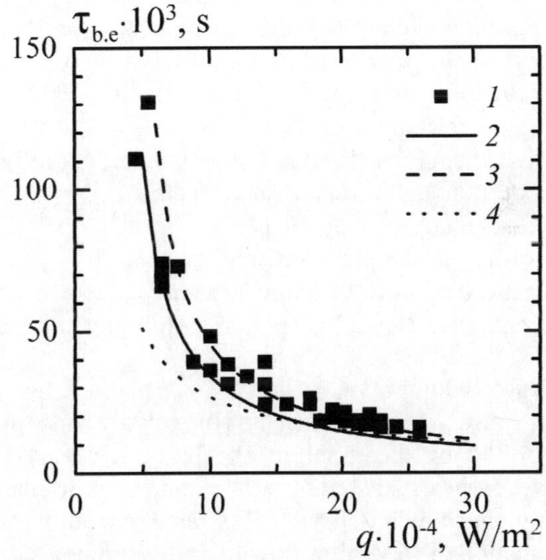

Figure 2. Comparison between experimental and calculated boiling expectation times with different $\Delta T_{b.e}$, $Re_{in} = 230$: *1* — Experimental data; *2* — $\Delta T_{b.e} = 26$ K; *3* — $\Delta T_{b.e} = 32$ K; 4 — $\Delta T_{b.e} = \Delta T_{b.e}(q)$.

The calculated boiling expectation times are in satisfactory agreement with the experimental data [2] found under the conditions of evolution of three-dimensional waves on a free surface of the film (Figure 2). The results of numerical simulation agree with experimental data because evolution of three-dimensional waves within the investigated range of Reynolds numbers does not lead to significant variation of the defining wave parameters that determines the contribution of the heat-transfer convective component.

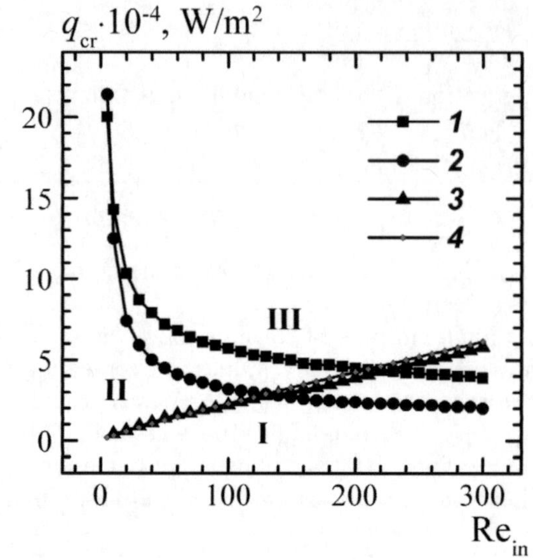

Figure 3. Regime map. *1,2* — Film decay on boiling-up, $\Delta T_{b.e} = 26$ K, $\Delta T_{b.e} = \Delta T_{b.e}(q)$, respectively; *3* — Surface draining on unsteady heat release; *4* — Calculation of critical heat flux for evaporation on steady heat release using (Pavlenko, Lel, 1997) model.

The regime map, which describes the different mechanisms of decay of the falling wavy film of liquid nitrogen, has been obtained by summing up the results of numerical simulation (Figure 3). Calculated curves 1, 2 and 3 divide the area of the investigated Reynolds numbers and heat flux densities into the set of regions. Region I is characterized by the absence of crisis. Region II describes the set of the problem parameters, where liquid film decay will be performed by total local evaporation without liquid boiling-up. Region III describes the set of parameters where the liquid film decay will be performed by the film boiling-up. Calculated critical heat flux densities of the total local evaporation (curve 3) agree satisfactorily with the values calculated for steady heat release conditions using the formula presented in [6] — curve 4. This is a consequence of the system Fourier number Fo > 1 which means that the time of temperature profile formation is less than the time of total local evaporation and critical heat flux densities in this regime are similar to critical heat flux densities in the case of steady heat release.

According to the presented regime map, it is possible to make a conclusion that boiling-up becomes the prevailing decay mechanism independently on heat flux density, when the inlet Reynolds number Re_{in} exceeds 130 – 220 depending on liquid and heater parameters.

Acknowledgments

The study was performed at Kutateladze Institute of Thermophysics SB RAS and was financially supported by the grant of the Russian Science Foundation (project No. 14-49-00010: blocks of unsteady heat transfer simulation and regime map with different mechanisms of development of the crisis phenomena) and the Russian Foundation for Basic Research grant (project No. 15-08-03221: simulation of wavy film flow hydrodynamics).

References

[1] Vorontsov E 1999 *Theor. Found. Chem. Eng.* **33** 117
[2] Pavlenko A, Surtaev A, Matsekh A 2007 *High Temp.* **45** 826
[3] Markides C, Mathie R, Charogiannis A 2016 *Int. J. Heat Mass Tr.* **93** 872
[4] Chernyavskiy A, Pavlenko A 2017 *Int. J. Heat Mass Tr.* **105** 648
[5] Skripov V et al. 1988 *Thermophysical Properties of Liquids in the Metastable (Superheated) State* (Gordon and Breach Science Publishers, New York)
[6] Pavlenko A, Lel' V 1997 *Rus. J. Eng. Therm.* **7** 177

Contact line motion over heated substrates with spatially non-uniform wetting properties

V S Ajaev[1,2] and E Ya Gatapova[2,3]

[1]Southern Methodist University, Dallas Texas 75275, USA
[2]Kutateladze Institute of Thermophysics SB RAS, Novosibirsk 630090, Russia
[3]Novosibirsk State University, Novosibirsk 630090, Russia

E-mail: ajaev@smu.edu

Abstract. We develop mathematical models of moving contact lines over flat heated solid surfaces with spatial variation of wetting properties under the conditions when evaporation is significant. The gas phase is assumed to be pure vapor and a lubrication-type framework is employed for describing viscous flow in the liquid. Substrate wetting properties are modeled through disjoining pressure, with the main focus on how the contact line motion changes near the regions of localized spatial variation of wetting properties.

1. Introduction

In heat transfer applications such as boiling or spray cooling, liquid-gas interfaces can come into contact with heated solid substrates [1-3]. Studies of solid surface non-uniformity and contact angle hysteresis [4] are important for many practical applications. The gas phase can be pure vapor or moist air, leading to different limiting mechanisms for local evaporation rates near contact lines. Several mathematical models of evaporation process from the liquid-gas interface in the contact line region have been developed over the past decades in order to describe coupling of evaporation and a number of other relevant physical phenomena such as capillarity, viscous liquid flow, London-van der Waals disjoining pressure, Marangoni stresses, diffusion and fluid flow in the gas phase, electrostatic interactions of interfaces, and vapor recoil. These studies address static and dynamic configurations for both triple lines, i.e. lines of contact of liquid, gas, and solid phases, and the so-called apparent contact lines, which are transition regions between macroscopic liquid-gas interfaces and ultra-thin films covering the solid substrate. The present study is focused on the latter case, illustrated in the sketch in Fig. 1; the gas phase is assumed to be pure vapor here and below. Local evaporation rate in this configuration has a maximum in the contact line region, while evaporation is completely suppressed in the ultra-thin film. Evaporative mass loss is compensated by the in-flow into the apparent contact line region.

The general framework for studies of the effect of evaporation on contact line motion for the case when the gas phase is pure vapor was provided by Ajaev et al. [5]. This paper deals with dynamics of vapor bubbles confined between two parallel walls. The interface shape is dominated by the capillary forces away from the walls, but the effects of viscous stresses, evaporation, and surface tension are all important near the walls. The apparent contact line model was used to define the moving boundary of macroscopically dry region of the wall. The main objective of the present study is to generalize the framework developed in Ajaev et al. [5] to include spatial variation of wetting properties of the

IOP Conf. Series: Journal of Physics: Conf. Series **925** (2017) 012016 doi:10.1088/1742-6596/925/1/012016

substrate via appropriate modification of the disjoining pressure terms in the equations which describe the local interface shape near the contact line. The substrate temperature is assumed to be fixed, although the same framework can in principle be used to describe contact line motion over substrate with spatial variation of temperature and/or heat flux.

2. Model formulation

We consider a geometric configuration of a meniscus moving over a solid surface with spatially non-uniform properties. The meniscus radius of curvature R away from the solid is constant (assuming that the appropriately defined Bond number is small enough for the effects of gravity to be negligible, which is typically a reasonable assumption for microsystems). Near the solid, the interface shape is affected by the viscous flow which in turn is coupled to the local evaporation rates from the interface. Evaporation here is assumed to be taking place in pure vapor environment. The solid substrate temperature is constant and higher than the vapor saturation temperature by the superheat value of ΔT.

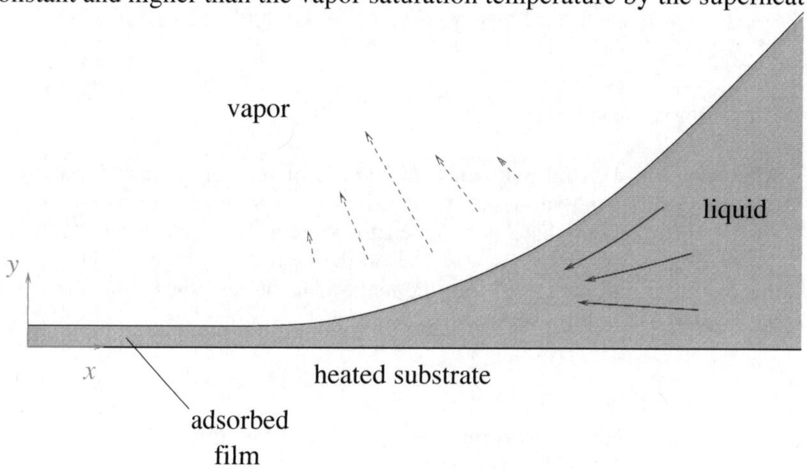

Figure 1. A sketch of geometric configuration of the apparent contact line region with solid substrates covered by ultra-thin adsorbed film.

The characteristic velocity U is dictated by evaporation and thus can be estimated from the mass conservation condition at the interface where evaporation takes place. For evaporating liquid of density ρ and thermal conductivity k we can therefore write

$$U = \frac{k\Delta T}{\rho \ell R} \tag{1}$$

where ℓ is the latent heat of vaporization. For the geometric configuration considered, it is convenient to define nondimensional variables using the capillary number, which in turn is defined by the following equation,

$$\mathrm{Ca} = \frac{\mu U}{\sigma} \tag{2}$$

Here μ is the dynamic viscosity of the evaporating liquid, σ is the surface tension at the liquid-vapor interface. The capillary number is one of the key parameters in our formulation and provides a measure of relative importance of viscous effects and surface tension. Estimating the capillary number based on Eq. (2) for several liquids used in heat transfer applications shows that the numerical value of this parameter is typically small. However, in the immediate vicinity of the contact line both viscous and surface tension effects can become important. To account for this, appropriate local Cartesian coordinates, shown in the sketch in Fig. 1, can be defined in terms of the capillary number by the following formulas,

$$x = \frac{x^*}{R\mathrm{Ca}^{1/6}}, \quad y = \frac{y^*}{R\mathrm{Ca}^{1/3}} \tag{3}$$

Here the asterisks denote dimensional variables. Note that for small capillary numbers, the horizontal length scale is much larger than the vertical one, so that using the lubrication-type approach is appropriate [6,7]. The details on the application of this approach to moving contact lines under the conditions when evaporation is taking place can be found in several recent works [5,8,9]. The key steps of the derivation in the present case are the same as described in e.g. Ketelaar and Ajaev [8] except that the disjoining pressure term is now a function of the local coordinate along the substrate. Also, for simplicity the disjoining pressure is assumed to only have one component, based on the London-van der Waals model of unbalanced intermolecular interactions in the thin film. The final evolution equation for the local film thickness h, scaled by the same characteristic value as the y-coordinate in Eq. (3) above, can then be written in the form

$$h_t + J + \frac{1}{3}\left[h^3\left(h_{xx} + \frac{\alpha f(x)}{h^3}\right)_x\right]_x = 0 \tag{4}$$

and the nondimensional evaporative flux J in this equation can be expressed in terms of the interface shape and various nondimensional parameters of the problem as follows,

$$J = \frac{-\delta\left(h_{xx} + \dfrac{\alpha f(x)}{h^3}\right) + T_0}{K + h} \tag{5}$$

Here the nondimensional parameter α is the characteristic scaled Hamaker constant which measures the relative value of the London- van der Waals disjoining pressure; note that the dimensional values of the Hamaker constant have been measured for a wide variety of liquids on different substrates, as discussed e.g. in Israelachvili [10]. In the expression for the mass flux, we also introduced nondimensional parameters K, δ. The first one is the measure of the non-equilibrium effects due to evaporation at the interface, while the second defines the change in local equilibrium interfacial temperature due to changes in curvature and effects of disjoining pressure in the liquid. In all simulations reported below we use the following values: $\alpha = 10^{-4}, K = 0.01,\ \delta = 10^{-4}$. The nondimensional quantity T_0 is the wall superheat scaled by the saturation temperature.

Numerical solution of the system defined by Eqs. (4)-(5) has been carried out using a finite-difference approach on the spatial interval [-5, 5], with the boundary conditions of $h(5) = 12$ (fixed value to avoid multiple solutions in the limit of spatially uniform substrate) and $h_{xx}(5) = 1$ to account for constant curvature of the interface sufficiently far away from the wall (note that in the lubrication-type framework the interfacial curvature is well-approximated by the second derivative). At the left end-point of the computational domain, the conditions of zero first and third derivatives are imposed. Spatially non-uniform mesh was used to provide better resolution in the contact line region; the mesh size in that region is decreased by a factor of 8 compared to the rest of the simulation domain. The results shown below correspond to the total number of points in the coarse mesh equal to 1,000. Time stepping is performed using the subsroutines from the DVODE package with the initial condition representing a parabola that satifies boundary conditions at the right endpoint of the domain and the ultra-thin film patched together. The thickness of the ultra-thin film is determined from the condition of the evaporative flux J in equation (5) (with $f(x)=1$) being equal to zero.

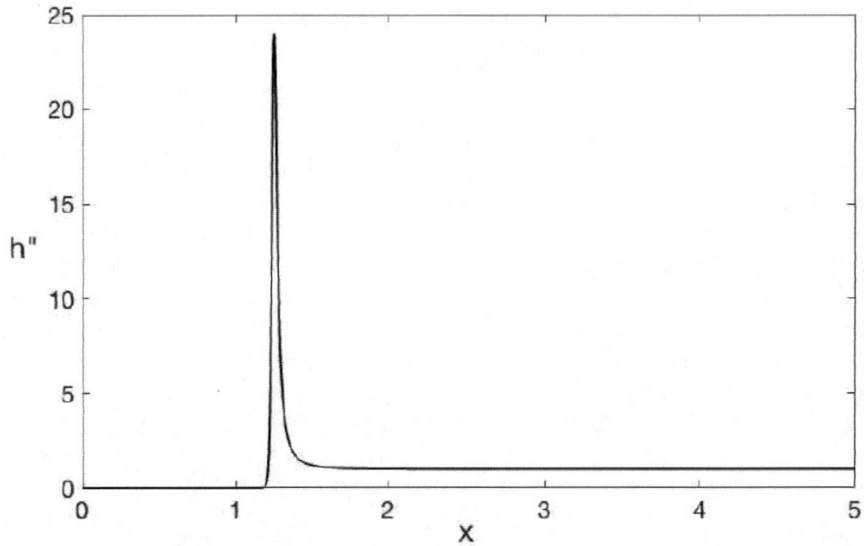

Figure 2. Interfacial curvature profile for the wall superheat $T_0 = 0.01$ and spatially uniform substrate wetting properties

3. Simulation results

Let us now discuss the numerical results. For all cases, the general outline of the procedure is as follows. We first find the steady-state solution corresponding to $T_0 = 0.01$ by running the code for sufficiently long time for the transients to decay. This situation corresponds to evaporation balancing the flow into the contact line region, meaning that the interface shape does not change with time. Curvature profiles for such steady state solutions typically have a well-defined maximum, as illustrated in Fig. 2. The scaled temperature is then instantaneously reduced by $\Delta T_0 = 0.006$, resulting in contact line motion to the left as suddenly weakened evaporation no longer compensates for the mass supply by the viscous flow. As the contact line moves, the interface profile also changes and these changes are recorded. The data on interface profiles at different moments in time is then used to track the motion of the contact line. Specifically, the position of the contact line is defined by the local maximum of curvature in the contact line region. For calculating this value, we first evaluate the curvature at every grid point and identify the point with the largest curvature value. Then, the local shape of the curve is approximated by a quadratic polynomial and the coordinate of the maximum of that polynomial graph is defined as the contact line location.

We first conduct simulations for the case when the substrate wetting properties are spatially uniform, so that $f(x) = 1$. The contact line location as a function of time for this case is shown by the dot-dashed line in Fig. 3, showing gradual slow-down as the new steady-state solution, corresponding to the smaller value of the wall superheat, is approached. The horizontal axis in Fig. 3 is the time measured from the moment of instantaneous change of the substrate temperature and scaled by the characteristic value of $R_0 Ca^{2/3}/U$; the maximum value of t in the plot is chosen such that the system is close to the new steady-state solution; we verified that this is indeed the case by running the simulation up to a much higher value of $t = 100$.

IOP Conf. Series: Journal of Physics: Conf. Series **925** (2017) 012016 doi:10.1088/1742-6596/925/1/012016

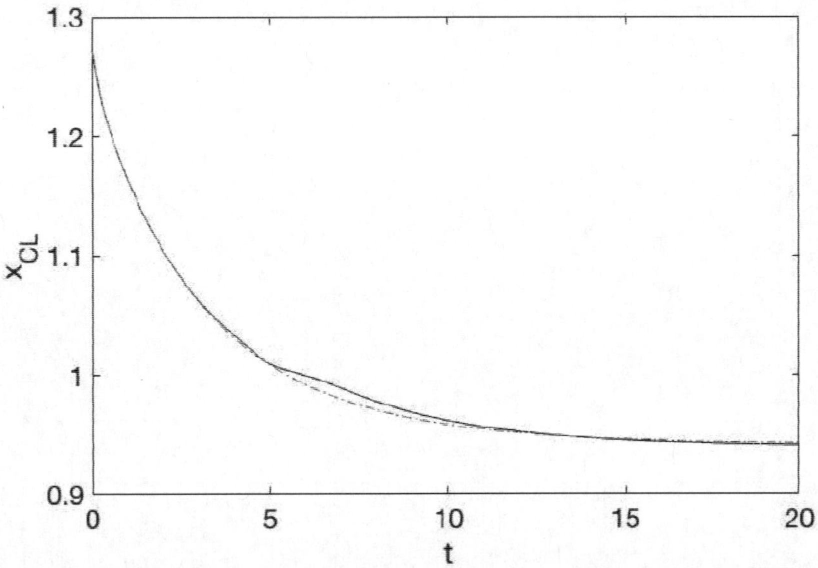

Figure 3. Evolution of the contact line position for spatially uniform substrate (dot-dashed line) and the case of an isolated defect (solid line) at $x_0 = 1$, as defined by Eq. (6).

Now consider the case when there is a localized region of variation of wetting properties, essentially a surface defect. This situation is modeled by a Gaussian profile so that the function *f(x)* appearing in the equations above is defined by

$$f(x) = 1 - \frac{1}{2}\exp\left(-\frac{(x-x_0)^2}{(\Delta x)^2}\right) \tag{6}$$

with $x_0 = 1, \Delta x = 0.01$. The contact line position for this case is shown as a function of time by the solid line in Fig. 3. Clearly, as the defect is approached, the motion of the contact line slows down. However, after the contact line passed over the region of spatial non-uniformity, its speed increases. The two phases of motion, the slow-down and speed-up, do not entirely compensate each other as compared to the motion without the defect: the contact line ends up slightly further to the left. However, for this dynamics it is essential that the contact line rapidly passes over the defect and into the region of spatially uniform disjoining pressure. Suppose this condition is violated due to the presence of a second Gaussian term in the formula for *f(x)*, centered at $x_1 = 0.95$ rather than 1, so that

$$f(x) = 1 - \frac{1}{2}\exp\left(-\frac{(x-x_0)^2}{(\Delta x)^2}\right) - \frac{1}{2}\exp\left(-\frac{(x-x_1)^2}{(\Delta x)^2}\right) \tag{7}$$

This situation is relevant for the case of patterned substrates when wetting properties variation is introduced in the form of a pattern, in contrast to a situation with an isolated random defect. The result for such modified configuration is shown in Fig. 4 and clearly corresponds to a significantly different behavior than seen in Fig. 3. The contact line slows down near $x = 1$ in a manner similar to the case of a single defect, but then instead of recovering and speeding up it slows down due to the interaction with the second localized region of the variation of wetting properties. The total contact line displacement at the point when the steady state is reached is in fact less than for uniform substrate. Thus, depending on the nature of wetting properties variation, the contact line can be displaced either more or less that for the case of uniform Hamaker constant. Our observation from many simulations similar to the ones generating Figs. 3, 4 is that the outcome is very sensitive to the contact line speed in the region where local variation of the Hamaker constant is imposed. For the result shown in Fig. 4, the contact line speed is much slower near $x = x_1$ than near $x = x_0$, so the dynamics is different.

IOP Conf. Series: Journal of Physics: Conf. Series **925** (2017) 012016 doi:10.1088/1742-6596/925/1/012016

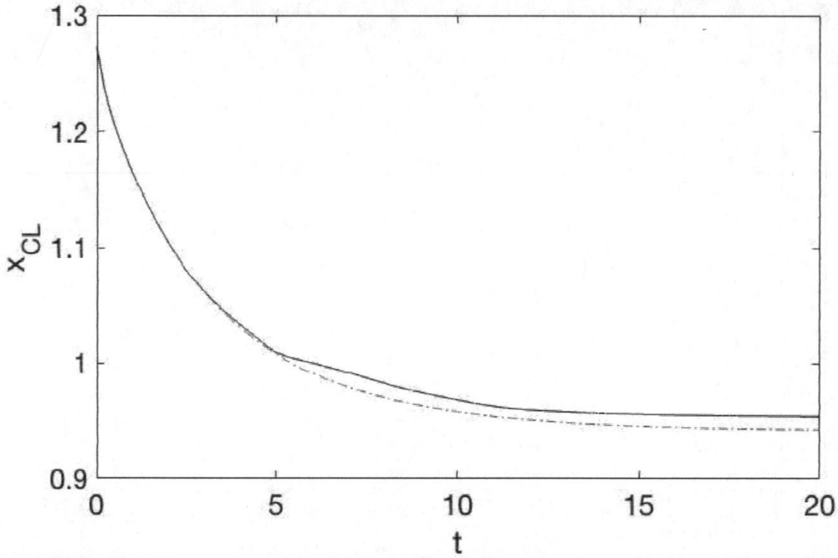

Figure 4. Evolution of the contact line position for spatially uniform substrate (dot-dashed line) and spatial variation of disjoining pressure modeled by two Gaussian profiles (solid line), one centered at $x_0 = 1$ and another one at $x_0 = 0.95$, as described by Eq. (7). Slow-down of the contact line is clearly seen

4. Conclusions

We carried out a numerical investigation of contact line motion over a susbtrate with spatially non-uniform wetting properties. The fully unsteady lubrication-type model is developed which allows us to track evolution of the interface as a function of time in response to substrate temperature changes. The non-uniformity of substrate properties is modeled through the spatial variation of the Hamaker constant in the formula for disjoining pressure. The contact line speed at the time when the region of spatial non-uniformity is approached is identified as the key parameter determining the contact line dynamics. We started by considering a Gaussian profile of local spatial variation of the Hamaker constant. If the contact line is moving relatively fast, its slow-down due to local change in disjoining pressure is compensated by the subsequent speed-up in the region when the disjoining pressure is uniform again. However, if another region of spatial non-uniformity is encountered afterwards, the contact line can slow down significantly and its total displacement in response to temperature change will be less than for the case of spatially uniform substrate properties.

Acknowledgements

This work was supported by the Russian Science Foundation (project No. 16-19-10675).

References

[1] Dhir V K 1998 Boiling heat transfer, *Annu. Rev. Fluid Mech.* **30** 365
[2] Stephan P and Kern J 2004 Evaluation of heat and mass transfer phenomena in nucleate boiling, *Int. J. Heat Fluid Flow* **25** 140
[3] Kim J 2007 Spray cooling heat transfer: The state of the art, *Int. J. Heat Fluid Flow* **28** 753
[4] Kirichenko E O and Gatapova E Ya 2016 *MATEC Web of Conferences* **72** 01045
[5] Ajaev V S, Morris S and Homsy G M 2002 *J. Colloid Interface Sci.* **254** 346
[6] Oron A, Davis S H and Bankoff S G 1997 *Rev. Mod. Phys.* **69** 931
[7] Craster R V and Matar O K 2009 *Rev. Mod. Phys.* **81** 1131
[8] Ketelaar C and Ajaev V S 2015 *Phys. Fluids* **27** 112110
[9] Murisic N and Kondic L 2011 *J. Fluid Mech.* **679** 219
[10] Israelachvili J N 2011 *3rd Ed. Academic Press* 704

IOP Conf. Series: Journal of Physics: Conf. Series **925** (2017) 012017 doi:10.1088/1742-6596/925/1/012017

Structure of gels layers with cells

B G Pokusaev[1], S P Karlov[1], A V Vyazmin[1,2], D A Nekrasov[1], N S Zakharov[1], D P Khramtsov[1], D A Skladnev[3] and D V Tyupa[3]

[1] Moscow Polytechnical University, Department of Chemical Engineering, ul. Staraya Basmannaya 21/4, Moscow, 105066, Russia

[2] Scientific Research Institute of Rubber Industry, poselok NIIRP, Sergiev Posad, 141312, Russia

[3] Federal Research Centre «Fundamentals of Biotechnology» RAS, Leninsky prospect, 33, build. 2, 119071, Moscow, Russia

E-mail: pokusaev2005@yandex.ru

Abstract. The structure of two-layer agarose gels containing yeast cells is investigated experimentally by spectrometry, to shed a light on the theoretical foundations for the development of bioreactors by the method of 3D bioprinting. Due to division, cells overcome the layer of the dispersion phase separating successively applied layers of the agarose gel. However a gel layer of 100 μm thick with a high concentration of silver nanoparticles completely excludes the infiltration of yeast cells through it. A special sort of agarose is suggested where the concentration of silver nanoparticles formed by cells from salt of silver can serve as an indicator of the state of the yeast cells in the volume of the gel.

1. Introduction

The idea of growing tissues and organs in vitro using stem cells is not new. However, in order to implement this idea, it is necessary to create special bioreactors which would be capable of maintaining the required temperature, pH-level, osmotic pressure, supplying cells with nutrients and oxygen, removing their metabolic products as well as fulfilling many other requirements, which provides necessary physiological conditions for immobilized cells [1]. The deviation from the optimal level of most of the listed factors leads to inhibition of metabolic activity, deterioration of the physiological state of cells and, ultimately, to a retirement of their growth [2-4].

The use of the method of 3D additive manufacturing to create such tissues and organs from stem cells seems to be very promising [5]. Currently gels are considered a high-potential structure-forming material for formation of artificial tissues from immobilized cells using the method of additive manufacturing. In the construction of 3D-bioreactors the gels can be used in the capacity of a neutral coat of gels in which biological microscopic objects are cultivated. They allow nutrients and oxygen to penetrate into the growing medium and remove metabolic products and carbon dioxide.

Agarose gel is a typical gel which is formed from the solution when the temperature is lowered. It is widely used in microbiology for growing of microbiological objects. Such gels have been thoroughly investigated (see, for example, [6]), including by the optical methods [7]. The following aspects have been studied: structure of links which are formed in the disperse phase of gel formation, physical-chemical properties of gels; also, technology of generation for various uses has been described [8]. Nevertheless, many properties of agarose gels which determine whether they may be

IOP Conf. Series: Journal of Physics: Conf. Series **925** (2017) 012017 doi:10.1088/1742-6596/925/1/012017

used for the creation of bioreactors with the help of additive manufacturing require additional studying.

Observation of biogenic nanoparticles growth (OBNG) methodology is used to assess the physiological state of living cells immobilized in an agarose gel [9, 10]. The idea is that within a few minutes only metabolically active cells are able to form nanoparticles of recover silver (Ag^o) from the solution of a silver salt in the liquid medium with cells. Respectively, the formation of nanoparticles is sharply inhibited with the weakening of the physiological state of the cells. In the present work, the OBNG methodology is used to analyze the state of living cells immobilized in agarose gels.

The aim here is to carry out an investigation the "thin" structure and adhesion between gel layers applied upon one another in presence of immobilized cells. The peculiarities of the displacement of yeast cells are studied in layered agarose gels intended for the formation of artificial biological structures using 3D-bioprinting technology. The directional restriction of growth of immobilized cells is considered due to the creation of layers of nanoparticles.

2. Materials and methods

In experiments related to immobilization in gels yeast cells *Yarrowia lipolytica* were used as model eukaryote biological object [11]. The cultivation of yeast immobilized cells in the gel is carried out under optimal temperature conditions - 28°C. Immobilization of cells is carried out by placing them into a gel of 1 mm^3 volume containing a 10% complete growth medium LB. A suspension of fresh yeast cells is introduced into an agarose gel at a temperature of 42°C with a concentration of 5×10^6 cells/mm^3 when studying the structure of gel layers with cells and 10^7 cells/mm^3 in experiments with nanoparticles. In all experiments, the yeast gel is placed on top of the sterile gel to ensure a better aeration of the cells.

An agarose-based gel "Chemapol" was used as a main gel during the experiments. It is known that gel properties can vary depending on their manufacturer. The gels used in the experiments were obtained by mixing agarose with distilled water with heating up to 90°C by convectional and UHF methods. Agarose gels with 0.6–1.5% of agarose are used in the experimental study of the layered structure of a gel with cells. The concentration of agarose gel in experiments with nanoparticles is 0.4% of agarose, both providing and immobilizing cells, and their volumetric proliferation. In these experiments, two types of agarose are used with differing in the level of electroendoosmosis (EEO).

For investigation two-layer gels system with immobilized cells the spectroscopic method is used. This method analyses the form of spectrum of transparent or reflected light of the sample. The growth and the displacement of yeast cells are investigated by recording the increase in the optical density of the samples in agarose gels of various densities using the optical fiber optical spectrometric system shown in figure 1.

Figure 1. Fiber-optic spectrometric system for recording changes occurring in multilayer gels used for evaluation of diffusion processes and the control of the of silver nanoparticles formation

IOP Conf. Series: Journal of Physics: Conf. Series **925** (2017) 012017 doi:10.1088/1742-6596/925/1/012017

The formation of biogenic silver nanoparticles Ag^0 is performed by the OBNG method. As a source of silver cations, a solution of silver nitrate salt $AgNO_3$ is used, introduced into a molten agarose gel (at a temperature of 42°C) immediately before its cooling at a final concentration of 5 μg/ml.

The registration of spectral properties and changes occurring in multilayer gels is carried out using an optical fiber spectrometric system based on equipment from OceanOptics (USA) for evaluation of diffusion processes and the control of the formation of silver nanoparticles. The system allows measurements at a 2 mm^2 point along the entire height of the cuvette.

3. Results and discussions

It was important to determine how gel of various thicknesses influences growth of cells in different conditions, practically in conditions of different aeration. Due to formation of layered gels with additive technologies, it is important to know the basic regularities in the behavior of immobilized microorganisms within the gel and in close proximity to border between gel layers, filled only with the dispersion medium. Here is the fundamental answer to the question of whether immobilized gels spontaneously spread through border between gel layers, formed of agarose gels, including different concentrations.

3.1. Gels layers with cells

Comparison photos of the two-layer gel system are presented, consisting of both sterile agarose gels with concentration of 0.6%, and the same system, when the upper layer contains encapsulated yeast cells (figure 2). Photos were taken 48 hours after the formation of two-layer systems. It is possible to notice that the gel with cells is much more turbid. Border between gel layers is also filled with the turbid substance and is blurred in the direction of the gel with the cells, whereas for a sterile system, it remains clear.

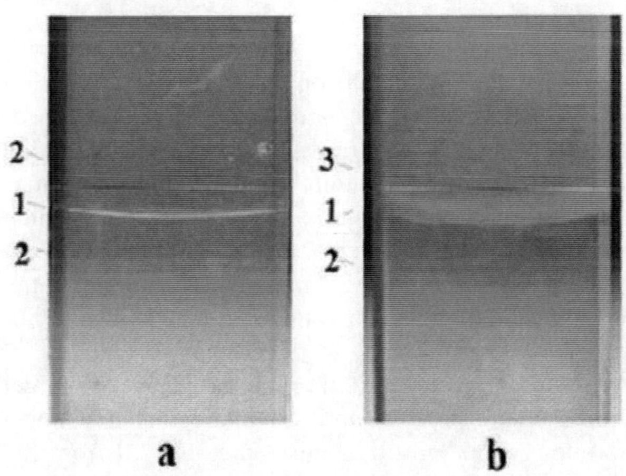

Figure 2. Comparison photos of the two-layer agarose gels without and with yeast cells. Notations: 1 – the interface of the gels; 2 – agarose gel with a concentration of 0.6%; 3 – the same gel with yeast cells

Using a specially developed thin cuvette by visual observation it was found that the density of the yeast growth remains unchanged in the volume of the agarose gel with a concentration of 0.6% to a depth of 11–12 mm. The density of yeast growth increased with continued incubation cuvettes up to 48 hours and its concentration is 10 times more compared with initial concentrations. It should be noted that during the cultivation of yeast in special cuvettes with layered top layer agarose gel with a concentration of 0.3% by weight (during modeling of usage of protection layer of liquid gel) the entire volume of the upper, the source of sterile liquid gel, also gradually filled with yeast cells. This indicates the ability of cells to overcome pre-formed boundary between gel layers in the direction of concentration changes for agarose gel of 0.6% to 0.3% by weight. This circumstance will require additional studies of the movement of specific living cells, designed specifically for 3D-bioprinting in the gels used for their cultivation, variety, and speed changing density.

Figure 3 shows the results of spectral sensing region adjacent to the boundary between gel layers for two-layer gel system formed by the gels with agarose concentration of 0.6% if the top layer of yeast cells (for gel cells distance is measured in the positive region). The measurements were carried out as one hour after the formation of such a system, and in 48 hours, which allows to trace the spectral changes caused by the condition of the gel and immobilized yeast cells. Intensity values of light transmission were treated to an appropriate amount of sterile gel of the same concentration, measured immediately after its formation. A layer of gel containing the cells has a lower ability for passing light compared with sterile gel, due to the increasing light scattering on yeast cells which size is about 10 μm.

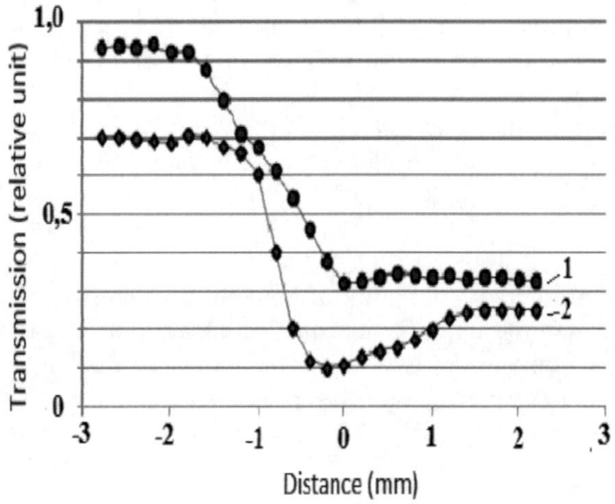

Figure 3. The time variation of the relative intensity of light transmittance of the layered gels with agarose concentration of 0.6% in the border areas of the distance to the boundary in the presence of the top layer of yeast cells. Visually observing the interface between the layers of the gels in the beginning of the experiment corresponds to a value of 0, with minus – without yeast (bottom layer), with a plus - with yeast (top layer). Notation of curves: 1 – within one hour after formation. 2 – after 2 days

We had some differences in the dependency of relative intensity of light transmitted from a distance to border between gel layers. In a two-layer gel system with cells one hour after its formation is practically absent the minimum intensity of transmitted light, which is detected for distinct two-layer sterile system. The reason for this is to that space between gel layers is filled by yeast cells in the process of formation. With increasing time of cell cultivation, such a minimum appears and becomes more and more noticeable, while its width increases significantly in the part of the layer, in which yeast cells were introduced originally. The behaviour of curve 2 in figure 3 resembles the change in concentration of a substance in a stationary environment when the diffusive mass transfers through the interface. It can be assumed that the mechanism of migration of cells from one environment to another via boundary between gel layers is presented.

It is important to note that the relative intensity of light transmittance through the layers of the gel after 48 hours it becomes lower than at the beginning of the process, and the gradient of its change from the sterile gel increases. The decrease in the intensity of light transmission explained from the standpoint of the self-seal dispersion medium in the gel when it is aging. This process takes place in both layers of the gel. However, the decrease in relative intensity for sterile gel is significantly more than that of the layer containing cells. This suggests that the observed pattern is caused by the penetration of yeast cells into sterile gel and the beginning of their reproduction and growth in it. Moreover, large gradient of relative intensity of light passing near the boundary shows that, initially, the yeast cells begin their reproduction and growth under sterile gel side boundary between gel layers.

3.2. Formation of nanoparticles in agarose gels by cells

For yeast cells grown in a complete growth medium the addition of a solution of silver nitrate $AgNO_3$ leads to the formation of silver nanoparticles 25–70 nm in size within 20–30 minutes, as it was previously found by [9, 10]. The formation of nanoparticles was observed both in photographs (with the transmission electron microscope) and by spectrometry (a specific peak close to the wavelength of the light 400 nm). In a sterile growth medium (that is, an environment that does not contain live yeast cells), nanoparticles were not formed at the same time, since the reduction of cations ceased at a stage not larger than clusters of silver atoms (up to 0.5–2 nm). Such clusters are detected using a transmission electron microscope, but they are not determined spectrometrically.

To study the formation of nanoparticles by cells in gels, the silver salt solution was introduced into a molten 0.4% agarose gel and into the same gel containing fresh yeast cells at a concentration of 10^7 cells/ml. Control cuvette with agarose gel did not contain cells. The cuvettes were incubated for 30 minutes at room temperature, after which the spectrum of both preparations was determined. As expected, a characteristic peak is observed indicating the presence of silver nanoparticles in cuvette with yeast cells. However, in a cuvette with a sterile agarose gel, such a peak is quite pronounced (see figure 4). The magnitude of the peak in the presence of yeast cells is higher than in the sterile gel. However, the difference is not significant as to evaluate the physiological state of the cells the gel using these data. Thus, it is established that silver nanoparticles are formed in a 0.4% agarose gel when the silver nitrate solution interacts with the gel itself.

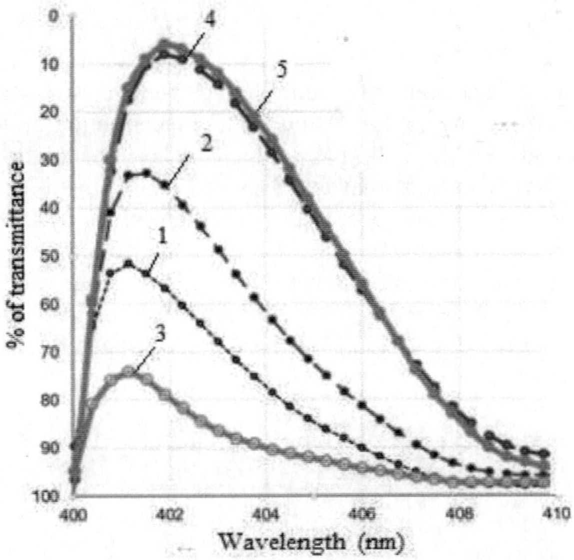

Figure 4. Transmission spectra during the formation of silver nanoparticles in agarose gels with different levels of EEO, including in the presence of yeast cells The composition of the samples: 1 - aqueous solution of $AgNO_3$, 2 - ordinary agarose gel with $AgNO_3$, 3 – low EEO agarose gel with $AgNO_3$, 4 - ordinary agarose gel with $AgNO_3$ and cells, 5 – low EEO agarose gel with $AgNO_3$ and cells

To prevent the formation of silver nanoparticles in agarose, another substance was used with a low level of EEO. Figure 4 shows that the formation of nanoparticles in the presence of agarose with low EEO is insignificant. In further experiments concerning the interaction of living cells and silver salts (as indicators of the physiological state of cells), only low EEO agarose was used.

When two-layer agarose gels are formed with agarose concentration of 0.4%, containing in the upper layer of the yeast cell, it is established that the yeast grows at an optimum temperature conditions pass through the interlayer boundary. After incubating the cuvette for 12 hours, the concentration of yeast cells in the upper layer increased approximately 10 times corresponding to approximately three divisions of a cell. Yeast cells also appear in the lower layer, but at a concentration much lower than in the upper layer. Hence the yeast cells pass through the boundary in double-layered gels as a result of their division.

In experiments on the evaluation of the interaction of agarose gels with different levels of EEO and a solution of silver nitrate, it is found that the formation of silver nanoparticles is much slower in low EEO agarose than in standard agarose. Formed nanoparticles show a high ability to diffuse through a 0.4% gel. The ability of the nanoparticles to pass through the gel makes it possible to use them to assess the state of the cells immobilized in the gel. In case of presence the agarose of fresh yeast cells and the silver salt solution in the upper layer of 0.4% low EEO gel, after the cultivation in the middle of the cuvette, two intermediate layers are formed. The upper one corresponds to the cell suspension on spectral characteristics; the lower one corresponds to the suspension of nanoparticles. The concentration of nanoparticles at high diffusion rate can be considered homogeneous over the volume of the gel. This allows us to obtain the information on the state of the cells providing the reduction of cations from the parameters of the nanoparticles. This assumption requires further confirmation.

The above experiment demonstrated that concentrations of silver nitrate up to 5 μg/ml are useful for assessing the physiological state of the active cells, since the cells retain the same ability to divide, as without any silver substance.

In the experiment, it was checked whether a layer of nanoparticles could prevent the propagation of yeast cell growth in layers of 0.4% gel from low EEO agarose. The mature agent of silver nanoparticles is mixed with a small volume of gel and is applied as a separator between two layers of the gel with cells and without cells. The thickness of the separation layer is 100 μm. Yeast capable of growing in the upper layer does not overcome the gel layer containing the preparation of silver nanoparticles. Thus, it can be suggested to use layers with silver nanoparticles to limit the growth of bioobjects in gels during the realization of biotechnology of 3D-bioprinting.

4. Conclusions

Experimentally for cells of the yeast *Yarrowia lipolytica*, the possibility of their spontaneous transition from one layer agarose gel to another through the surface between gel layers filled dispersive medium was determined. This phenomenon occurs both when gel layers are of the same concentration and when concentration is drastically changes. It is found that the cells pass from one layer of gel to another through the boundary as a result of their division. Note that a gel layer of 100 μm thick containing concentrated silver nanoparticles does not allow cells to pass through themselves by the division.

It is observed that the addition of silver nitrate with a concentration lower than 5 μg/ml can be used as an indicator of the state of cells in the gel volume. Healthy and active yeast cells restore silver to form nanoparticles. Such nanoparticles readily diffuse through a 0.4% agarose gel. Their concentration outside the gel revel the state of the yeast cells inside the gel.

Acknowledgments

This work was supported by the Russian Science Foundation (project no. 15-19-00177).

References

[1] Rodrigues C A V, Fernandes T G, Diogo M M, da Silva C L and Cabral J M S 2011 *Biotech. Adv.* **29** 815
[2] Ott H C, Matthiesen T S, Goh S K, Black L D, Kren S M and Netoff T I 2008 *Nat. Med.* **14** 213
[3] Placzek M R, Chung I M, Macedo H M, Ismail S, Mortera B T and Lim M 2009 *J. R. Soc. Interface* **6** 209
[4] Lee M K, Rich M H, Baek K, Lee J and Kong H 2015 *Sci. Rep. (Nature)* **5** 8948
[5] Marga F, Jakab K, Khatiwala C, Shepherd B, Dorfman S, Hubbard B, Colbert S and Gabor F 2012 *Biofabr.* **4** 02200
[6] Duckworth M and Yaphe W 1971 *Carbohyd. Res.* **16** 189
[7] Rees D A, Scott W E and Williamson F B 1970 *Nature* **227** 390
[8] Santos G A 1990 *A Manual for the Processing of Agar from Gracilaria* (Manila: ASEAN/UNDP/FAO Regional Small-Scale Coastal Fisheries Development Project)
[9] Sorokin V V, Skladnev D A, Volkov V V, Tereshchenko E Y, Mulukin A L and Gal'chenko V F 2013 *Dokl. Biol. Sci.* **452** 325
[10] Skladnev D A, Sorokin V V and Gal'chenko V F 2017 *Microbiology* **86** 355
[11] Scioli C and Vollaro L 1997 *Water Res.* **31** 2520

IOP Conf. Series: Journal of Physics: Conf. Series **925** (2017) 012018 doi:10.1088/1742-6596/925/1/012018

Synthesis of silicon oxide microropes on the copper substrate with SiO_2 interlayer

E Baranov, S Khmel, A Zamchiy and E Shatskiy

Kutateladze Institute of Thermophysics SB RAS, 630090 Novosibirsk Ac. Lavrentiev ave. 1, Russia

E-mail: itpbaranov@gmail.com

Abstract. Nanostructuring of the surface is a promising technology for the processes of boiling. In this paper, we synthesized array of "microropes" from silicon oxide nanowires on the copper substrate with a silicon oxide intermediate layer by gas-jet electron beam plasma CVD method. The morphology for the synthesis time of 2 minutes 30 seconds and 5 minutes was obtained. The water droplet on the silicon oxide nanowires shows the measured contact angles 14° and 10° for deposition times of 5 min and 2 min 30 sec, respectively.

1. Introduction

Nanostructuring of the surface, which is used for the processes of boiling, is a promising technology by Surtaev and Serdyukov (2016). Most often, monocrystalline silicon wafers are used for nanostructuring. But copper heaters can be more promising and these heaters with silicon nanostructures are a very interesting system. In addition, one of the important properties of nanomodification is to change of the surface wettability.

In this paper, we synthesized array of "microropes" from silicon oxide nanowires on the copper substrate with a silicon oxide (SiO_2) intermediate layer. The morphology for the synthesis time 2 minutes 30 seconds and 5 minutes was obtained. The contact angle for the nanostructures was measured.

2. Experiment

2.1. Synthesis of the silicon oxide nanowires array

The silicon oxide nanowires were synthesized from mixture monosilane-argon (37 sccm) with gas diluent hydrogen (386 sccm) by gas-jet electron beam plasma CVD method by Zamchiy and Khmel (2014). Oxygen (3 sccm) was supplied directly into the vacuum chamber. The synthesis was carried out on the copper substrate with/without SiO_2 interlayer and with tin film mass thickness of about 60 nm. The substrate is copper core with a round head with diameter d=5 mm. The SiO_2 film was deposited by magnetron sputtering and the tin film was deposited by thermal vacuum deposition. The process of nanowire synthesis on the substrate with the catalyst consists of three stages: heating up to operating temperature (400°C), treatment of hydrogen plasma and the actual growth of nanowires by vapor–liquid–solid (VLS) mechanism. The synthesis temperature was 330°C.

IOP Conf. Series: Journal of Physics: Conf. Series **925** (2017) 012018 doi:10.1088/1742-6596/925/1/012018

2.2. Measuring techniques

The surface morphology was determined by the methods of scanning electron microscopy (SEM) using the JEOL JSM-6700F microscope.

The contact angle of the sample with water was measured by DSA-100 KRUSS device with high-precision system of liquid supply with a minimal dosing step of 0.1 μl.

3. Results and Discussion

Figure 1 shows the SEM image of the tin film on the copper substrate. It is seen that the tin wets the surface. Individual particles with a typical size of 500 nm do not have a fixed shape. After heating and treatment with hydrogen plasma, the surface morphology changes dramatically, no catalyst particles are observed in the SEM image. We assume that the heating of the tin above the melting temperature in combination with hydrogen plasma treatment causes chemical reactions between the copper substrate and the tin and leads to the formation of intermetallic compounds (Yin and Chauhan, 2008). Figure 2 shows a SEM image of the copper substrate without SiO_2 interlayer after synthesis. Synthesis of the array of "microropes" from silicon oxide nanowires does not occur on this catalyst.

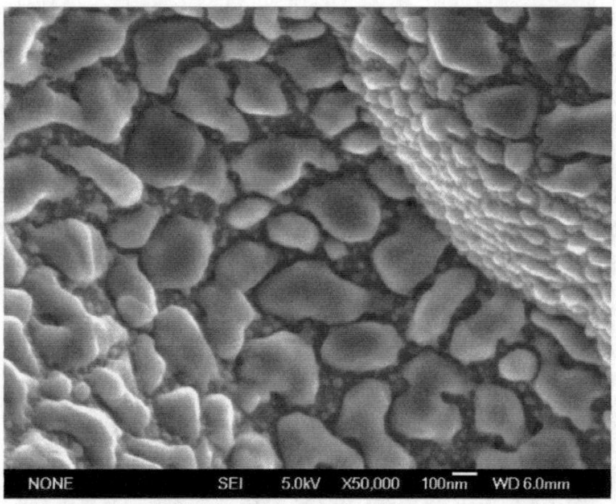

Figure 1. SEM image of tin film on the copper substrate without SiO_2 interlayer.

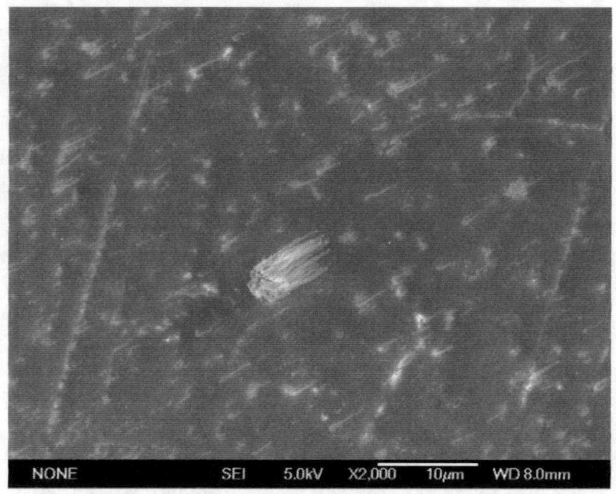

Figure 2. SEM image of the copper substrate without SiO_2 interlayer after synthesis.

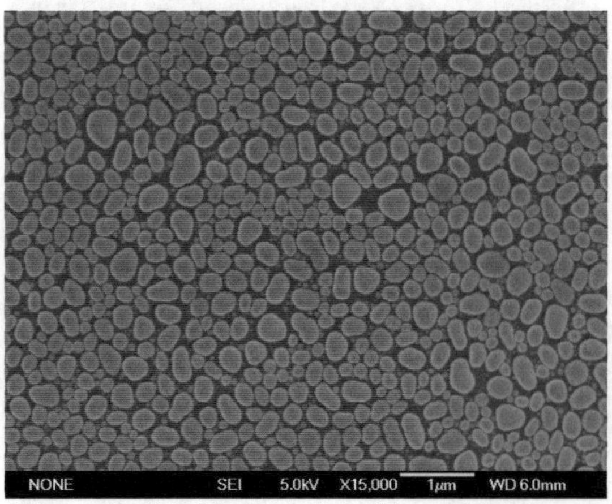

Figure 3. SEM image of tin film on the copper substrate with SiO_2 interlayer.

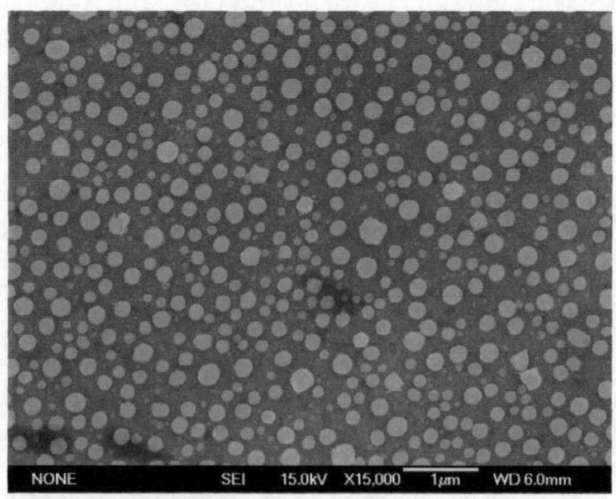

Figure 4. SEM image of tin film on the copper substrate with SiO_2 interlayer after heating and treatment with hydrogen plasma.

To solve this problem, it was decided to deposit the intermediate layer, which was supposed to have good adhesion with copper, and which is not wetted by tin. This intermediate layer was a silicon oxide with a thickness of 100 nm.

Figure 3 shows the SEM image of the tin film on the copper substrate with SiO_2 interlayer. Particles of tin catalyst have clear boundaries. Particle shape is round and oval. Similar morphology of the catalyst is suitable for synthesis of nanowires of silicon oxide (Khmel and Baranov, 2016).

Figure 4 shows the SEM image of the tin film on the SiO_2 interlayer after heating and hydrogen plasma treatment. The treatment promotes reduction of oxide layers on the surface of tin particles. Since the temperature is above the melting point of tin and since the particles are liquid, they become of a regular spherical shape. The particle size distribution is within 400–700 nm. X-ray EDS analysis showed the oxygen content in the catalyst particle to be about 3%.

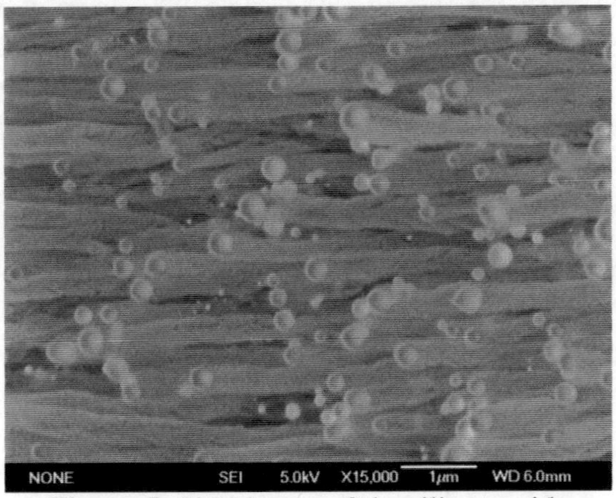

Figure 5. SEM image of the silicon oxide nanowires array on the copper substrate with SiO_2 interlayer for deposition time 5 min.

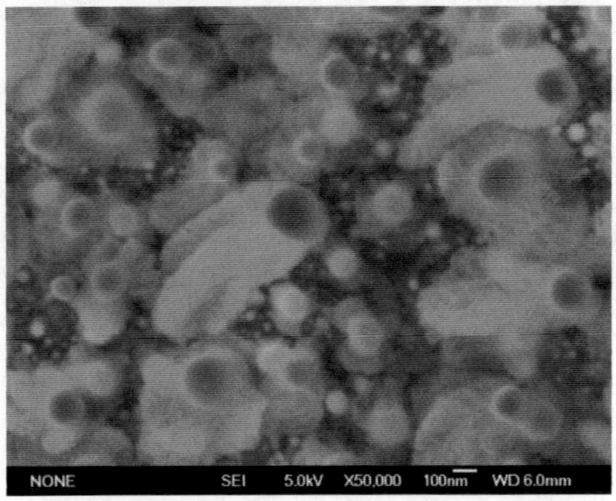

Figure 6. SEM image of the silicon oxide nanowires array on the copper substrate with SiO_2 interlayer for deposition time 2 min 30 sec.

The SEM image in Fig. 5 shows the silicon oxide nanowires array for deposition time of 5 min. Each microrope consists of numerous nanowires with average diameter of about 15 nm. Nanowires bunch is on the bottom of the catalyst particle, whereas the nanowires are absent at the top of it. Length of microrope was about 10 micrometer. Reducing the time of synthesis to 2 min 30 sec resulted in a decrease of the length of the microrope and change of the morphology. As can be seen in the Fig. 6, the length of the microrope has decreased to 1 micrometer. For single particles, the microrope is divided into several separate tails. Most likely, as the length of the microrope increases, they begin to act on each other forming an ordered array.

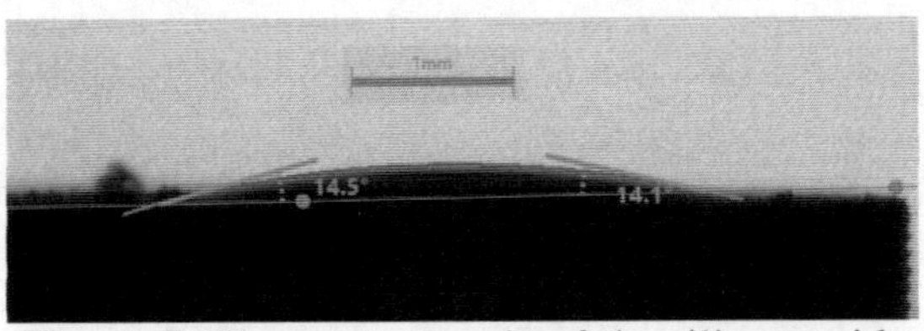

Figure 7. The contact angle of the silicon oxide nanowires grown on the copper substrate with SiO_2 interlayer for deposition time 5 min.

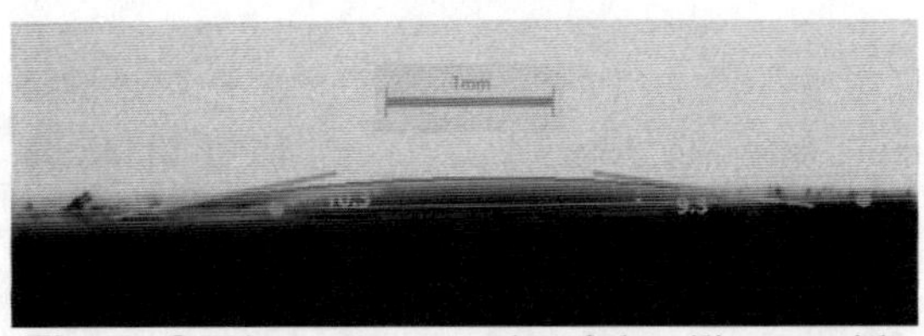

Figure 8. The contact angle of the silicon oxide nanowires grown on the copper substrate with SiO_2 interlayer for deposition time 2 min 30 sec.

Analysis of the photograph of the water droplet on the silicon oxide nanowires grown on the copper substrate with SiO2 interlayer shows the measured contact angle were for deposition time 5 min - 14° (Fig. 7) and for deposition time 2 min 30 sec - 10° (Fig. 8). This confirms that the surface of the silicon oxide nanowires array, synthesized in this work, is hydrophilic in nature (Baranov and Zamchiy, 2016). In addition, it was observed that the droplets on the surface have a very large imbibition layer. We assume that water penetrates easily into grooves constituting the relief in the case of Cassie impregnating wetting.

4. Conclusion

In this paper, the array of "microropes" from silicon oxide nanowires on the copper substrate by gas-jet electron beam plasma CVD method were synthesized. The silicon oxide intermediate layer with a thickness of 100 nm was deposited to solve the problem with wetting the copper substrate by tin. The morphology for the synthesis times of 2 minutes 30 seconds and 5 minutes was obtained. The water droplet on the silicon oxide nanowires, grown on the copper substrate with the silicon oxide interlayer, shows the measured contact angles of 14° and 10° for deposition times of 5 min and 2 min 30 sec, respectively.

Acknowledgements

The study was carried out by the grant of Russian Science Foundation (agreement no. 15-19-30038). The authors thank Ekaterina Kirichenko for the contact angles measurements.

References

[1] A. Surtaev, V. Serdyukov, A. Pavlenko 2016 *Nanotechnologies Russ.* **11** 696
[2] A. Zamchiy, E. Baranov and S. Khmel 2014 *Physica status solidi (c)* **11** 1397
[3] L. Yin, A. Chauhan, T.J. Singler 2008 *Mater. Sci. Eng. A.* **495** 80
[4] S. Khmel, E. Baranov, A. Zaikovskii, A. Zamchiy, E. Maximovskiy, D. Gulyaev, K. Zhuravlev, 2016 *Phys. Status Solidi. (a)* **213** 1774
[5] S. Khmel, E. Baranov, A. Zamchiy, E. Maximovskiy, D. Gulyaev, K. Zhuravlev 2016 *Phys. Status Solidi. (a)* **213** 1790

Comparison study on the calculation formula of evaporation mass flux through the plane vapour-liquid interface

L Zhang, Y R Li, L Q Zhou and C M Wu

Key Laboratory of Low-grade Energy Utilization Technologies and Systems of Ministry of Education, College of Power Engineering, Chongqing University, 174 Shazhengjie, Shapingba, Chongqing 400044, China

E-mail: liyourong@cqu.edu.cn

Abstract. In order to understand the influence of various factors on the evaporation rate on the vapor-liquid interface, the evaporation process of water in pure steam environment was calculated based on the statistical rate theory (SRT), and the results were compared with those from the traditional Hertz-Knudsen equation. It is found that the evaporation rate on the vapor-liquid interface increases with the increase of evaporation temperature and evaporation temperature difference and the decrease of vapor pressure. When the steam is in a superheated state, even if the temperature of the liquid phase is lower than that of the vapor phase, the evaporation may also occur on the vapor-liquid interface; at this time, the absolute value of the critical temperature difference for occurring evaporation decreases with the increase of vapor pressure. When the evaporation temperature difference is smaller, the theoretical calculation results based on the SRT are basically the same as the predicated results from the Hertz-Knudsen equation; but the deviation between them increases with the increase of temperature difference.

1. Introduction

The evaporation phenomenon exists widely in daily life and industrial production process, such as, spray drying, electronic cooling, spray printing, DNA detection and protein chip technology and so on, in which the calculation and control of evaporation rate has a very important influence on many industrial production process, and has already attracted the attention of many scholars [1-3].

Many factors will affect the interfacial evaporation rate, including gas composition, temperature, pressure and flow velocity and so on. When the evaporation of liquid is in the static pure steam environment, the evaporation mass flux j_m can be calculated according to the theory of Classical Kinetic Theory (CKT) [4-7], in which Hertz-Knudsen-Schrage equation or Hertz-Knudsen equation is used commonly

$$j_{\mathrm{m}} = \alpha\sqrt{\frac{M}{2\pi R}}\left[\frac{P_{\mathrm{s}}(T_{\mathrm{l}})}{\sqrt{T_{\mathrm{l}}}} - \frac{P^{\mathrm{V}}}{\sqrt{T_{\mathrm{v}}}}\right] \tag{1}$$

In the formula, M is the medium molar mass, R the universal gas constant, R=8.314 J/(mol·K), P_s the saturation pressure corresponding to interfacial liquid-phase temperature T_l, P_v and T_v as the vapor pressure and the vapor temperature respectively, and α the evaporation adaptation coefficient determined by the experiment.

The equation (1) shows that the relationship of the evaporation flux and the temperature difference between two sides of vapor-liquid interface is nonlinear. If the vapor and the liquid are in equilibrium state, T_l=T_v, then the evaporation flux j_m=0, and no evaporation occurs. The Hertz-Knudsen equation (1) is linearized around equilibrium, and becomes the following form [8-9]

$$j_m = \alpha\rho_v h_{fg}\sqrt{\frac{M}{2\pi RT_v^3}}(T_l - T_v)=\alpha\rho_v h_{fg}\sqrt{\frac{M}{2\pi RT_v^3}}\Delta T \quad (2)$$

in which, ρ_v is the vapor density, h_{fg} the latent heat of vaporization, ΔT the temperature difference between the liquid and vapor phases, namely, ΔT=T_l–T_v.

It should be noted that both the equations (1) and (2) are the calculation formulas of vapor-liquid interfacial evaporation flux on basis of the experiment, in which the evaporation coefficient α is constant and is determined by the experiment. Therefore, Ward and Fang [10] derived the evaporation mass flux formula according to the Statistical Rate Theory (SRT) in 1999, which was certified by a series of experimental results [11-14]. Compared with the classical Hertz-Knudsen equation, there is no empirical constant in this expression, which not only can calculate the interfacial evaporation rate more accurately, but also can make up for the deficiencies of the classical kinetic theory that cannot reflect the interfacial non-equilibrium effect.

Based on the interfacial evaporating statistical rate theory and taking water as the working medium, this paper analyzes the effects of various non-equilibrium thermodynamic factors on the interfacial evaporation mass flux at the vapor-liquid interface. The results are compared with those of the classical Hertz-Knudsen equation.

2. Evaporation statistical rate theory

Take the single component steady evaporation system with the liquid phase, the vapor phase and the interface, as shown in figure 1. The temperatures of vapor phase and liquid phase in the vapor-liquid interface were T_I^V and T_I^L, the vapor pressure P^V, the saturation pressure $P_s(T_I^L)$ corresponding to the liquid phase temperature. To maintain steady state evaporation, the liquid is continually supplied from the bottom part, and the steam from evaporating is continuously removed from the upper part. The liquid supplement rate should be equal to the evaporation rate to maintain the vapor-liquid interfacial position. Simultaneously, the liquid phase is touched with the constant temperature heat source T^L, and all the heat for vaporing comes from the heat source.

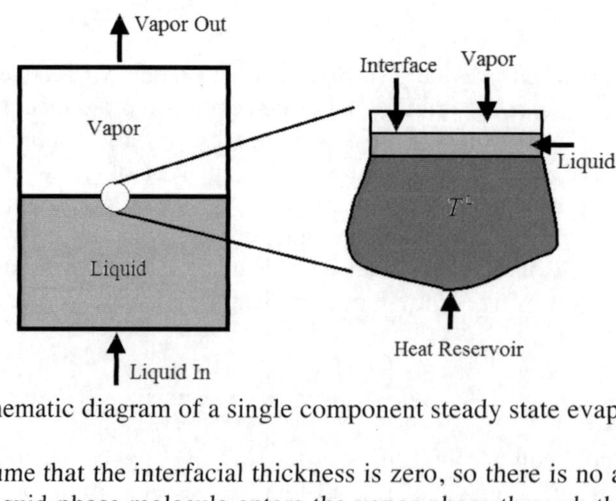

Figure 1. Schematic diagram of a single component steady state evaporation system.

For simplicity, assume that the interfacial thickness is zero, so there is no adsorption process at the interface. Namely, a liquid-phase molecule enters the vapor phase through the interface and does not stay at the interface, and vice versa. Meanwhile, each phase of vapor and liquid is homogeneous and all the intensive parameters are constant. According to the statistical rate theory, the evaporation mass rate at the vapor-liquid interface can be deduced [10-14]:

$$j_m = 2m_w K_e \sinh\left(\frac{\Delta s_{LV}}{k_b}\right) \tag{3}$$

$$K_e = \frac{P_s\left(T_I^L\right)}{\sqrt{2\pi m k_b T_I^L}} \exp\left\{\frac{v_f\left(T_I^L\right)}{v_g\left(T_I^V\right)}\left[\frac{P^L}{P_s\left(T_I^L\right)} - 1\right]\right\} \tag{4}$$

$$\frac{\Delta s_{LV}}{k_b} = 4\left(1 - \frac{T_I^V}{T_I^L}\right) + \ln\left[\left(\frac{T_I^V}{T_I^L}\right)^4 \frac{P_s\left(T_I^L\right)}{P_I^V}\right] + \ln\left[\frac{q_{vib}\left(T_I^V\right)}{q_{vib}\left(T_I^L\right)}\right] + \left(\frac{1}{T_I^V} - \frac{1}{T_I^L}\right)\sum_{i=1}^{3n-6}\left(\frac{\theta_i}{2} + \frac{\theta_i}{e^{\theta_i/T_I^V} - 1}\right) + \frac{v_f\left(T_I^L\right)}{k_b T_I^L}\left[P^L \quad P_s\left(T_I^L\right)\right] \tag{5}$$

$$q_{vib}(T) = \prod_{i=1}^{3n-6} \frac{e^{-\theta_i/2T}}{1 - e^{-\theta_i/T}}, \quad \theta_i = \frac{h\omega_i}{2\pi k_b} \tag{6}$$

The liquid phase pressure P^L under equilibrium state should satisfy the following relation

$$P^L = P_I^V + \frac{2\gamma^{LV}\left(T_I^L\right)}{r_I} \tag{7}$$

Above all, Δs^{LV} indicates the entropy change caused by liquid-phase molecules into the vapor-phase space, k_b is Boltzmann constant, k_b=1.38×10^{-23} J/K, and h is Planck constant, h=6.626×10^{-34} J·s; m_w is for the molecular mass of evaporation, n for the atomic number, v_f and v_g for the specific volumes of saturated liquid molecules and gas molecular, γ^{LV} as the surface tension of vapor-liquid interface, θ_i as the characteristic temperature, ω_i as the phonon frequency, and q_{vib} as the vibrational partition function. If the evaporation interface is flat, the curvature radius of the interface is r_I=0, and the equation (7) becomes $P^L = P_I^V$.

3. Results and discussion

The working medium is water, the atomic number is n=3, and the molar mass is M=0.018015kg/mol. For the sake of simplicity, assume that the evaporating vapor-liquid interface is flat and the thermodynamic properties of water and steam are calculated according to the method in the literature [15]. Vibrational frequency of covalent bond of water molecules includes symmetric stretching vibration frequency, anti-symmetric stretching vibration frequency and flexural vibration frequency. Stretching vibration frequency is usually about 2 times of flexural vibration frequency. The values for three frequencies ω_i are ω_1=3.651×10^6×2.9979×10^8=1.0945×10^{15} Hz, ω_2=3.756×10^6×2.9979×10^8 1.1260×10^{15} Hz and ω_3=1.590×10^6×2.9979×10^8=0.4767×10^{15} Hz, respectively [14].

Figure 2 shows the variation of vapor flux with the liquid-phase temperature T_l and the evaporation temperature difference $\Delta T = T_l - T_v$. When the steam is saturated, if the liquid-phase temperature is higher than the steam temperature, the liquid is in the super-saturated state, and the saturated pressure corresponding to the liquid temperature must be higher than the vapor pressure. Therefore, the evaporation will occur at the vapor-liquid interface. When both the vapor and the liquid are in the saturated state, the Clausius-Clapeyron equation is described by

$$\frac{dP}{dT} = \frac{\rho_v h_{fg}}{T} \tag{8}$$

Thus we can see that under the given evaporation temperature difference, the vapor density ρ_v increases rapidly with the increase of temperature, which makes the difference between the saturation pressure and the vapor pressure increase, so the interfacial evaporation rate increased rapidly, as shown in figure 2 (a). At a given liquid-phase temperature, the value for the right-hand side of equation (8) is constant, so the interfacial evaporation flux increases linearly with the increase of evaporation temperature difference, as shown in figure 2 (b). Special attention should be paid to that when the steam is saturated, if the liquid-phase temperature is lower than the steam temperature $\Delta T<0$, the liquid phase must be in a supercooled state, and not the evaporation process but the condensation process occurs at the vapor-liquid interface, so the interfacial evaporation flux is negative value.

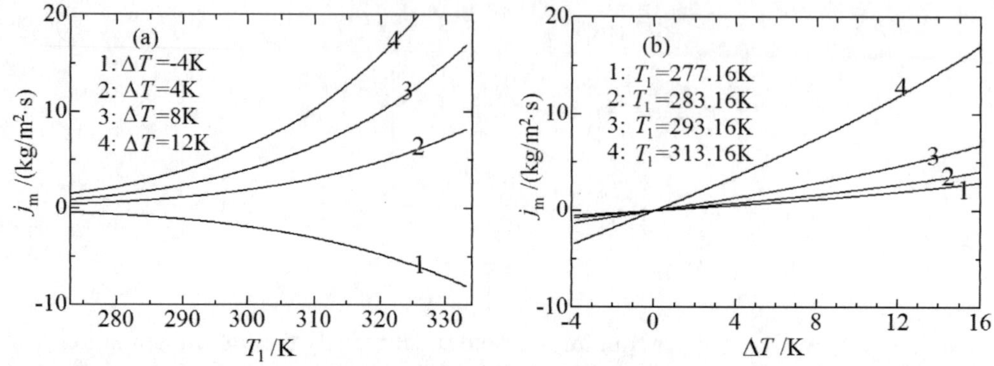

Figure 2. Variation of evaporation mass flux with the liquid-phase temperature (a) and evaporation temperature difference (b) in saturated steam environment.

In the experimental studies on vapor-liquid interfacial evaporation phenomenon [11-14], the continuous pumping on the steam chamber is to maintain a constant pressure. At this time, the steam is usually in a superheat state. When the steam is superheated, a new parameter η is introduced to describe the superheat degree and is defined as the ratio of steam pressure and saturation pressure corresponding to steam temperature, i.e.

$$\eta = P_{\mathrm{I}}^{\mathrm{V}} / P_{\mathrm{s}}^{\mathrm{V}}\left(T_{\mathrm{I}}^{\mathrm{V}}\right) \tag{9}$$

Obviously, η is smaller, the superheat is higher; $\eta=1$ means the saturated steam.

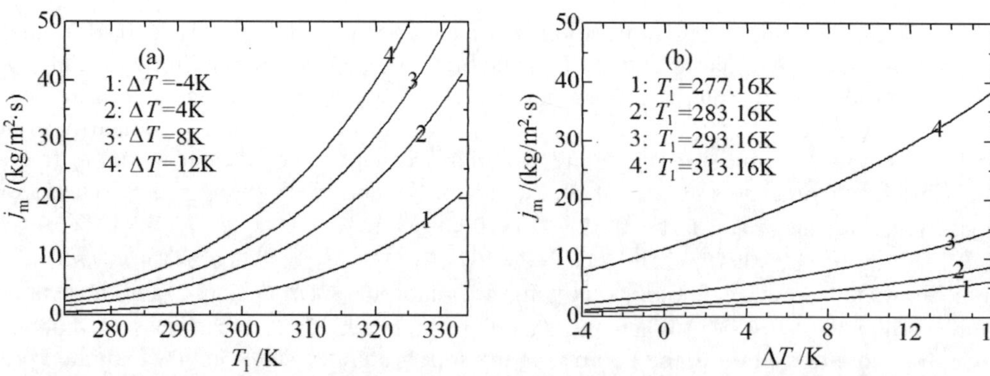

Figure 3. Variation of evaporation mass flux with the liquid-phase temperature (a) and temperature difference (b) in superheated steam environment at $\eta=0.5$

Figure 3 gives the variation of evaporation mass flux with the liquid-phase temperature and the evaporation temperature difference in superheated steam environment at $\eta=0.5$. Overall, the variation of mass flux in the superheated steam is similar to that in the saturated vapor. However, there are three differences. (1) Because the steam is superheated, the pressure dropped, the vapor-phase chemical potential decreased, the chemical potential difference between liquid phase and vapor phase as the vaporing driving force increased, so the interfacial evaporation mass flux increased. (2) The chemical potential of the pure substance is not only related to the pressure, but also depends on the temperature. The lower the pressure is and the higher the temperature is, the lower the chemical potential is. Therefore, when the steam is in the superheated state, even if its temperature is higher than the liquid-phase temperature $\Delta T=T_l-T_v<0$, it also ensures that the chemical potential of the vapor phase is lower than that of the liquid phase. Therefore, the evaporation process can still be carried out and evaporation mass flux is still positive value, which has been confirmed by a large number of experimental results [11-14]. (3) Due to the superheated steam, the Clausius-Clapeyron equation is no longer applicable, and the evaporation mass flux no longer increases linearly with the increase of evaporation temperature difference, as shown in figure 3 (b).

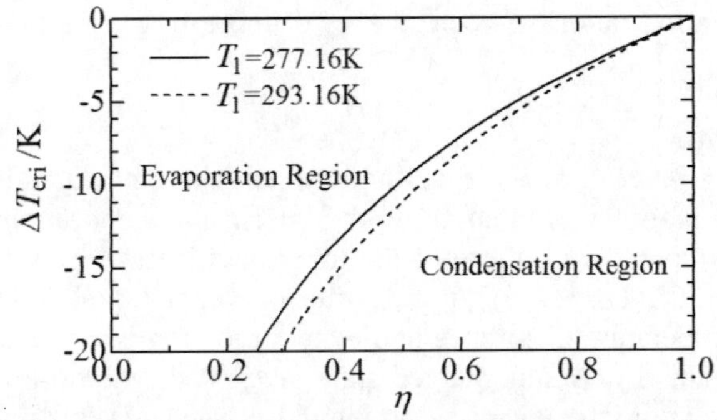

Figure 4. Critical conditions of interfacial evaporation.

From figure 2, when the steam is saturated, if the liquid-phase temperature is lower than the steam temperature, $\Delta T<0$, not the evaporation but the condensation occurs on the interface. When the steam is superheating, whether the evaporation process occurs or not depends on the two phase temperature difference and the vapor-phase pressure. Obviously, when the liquid-phase temperature is low enough, the condensation process will occur. Figure 4 shows the critical conditions whether the evaporation process can occur on the interface under two typical liquid-phase temperatures. The left-upper part is the evaporation region, and the right-lower part is the condensation region. The absolute value of the critical temperature difference in the evaporation process decreases with the increase of the vapor-phase pressure. At $\eta=1$, the steam is saturated and the vapor and the liquid are in equilibrium state, therefore, $\Delta T_{cri}=0$.

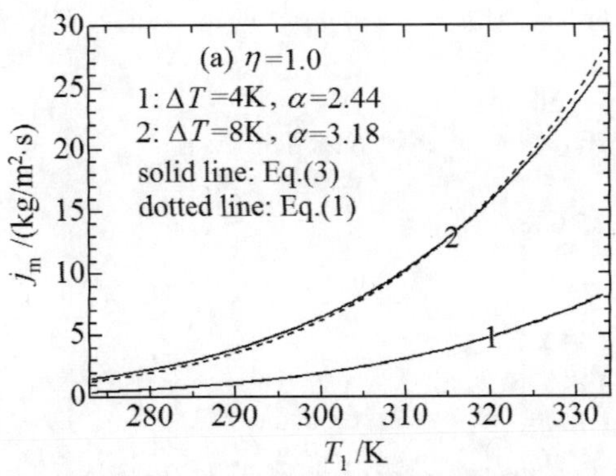

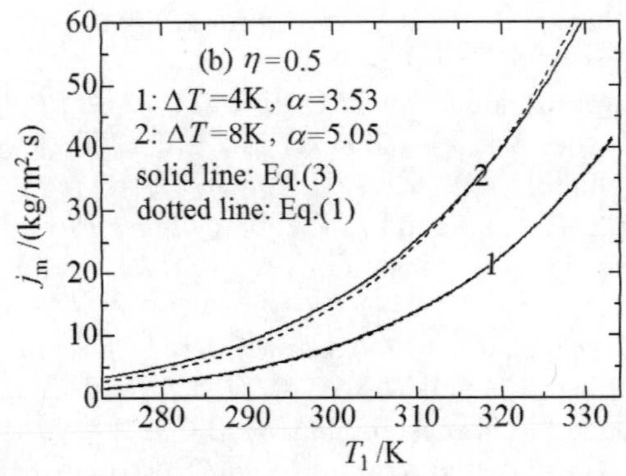

Figure 5. Comparison of calculating results between SRT theory and Hertz-Knudsen equation for vapor-liquid interfacial evaporation rate.

Compared with the traditional Hertz-Knudsen equation, the greatest merit of the SRT theory for calculating the evaporation rate on the vapor-liquid interface is that it does not require any empirical constant or coefficient, and is a pure theoretical calculation formula. The comparison between the two calculation results of evaporation mass flux for water is shown in figure 5. Here, the evaporation adaptation coefficient α in equation (1) is fittingly determined according to the calculation results of the SRT. From the figure, when the steam is saturated and the evaporation temperature difference is small, the calculated results of SRT theory are very close to those of the Hertz-Knudsen equation, which is in the vicinity of the equilibrium state. As long as the suitable evaporation coefficient is selected, both calculation results are basically the same. However, when the temperature difference is larger, such as, ΔT=8K, the difference between the calculated results increases, and the increasing speed of the evaporation rate predicted by SRT theory with the liquid-phase temperature is slower than that of the Hertz-Knudsen equation. On the other hand, when the vapor is in a superheat state, the difference of the evaporation rate between the two indicating results increases, as shown in figure 5 (b).

4. Results and discussion

The evaporation process of water in pure steam environment was calculated by adopting the statistical rate theory (SRT), and the main factors that affect the interfacial evaporation rate were analyzed. The calculation results were compared with those of the traditional Hertz-Knudsen equation, and the main conclusions were as follows: (1) the evaporation rate of vapor-liquid interface increases with the increase of evaporation temperature difference and evaporation temperature and the decrease of vapor pressure; (2) when the steam is in a superheated state, even if the liquid-phase temperature is lower than the vapor-phase temperature, the evaporation may also occur. At this point, the absolute value of the critical temperature difference decreases with the increase of vapor pressure; (3) when the evaporation temperature difference is smaller, the calculation results for the SRT theory and the Hertz-Knudsen equation are basically the same, but the deviation increases with the increase of temperature difference.

Acknowledgement

This work is supported by National Natural Science Foundation of China (Grant No. 11532015).

References

[1] Manokar A M, Murugavel K K and Esakkimuthu G 2014 *Rene. Sust. Energ. Rev.* **38** 309
[2] Erbil H Y 2012 *Adv. Colloid Interface* **170** 67
[3] Davoust L and Theisen J 2012 *Procedia Engineering* **47** 1
[4] Marek R and Straub J 2001 *Int. J. Heat Mass Transf.* **44** 39
[5] Schrage R A 1953 *Theoretical Study of Interphase Mass Transfer* (New York: Columbia University Press)
[6] Barrett J and Clement C 1992 *J. Colloid Interf. Sci.* **150** 352
[7] Kucherov R Ya and Rikenglaz L E 1960 *Doklady Akademii Nauk SSSR* **133** 735
[8] Palmer H J 1976 *J. Fluid Mech.* **75** 487
[9] Ji Y, Liu Q S and Liu R 2008 *Chinese Phys. Lett.* **25** 608
[10] Ward C A and Fang G 1999 *Phys. Rev. E* **59** 429
[11] Fang G and Ward C A 1999 *Phys. Rev. E* **59** 443
[12] Kapoor A and Elliott J A W 2010 *J Phys. Chem. B* **114** 15052
[13] Persad A H and Ward C A 2010 *J Phys. Chem. B* **114** 6107
[14] Duan F, Thompson I and Ward C A 2008 *J Phys. Chem. B* **112** 8605
[15] Wagner W and Kruse A 1998 *Properties of Water and Steam* (Berlin: Springer-Verlag)

IOP Conf. Series: Journal of Physics: Conf. Series **925** (2017) 012020 doi:10.1088/1742-6596/925/1/012020

Investigation of high frequency external perturbation effects on flow in a T-shape microchannel by μLIF technique

A Yu Kravtsova[1,2], **Yu E Meshalkin**[1,2] **and A V Bilsky**[1,2]

[1]Kutateladze Institute of Thermophysics, Siberian Branch of the Russian Academy of Sciences, 1 Lavrentyev Ave., Novosibirsk 630090, Russia
[2]Department of Scientific Research, Novosibirsk State University, 2 Pirogova Str., Novosibirsk 630090, Russia

E-mail: Kravtsova.Alya@gmail.com

Abstract. Investigation of high frequency external perturbation effect on flow inside T-shape microchannel was examined. In-phase pulsations of different frequencies were added to both inlets of the T-shaped microchannel to study mixing by means of Micro Laser Induced Fluorescence (μLIF) technique. For all flow regimes studied, mixing enhancement was obtained. Significant enhancement can be achieved at the beginning of the outlet channel operating in steady asymmetric regime (Re=186) by forcing at certain frequency ranges (f = 500Hz, f = 800Hz). Mixing suppression was also observed for two flow regimes (Re = 400, f = 1000Hz) and (Re = 120, f = 700Hz).

1. Introduction

Mixing in microchannels is important process for various biomedical and chemical applications. Microchannels usually operate in laminar or stratified flow conditions when mixing is governed by diffusion, which gives low mixing rate comparing to turbulent mixing. Therefore mixing enhancement in micromixers is an important topic. T-shaped micromixer has a simple and robust geometry. Flow regimes in T-shaped micromixer for different Reynolds numbers have been extensively studied previously. Engler et al. [1] showed that vertical flow arises in the microchannel because of Dean vortices. The critical Reynolds number, when Dean vortices lose their symmetry is 150. Mixing efficiency of two liquids within the range of Reynolds numbers from 50 to 1400 is investigated in [2]. Gobert et al. [3] numerically examined unsteady periodic regime. Mixing in a T-shaped microchannel for Reynolds numbers from 100 to 500 was done in [4]. The authors obtained the velocity and concentration fields using μPIV and μLIF techniques. The mixing efficiency was found to be highest for moderate Reynolds numbers (Re = 240-400) with S-shaped vortex flow and unsteady periodic flow regimes.

In order to study possibility to control and enhance mixing, a few authors investigated pulsed flow as inlet condition in T-shaped active micromixer. W B Mao and J L Xu [5] showed that mixing at low Reynolds numbers can be enhanced by pulsed flow with high amplitude. The higher the frequency the more enhancements can be achieved [6]. Ma et al. [7] found that there are optimal pulsation frequencies. Phase shifting of the pulsations will affect mixing as well [8]. All studies cited here used pulsed flow with high amplitude pulsing to get stratified flow in outlet channel. Contrary, in our paper we studied low amplitude perturbation, which can be enhanced by natural instability of the flow at

IOP Conf. Series: Journal of Physics: Conf. Series **925** (2017) 012020 doi:10.1088/1742-6596/925/1/012020

moderate Reynolds numbers, and lead to mixing enhancement with less energy consumption needed to force the flow.

2. Experimental setup and conditions

Experiments were carried out on the microchannel setup equipped with syringe KD Scientific pump with two outlets. Carl Zeiss Axio Observer.Z1 microscope with a number of lenses (figure 1) and filter set were used to get fluorescence images. To create the external disturbances of the flow with controlled frequencies, special excitation system based on the piezoelectric actuator was integrated in a hydrodynamic loop. Piezoelectric actuator with maximum motion 150 μm was preloaded by 1kN force and was driven by piezo drive with sine signal.

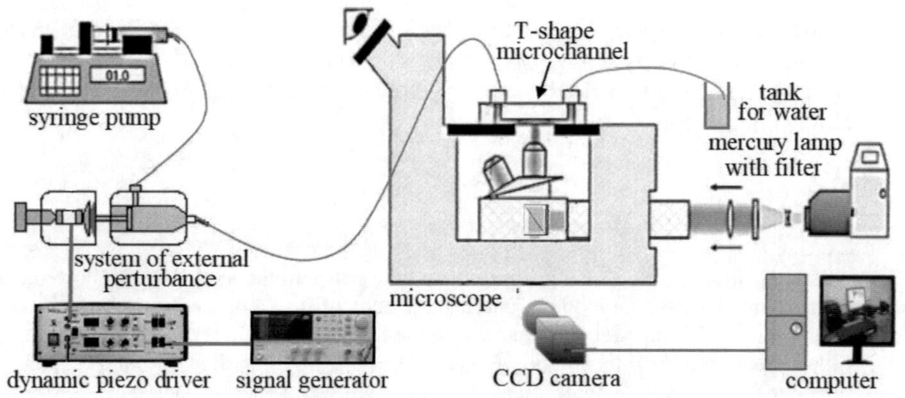

Figure 1. Sketch of the experimental setup

A T-shaped microchannel made of optically transparent material SU-8 with dimensions of 120×120×240 μm (height, width of the inlet channel and the width of the outlet channel) was mounted on the microscope stage. The variation of flow regime in microchannel is defined by Reynolds number, $\mathrm{Re} = U_0 D_h / \nu$ where U_0 is the mean flow rate velocity in outlet channel, $D_h = 4S_{mix} / P_{mix} = 160$ μm is hydraulic diameter, S_{mix} and P_{mix} are an cross-section area and a perimeter of mixing channel, correspondingly, ν is the kinematic viscosity of the flow. The experiments were carried out at Reynolds numbers Re = 120, 186, 300, and 400. All experiments were done with flow rates equal for both inlet channels. The high frequency external perturbation is generated in the same phase on both inlets of the microchannel with the frequency f up to 1000 Hz. The piezoelectric actuator motion conforms sinusoidal waveforms (U^+) from signal generator intensified by dynamic piezo driver. Amplitude is fixed to 100 μm, which corresponds to $3.14 \cdot 10^{-5}$ m^3/s of squeezing volume for $f = 1000$ Hz. The schematic of the active mixing on microchannel are showed in figure 2. The working liquid was distilled water.

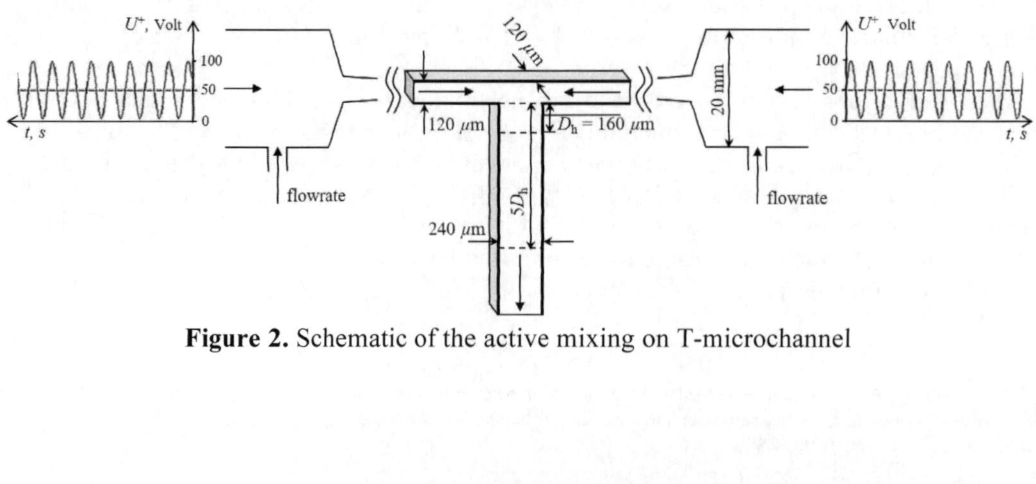

Figure 2. Schematic of the active mixing on T-microchannel

3. Measurement technique

To analyze flow patterns and a mixing efficiency in outlet channel Micro Laser Induced Fluorescence (μLIF) inverted in microscope was used. One inlet of microchannel was fed by the distilled water and the second inlet was fed by solution of distilled water with Rhodamine 6G dye. The concentration of the solution was 15mg/l. A mercury lamp was used for flow illumination. Band pass filter with a maximum transmission at 532 nm and beam splitter with 560 nm cutoff frequency were used to illuminate and register intensity of the fluorescent dye. Images from CCD camera (10 bits, matrix resolution 2048x2048 pixels) were processed in Actual Flow software to get intensity calibration and to calculate concentration fields.

4. Results

Figures 3-6 show time-averaged concentration fields for different Reynolds numbers and various perturbation frequencies together with the mixing efficiency calculated in outlet channel on distance equal to one and five hydraulic diameters D_h far from the inlet. In the concentration fields, the red color corresponds to liquid with Rhodamine, the blue one - to distilled water. The green color corresponds to concentration range c = 6.1÷8.9 mg/l and shows the area of complete mixing. Frequency f = 0 Hz corresponds to undisturbed flow case (figures 3-6 a).

In conducted experiments, the mixing efficiency was calculated using a concentration profiles measured at distances equal to D_h and $5D_h$ from inlet channel (figures 3-6). The estimate of mixing efficiency was made using Dankwerts segregation intensity I_M [9]: $I_M = 1 - \sigma/\sigma_0$, where $\sigma^2 = \frac{1}{N}\sum_{i=1}^{N}(c_i - \bar{c})^2$, $\bar{c}$ - is the average value of the concentration, $\sigma_0^2 = \bar{c}(c_{max} - \bar{c})$ - is the maximal root-mean-square deviation for mixing of liquid with concentration 0 and c_{max}. Thus, the liquids are completely mixed at $I_M = 1$, while segregated at $I_M = 0$.

The first series of experiments are presented for the stationary vortex flow regime Re = 120 (figure 2). For this regime, increase of the external perturbation frequency to 700 Hz leads to thinning of interface between liquids (figure 3b). However, for f = 1000 Hz the interface width increases (figure 3c) from 10% of the channel width at the distance $1D_h$ to 26% at distance $3D_h$. One can see also the insignificant decrease of interface width further downstream. In figure 2d the mixing efficiency increases within a range of frequencies from 0 Hz to 500 Hz at a distance equal to D_h from the entrance by 5%, and at a distance of $5D_h$ - by 9%. Here in after the change in mixing efficiency is expressed in % and presented in comparison with undisturbed case in the same cross-section. As the frequency increases further up to 675 Hz, the mixing efficiency decreases by 10.6% at a distance of D_h, and by 5% at a distance of $5D_h$ comparing to the undisturbed flow. Further increasing the frequency leads to growth of the mixing efficiency by 16% at a distance of $5D_h$, as well as by 9% and 6% at a distance of D_h for frequencies 800 and 1000 Hz, respectively. According to the visualization of mixing patterns efficiency increases significantly.

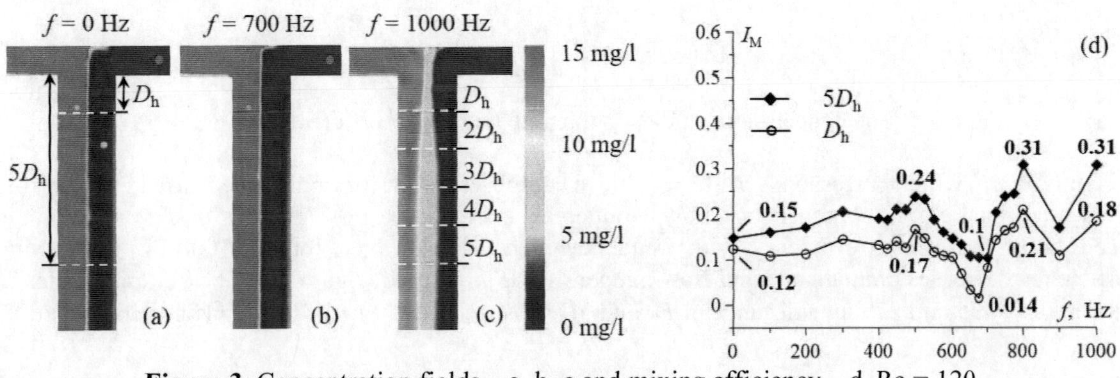

Figure 3. Concentration fields – a, b, c and mixing efficiency – d, Re = 120

IOP Conf. Series: Journal of Physics: Conf. Series **925** (2017) 012020 doi:10.1088/1742-6596/925/1/012020

For stationary asymmetric vortex flow regime Re = 186 f = 0 Hz (figure 4a) one can observe that the flow in outlet channel is structured. However, the plots of concentration fields sharply change in case of the external perturbations corresponding to 500 and 800 Hz (figure 4b,c).

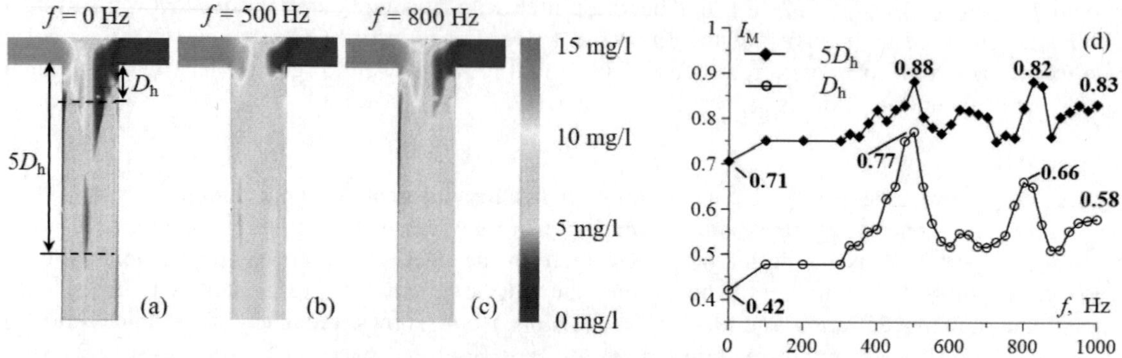

Figure 4. The concentration fields – a, b, c and the mixing efficiency – d, Re = 186

Concentration fields become homogeneous. We can suppose that it happened due to Kelvin-Helmholtz instability in liquid layers of jets interacting at the entrance of the outlet channel. The mixing efficiency increases by 35 and 17% with frequency raise to 500 Hz at a distance of D_h and $5D_h$, correspondingly (figure 4d). A further increase of the external perturbation frequency up to 650 Hz leads to a decrease in the mixing efficiency, however it is greater by 12% comparing to the case of undisturbed flow. The mixing efficiency increases by 24% for the frequency growth up to 800 Hz at a distance of D_h and by 11% at $5D_h$. Within the frequency range from 800 to 900 Hz the mixing efficiency decreases to that for 700 Hz. Next, the mixing efficiency increases by 6% for frequencies up to 1000 Hz.

For the periodic unsteady flow regime Re = 300 (figure 5), the mixing efficiency changes insignificantly, namely up to 3%.

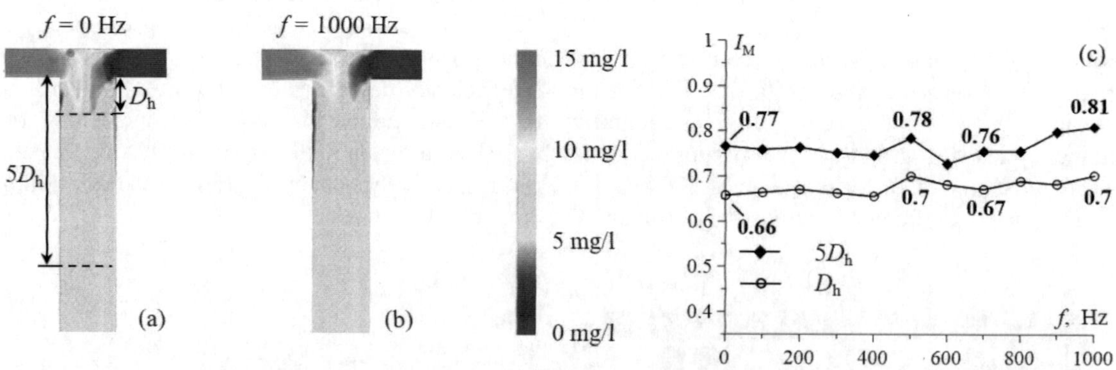

Figure 5. The concentration fields – a, b and the mixing efficiency – c, Re = 300

In figure 6, which corresponds to Re = 400, we can observe that for undisturbed case (f = 0 Hz) the concentration field is homogeneous in the outlet channel (figure 6a). We can observe monotonic increase of the mixing efficiency with frequency increase. However, for f = 1000 Hz the mixing efficiency decreases dramatically and flow happens to be structured (figure 6b). The mixing efficiency decreases by 35 and 27% at a distance of D_h and $5D_h$, respectively, for f = 1000 Hz (figure 6c).

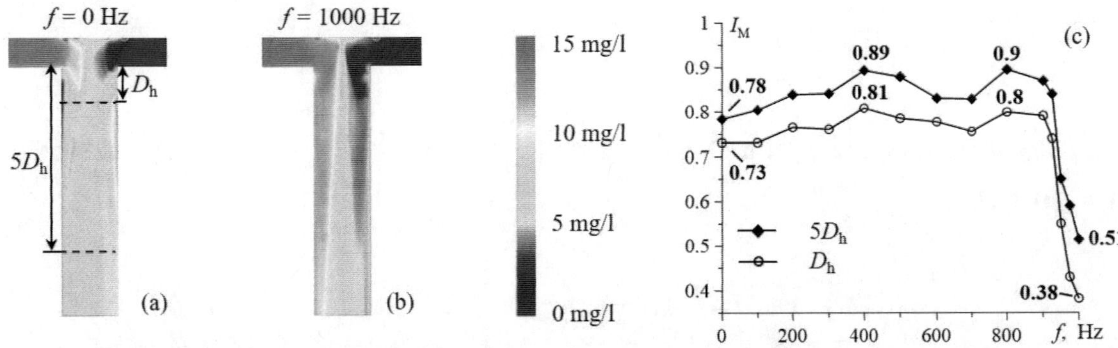

Figure 6. The concentration fields – a, b and the mixing efficiency – c, Re = 400

5. Conclusions

The μLIF technique was applied to measure concentration fields and mixing efficiency inside T-shape microchannel for different Reynolds numbers disturbed by external excitation with various frequencies. For all flow regimes studied, mixing enhancement was obtained. For stationary symmetric vortex flow (Re = 120) mixing can be increased up to two times at the excitation frequency f = 800 Hz. Significant enhancement can be achieved at the beginning of the outlet channel operating in steady asymmetric regime (Re = 186) by forcing within certain frequency ranges (f = 500 Hz, f = 800 Hz). Nevertheless, mixing suppression was also observed for two flow regimes (Re = 400, f = 1000 Hz) and (Re = 120, f = 700 Hz). Non-monotonical response of the flow to the different perturbation frequencies contradicts to previous studies of pulsed flows and means that there are preferable frequency ranges for the flow that probably correspond to the natural instability frequencies.

Acknowledgments

The research was funded by a grant of the Russian Scientific Fund (Project No. 16-19-10519) through Kutateladze Institute of Thermophysics SB RAS.

6. References

[1] Engler M, Kockmann N, Kiefer T and Woias P 2004 *Chemical Engineering Journal* **101** 315
[2] Wong S H, Ward M C L and Wharton C W 2004 *Sensors and Actuators B* **100** 359–79
[3] Gobert C, Schwert F and Manhart M 2006 *Proc. ASME Joint U.S.-European Fluids Eng. Summer Meeting*, (Miami, No. FEDSM2006-98035) 1053–62
[4] Minakov A, Yagodnitsyna A, Lobasov A, Rudyak V and Bilsky A. 2013 *Houille Blanche* **5** 12–21
[5] Mao W B and Xu J L 2009 *Int. J. Heat Mass Transf.* **52** 5258–61
[6] Glasgow I, Lieber S and Aubry N 2004 *Anal. Chem.* **76** 4825–32
[7] Ma Y, Sun C-P, Fields M, Li Y, Haake D A, Churchil B M and Ho C-M 2008 *J. Micromechanics Microengineering* **18** 45015
[8] Goullet A, Glasgow I and Aubry N. 2005 *Discret. Contin. Dyn. Syst.* **2005** 327–36
[9] Dankwerts P V 1952 *Appl. Sci. Res.* 279–96

IOP Conf. Series: Journal of Physics: Conf. Series **925** (2017) 012021 doi:10.1088/1742-6596/925/1/012021

Numerical investigation of thin film of polar liquid with added surfactant

V Y Gordeeva[1] and A V Lyushnin[2]

[1]Lysva branch, Perm National Research Polytechnic University, Perm territory, Lysva, 618910, Russia.
[2]Perm State Humanitarian-Pedagogical University, Perm, 614990, Russia

E-mail: varynka@gmail.com

Abstract. The thin film of polar liquid with an added surfactant is investigated numerically in this paper. The evolution equations for film thickness and surface concentrations were solved using the semi-implicit Crank-Nikolson scheme. A few profiles on the liquid film developing from an ellipse-shaped drop were received. It was confirmed that the developing film divides into two coexisting films with predictable thickness. It was discovered that this pecularity of the polar liquid is valid only in little range of vapor pressure, which corresponds to the disjoining pressure. It was found that the surfactant desorbed on the gas-liquid interface does not effect to the thickness of the film while the surfactant desorbed on the substrate does effect. It was also found that the stable thickness of the film grows with absolute value of the vapor pressure in stated little range.

1. Introduction

Thin films could be used in production of electronic devices, in biochemical investigations, and in various technological applications like coatings and paintings. One of ways to produce the technological thin film is to evaporate a liquid film with dissolved matter. Here, the behavior of the liquid solvent should be considered because the evaporating liquid can produce some patterns of the dissolved matters which is not always desired.

There are many experimental and theoretical papers dedicated to investigations of thin films [1; 2]. Only few theoretical papers consider the disjoining pressure [3] because properties connected to the disjoining pressure have their effect only in case of thin films with thickness around 100 Å.

Polar liquids have a specialty to “find” some special stable thickness compared to distance of dipole interaction. It corresponds to its disjoining pressure which consists not only of van der Waals interaction but also of electric interaction between interfaces. Disjoining pressure is an additional pressure in the thin film which was described by Deryagin [4]. It consists of van der Waals interaction, electrostatic interaction in polar liquids, capillary osmosis determined by a surfactant desorbed on substrate, and term of shape.

It was showed [5; 6] that little evaporating drops of water change their shape and decrease in thickness until they "found" some stable value of thickness, so that liquid layer divides into two parts: “thin” and “thick”. After that, the film saves this thickness, and volume of the evaporating liquid reduces in account of area of “thicker” part. The goal of this paper is to find some way to control thickness of thin film of the polar liquid.

2. Numerical simulation

The thin layer of polar liquid is investigated which is put on a solid substrate and bordered from above with a free-deformable gas-liquid interface. A surfactant is added to the liquid which can be desorbed on both interfaces. The disjoining pressure is considered in the mechanics of the thin film. Evolution equations received in paper [7] were used for simulation of the system: a lubrication equation for the liquid film, and two diffusion equations considering adsorbtion-desorbtion processes for surface concentration on both interfaces:

$$\partial_T H - \partial_X\left[H^2\left(\frac{H}{3}+\frac{\beta_0}{2}\right)\partial_X(P+\Phi)\right] - \frac{Ma}{2}\partial_X\left[\left(H^2+\beta_0 H\right)\partial_X\Gamma_A\right] + \Omega\left[-\frac{1}{H^3}+e^{-\chi H}-\partial_X^2 H + W_1 + W_0\ln(\gamma-\Gamma_S) - S = 0\right], \tag{1}$$

$$\partial_T\Gamma_A + \partial_X(U\Gamma_A) + \Gamma_A\partial_X H\,\partial_Z U - \frac{\partial_X^2\Gamma_A}{Pe_A} - K_1\Gamma_A(1-\Gamma_A) + K_2\Gamma_A = 0, \tag{2}$$

$$\partial_T\Gamma_S + \partial_X(U\Gamma_S) - \frac{\partial_X^2\Gamma_S}{Pe_S} - K_3\Gamma_S\left(1-\frac{\Gamma_S}{\gamma}\right) + K_4\Gamma_S = 0 \tag{3}$$

with the velocity equation

$$U = \left(\frac{Z^2}{2} - H(Z+\beta_0)\right)\partial_X(P+\Phi) - (Z+\beta_0)Ma\,\partial_X\Gamma_A \tag{4}$$

and the added pressure

$$P+\Phi = -Ca\,\partial_X^2 H - \frac{1}{H^3} + e^{-\chi H} + W_1 + W_0\ln(\gamma-\Gamma_S) \tag{5}$$

The following dimensionless governing parameters are used in the evolution equations:

- Marangoni number $Ma = \frac{a\Gamma_{A\infty}}{\sigma_0}$,
- adsorption parameters $K_{1,3} = \frac{k_{1,3}}{\varepsilon U_0}$ with subscript 1 for gas-liquid interface and 3 for solid-liquid interface,
- desorption parameters $K_{2,4} = \frac{k_{2,4}h_0}{\varepsilon U_0}$ with subscript 2 for gas-liquid interface and 4 for solid-liquid interface,
- capillary number $Ca = \frac{\varepsilon^3\sigma_0}{\mu U_0}$,
- Peclet numbers $Pe_{A,S} = \frac{h_0 U_0}{\varepsilon D_{A,S}}$,
- parameter of slipping along the substrate $\beta_0 = \frac{\beta}{h_0}$,
- evaporation parameter $\Omega = \upsilon\sigma_0\mu\left(\frac{6\pi}{A\sqrt{S^{PN}}}\right)^{2/3}$
- vapor pressure $S = \frac{\rho\varphi_v}{S^{PN}}$

IOP Conf. Series: Journal of Physics: Conf. Series **925** (2017) 012021 doi:10.1088/1742-6596/925/1/012021

where subscripts A and S mean the liquid-gas and liquid-solid interface correspondingly, μ is the liquid viscosity, D_i is the surface diffusion, ϱ and ϕ_v are the pressure and chemical potential of the vapor, and υ is the latent heat of vaporization.

The semi-implicit Crank-Nikolson scheme was used for linearizing of the equations. The equations were discretized by finite difference method.

3. Results and discussion

First we proved that the drop of polar liquid with initial shape of ellipse takes a form showed in paper [5] and finds some stable thickness. This simulation has been made for clear liquid without surfactant: $\Gamma_A = \Gamma_S = 0$. Another parameters used: $Ca = 0.001$, $Ma = 1$, $\chi = 1.085$, $\varepsilon = 0.001$, $\beta = 0.001$, $Pe_A = 5$, $Pe_S = 100$, $\Omega = 1$, $W_0 = 0.03$, $W_1 = -0.02$, $K_1 = 1$, $K_2 = 0.9$, $K_3 = 0$, $K_4 = 0$, $\gamma = 1$; dimensionless time interval $dt = 0.01$. Here, only parameters of evaporation and vapor pressure can influence. Results of the simulations at various values of vapor pressure S are showed in figure 1. One can see that the velocity of drop evaporation depend on the vapor pressure: the higher absolute value of the vapor pressure is, the quicker the drop evaporates.

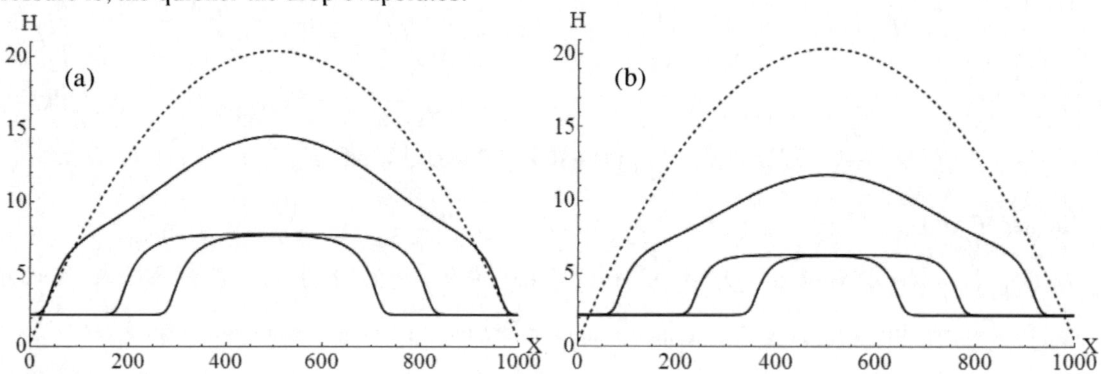

Figure 1. Evolution of drop of a polar liquid at various values of the vapor pressure and in various moment of simulation: (a) $S = -0.022$, solid lines at moment 0.3 10^4, 12 10^4, 20 10^4 of dimensionless time; (b) $S = -0.023$, solid lines at moment 0.3 10^4, 1 10^4, 1.5 10^4 of dimensionless time. The dashed line shows initial form of the drop.

Figure 2 shows that the vapor pressure also determines the higher stable thickness. We can conclude that the stable thicknesses can be predicted from the disjoining pressure equated with the vapor pressure because the evaporating drop comes to balance when the disjoining pressure becomes equal to the vapor pressure.

Earlier investigations showed that diffusivity of the surfactant on the liquid-gas interface can influence on stability of the film, so the faster the surfactant diffuses along the interface the more stable the interface is. Meanwhile, the solutal Marangoni instability arises only with anti-surfactant [7; 8] which increases surface tension. So, for goal of this paper, we simulated high-diffusive surfactants which decrease the surface tension but is present only on the gas-liquid interface. Simulation showed that the presence on the surfactant only on the gas-liquid interface does not effect to evolution of the film. After that, we simulated the liquid drop with the similar surfactant but desorbed also on the solid-liquid interface. As it was marked in [4], one of addends of the disjoining pressure is capillary osmosis determined by the surfactant concentrated on substrate below the liquid film. This idea was realized in equation (5). Result of the simulation is shown in figure 3.

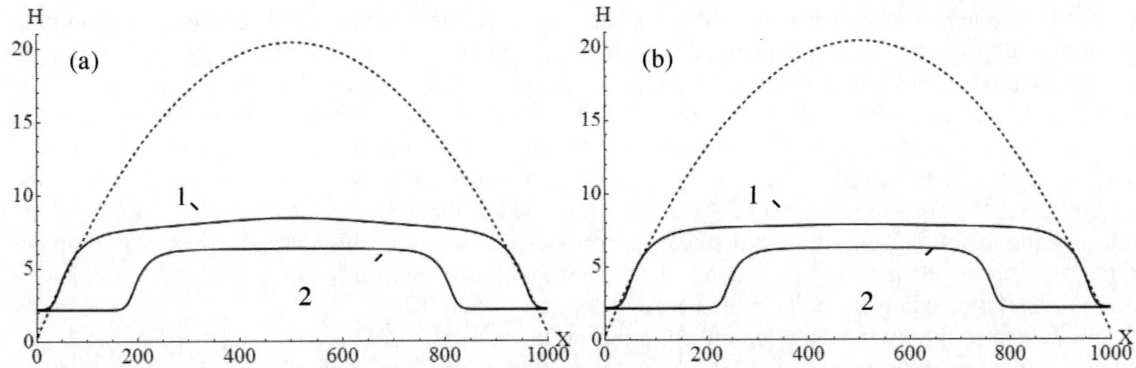

Figure 2. Comparison of liquid drop evolution at various vapor pressure: (a) after 0.8 10^4 of dimensionless time; (b) after 1 10^4 of dimensionless time. 1 means the layer at S = -0.022, and 2 means the layer at S = -0.023.

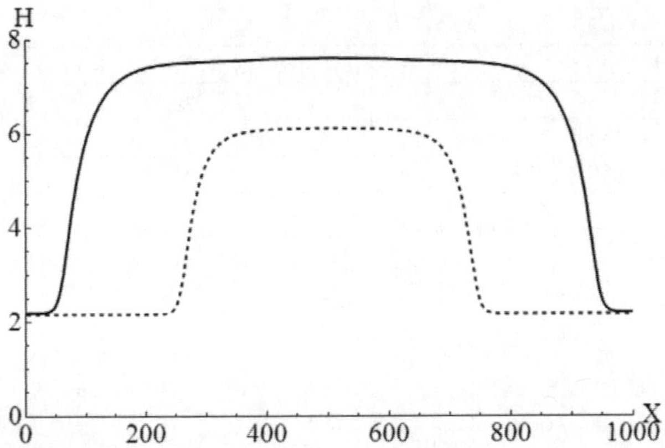

Figure 3. Simulation result for the drops with the surfactant desorbed on both interfaces (Γ_S = 0.15). The solid line is for vapor pressure S = -0.027, the dashed line is for S = -0.028.

Comparing figure 3 with figure 1 we can see that the values of stable thickness, both for "thick" and "thin" parts, do not change after addition of the surfactant on the substrate, but the vapor pressure does: the same values of thickness were received at another vapor pressure.

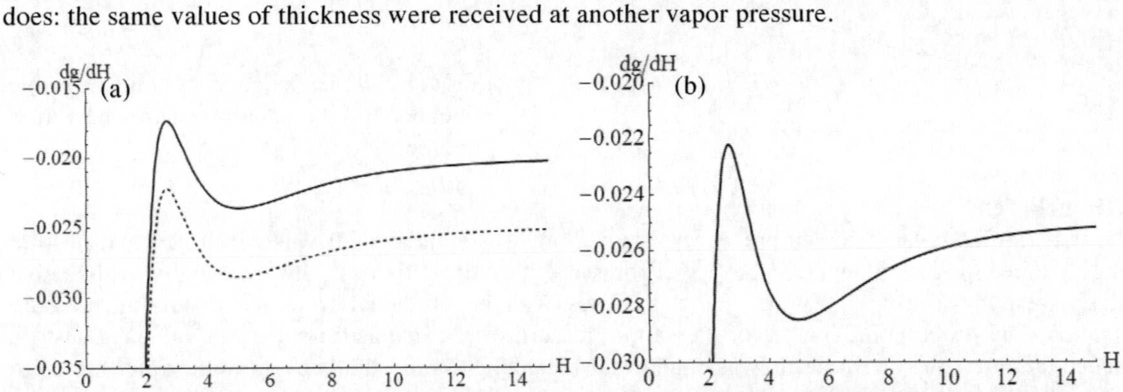

Figure 4. A chemical potential of a polar liquid in dependence on the film thickness. The plots (a) and (b) are received from equation (5) and reproduced in various scales. The dashed line is for Γ_S = 0.15 and the solid line is for Γ_S = 0. Plot (b) is for Γ_S = 0.15.

The possibility to predict the film thickness is valid only for polar liquids because their disjoining pressure is non-monotonically connected with the film thickness as it is shown in figure 4. It can be also understood from figure 4 that the surfactant desorbed on the substrate doesn't change dependence of disjoining pressure on the thickness but only reduces the values of the disjoining pressure. One can find that there is a little range of vapor pressure values where the two stable thicknesses can exist simultaneously: for example, from -0.028 to -0.025 in dimensionless units at concentration of the surfactant on the substrate Γ_S = 0.15. It is the range where three values of thickness coexist at one value on the disjoining pressure, two stable thicknesses and one unstable between them ("three-point range"). If the vapor pressure is outside of the range, the drop will have only one stable thickness, around 2.2 of dimensionless units, which corresponds to 1.1 nm [6].

In order to find how the surface concentration on the substrate effects to the three-point range, we calculated the ranges at various values of the concentration. Results of the calculation are present in table 1.

Table 1. "Three-point ranges" of disjoining pressure at various surface concentrations.

Suraface concentration at the substrate, Γ_S, dimensionless units	Min disjoining pressure dg/dH	Max disjoining pressure dg/dH
0.04	-0.02472	-0.02142
0.05	-0.02503	-0.02173
0.06	-0.02535	-0.02205
0.07	-0.02567	-0.02237
0.08	-0.02599	-0.02269

Further, we have to find how the higher stable thickness depends on the vapor pressure in stated range. Plot of this dependence is showed in figure 5. We can see from the figure 5 that the higher stable thickness grows while the vapor pressure reduces in its absolute value.

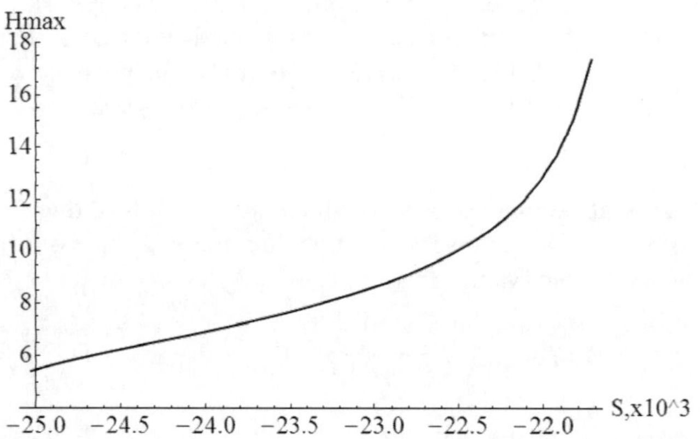

Figure 5. Dependence of higher stable thickness on the vapor pressure, the values are received for Γ_S = 0.05.

4. Conclusion

The thin film of polar liquid in presence of surfactant was simulated. It was confirmed that in little range ("three point range") of the vapor pressure the film divides into two parts with stable thicknesses. It was found that the stable thickness can be predicted from the disjoining pressure equation. The "three-point" ranges of the vapor pressure were calculated for various values of surface concentration on the substrate. It was found that the higher stable thickness grows while the vapor pressure reduces.

It can be concluded that the experimentalist could control the thickness of the film of polar liquid by change of vapor pressure in special experimental setup, or by putting some low-dissolving surfactant on the substrate under the liquid.

References

[1] Oron A, Davis S H and Bankoff S G 1997 Long-scale evolution of thin liquid films *Rev. of modern physics* **69(3)** 931.
[2] Craster R V and Matar O K 2009 Dynamics and stability of thin liquid films *Rev. of modern physics* **81(3)** 1131.
[3] Neto C, Jacobs K, Seemann R, Blossey R, Becker J and Grün G 2003 Satellite hole formation during dewetting: experiment and simulation *J. of Physics: Cond. Matter* **15(19)** 3355.
[4] Deryagin B V and Churaev N V 1987 *Langmuir* **3.5** 607–12.
[5] Leizerson I, Lipson S G and Lyushnin A V 2003 *Nature* **422** 395–6.
[6] Lyushnin A V, Golovin A A and Pismen L M 2002 Fingering instability of thin evaporating liquid films *Phys. Rev.* E **65(2)** 021602.
[7] Gordeeva V Y and Lyushnin A V 2017 Dynamics of an evaporating thin film of polar liquid with solutocapillary Marangoni effect and capillary osmosis. *Surf. & Coat. Technology* **320** 531–5.
[8] Gordeeva V Y and Lyushnin A V 2017 Numerical simulation and stability analysis of solutocapillary effect in ultrathin films *The Eur. Phys. J. Special Topics* **226** 1219–28.

IOP Conf. Series: Journal of Physics: Conf. Series **925** (2017) 012022 doi:10.1088/1742-6596/925/1/012022

Features of liquid fuel burning in a narrow channel

V V Zamaschikov[1,3] and E A Chinnov[2,3]

[1]Voevodsky Institute of Chemical Kinetics and Combustion, SB RAS, 630090, Institutskaya str. 3, Novosibirsk, Russia
[2] Kutateladze Institute of Thermophysics SB RAS, 630090, Lavrentied ave. 1, Novosibirsk, Russia
[3] Novosibirsk State University, 630090, Pirogov str. 2, Novosibirsk, Russia

E-mail: albor@kinetics.nsc.ru

Abstract. The paper deals with experimental investigations of the surface perturbation effects on the flame propagation velocity along the surface of the liquid. It is revealed that at low velocities the effect is quite significant. Under certain conditions the flame even stabilizes for some time. The experiments were conducted to elucidate the mechanism of flame propagation. However, it was impossible to determine expressly the flame propagation mechanism along the surface, because this requires visualization of gas flows.

1. Introduction

Making the miniature power sources, which use the energy released during hydrocarbon combustion, is an urgent task since high intensity of heat transfer in microsystems allows creating the effective heat exchangers that can be useful for development, for example, of thermoelectric converters. The process of flame propagation over the surface of liquid on the bottom of a cylindrical tube was investigated in [1]. At that, liquid heating was observed under the flame front. In more detail, the processes occurring in liquid during flame propagation are studied in [2]. The authors observed the formation of a vortex in liquid that is ahead of the flame front and plays a large role in the mechanism of flame propagation. It is believed [2] that the cause of the vortex generation is the movement of liquid under the impact of uncompensated surface tension forces. These forces arise due to the temperature gradient along the liquid surface, resulted from the heating of the liquid by combustion products.

Experimental studies of liquid combustion in horizontal channels have shown [1, 3] that the flame velocity can vary in a wide range from millimeters per second to several meters per second. The estimates made in [3], have shown that at flame propagation velocities of the order of several meters per second, movement of the liquid due to thermocapillary forces is unlikely. In this regard, the question arises about the mechanism of formation of the combustible mixture in the flame front. In [3] the evaluations were carried out for a fairly thick layer of liquid. However, it can be assumed that due to the high temperature and high flame propagation velocity under the combustion front, only a very thin layer of liquid is able to warm up. The viscosity of liquid decreases with increasing temperature. Therefore, because of the resulting thermocapillary force, the upper liquid layer begins to slide over the cold and therefore more viscous portion of liquid. For this case, the estimates given in [3] are unjust, and thermocapillary mechanism may still be implemented. Heating of the fluid ahead of the combustion front leads to an increase of the 1-butanol vapor flows into the oxidizer that leads to the formation of a combustible gas mixture.

Another possible mechanism of formation of the combustible mixture may be the following. Due to the large temperature difference between the combustion products and liquid (more than 1000 K) when near-surface liquid layer is heated (behind the combustion front, where the combustible products are situated), fuel vapors flow is formed. The stream flows around the flame, so that part of the stream is directed to the area ahead of the flame containing the oxidizer. Occurring turbulent mixing results in formation of the combustible mixture. Due to the rapid mixing, flame propagation velocity may be much higher than in the case of heat transfer through the liquid to the pre-flame area. The aim of this paper is to try to figure out the real mechanism of formation of the combustible gas mixture ahead of the combustion front.

2. Experimental setup

The experiments were carried out in a flat channel with the width of 4 mm and height of 42 mm (Fig. 1). Side walls were made of quartz plates.

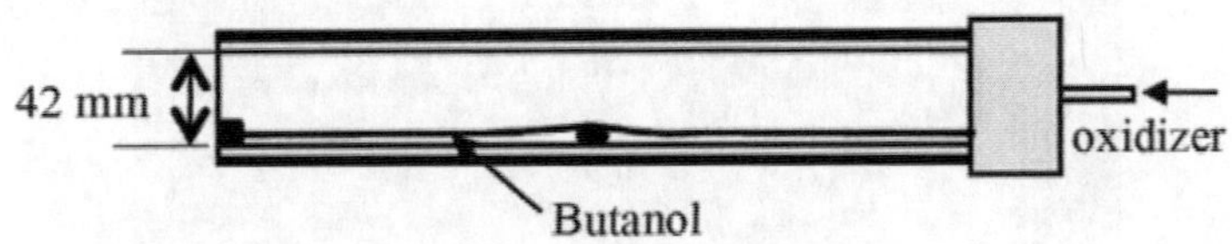

Figure 1. Setup.

In contrast to previous works, in this study, the bottom (lower part) of the channel was not a wide part [3], but a narrow one, and this allowed observing the combustion front and processes occurring in liquid. The oxidizer was fed from one side, while the other side was free. The barrier, preventing the leakage of fluid, was installed below the open side.

When air was used as an oxidizer, the flame entered to the channel and quenched. To exclude extinguishing, air was enriched with oxygen. To influence the processes occurring in liquid and combustion, a barrier was mounted on the channel bottom (Fig.1).

3. Experimental results

In contrast to [3], in this case, the flame velocity did not depend on oxidant flow rate, at least up to the flow rate of 5 liters per minute, which agrees with the data obtained by other investigators at flame propagation over the liquid surface in sufficiently large volumes [2]. This can be explained by the fact that the effect of oxidant flow on combustion should increase with a decrease in the free space above the flame. As expected, the velocity depends on the proportion of oxygen in the oxidant mixture. With an increase in oxygen proportion, the velocity increases almost linearly.

At flame velocities higher than 11 cm/s, there is no visual disturbance of the liquid surface (formation of a roll ahead of the front). However, the region of liquid heating is observed (Fig. 2 a). This region expands with a distance from the combustion front. The fact that in this area the temperature actually increases is shown by thermocouple measurements. A 25-μm chromel-alumel thermocouple was used (it can be seen in Fig. 2). As the flame velocity increases with oxygen enrichment of air, the heating region observed under the front becomes thinner, and eventually it cannot be distinguished (Fig. 2b). To understand the mechanism of flame propagation, it is important to determine where the heating region is relative to the leading point of combustion wave: is it ahead of this point or not.

However, thermocouple measurements did not allow this, since, first, when the thermocouple approaches the surface, the latter is bent, and second, the near-surface heated layer can be so thin that the thermocouple does not measure the actual temperature profile near the liquid surface. It should be noted that when measuring the gas temperature, i.e., in the case, when the thermocouple was located above the surface, the temperature profile obtained by the thermocouple was considerably shifted relative to the combustion front because of thermocouple inertia.

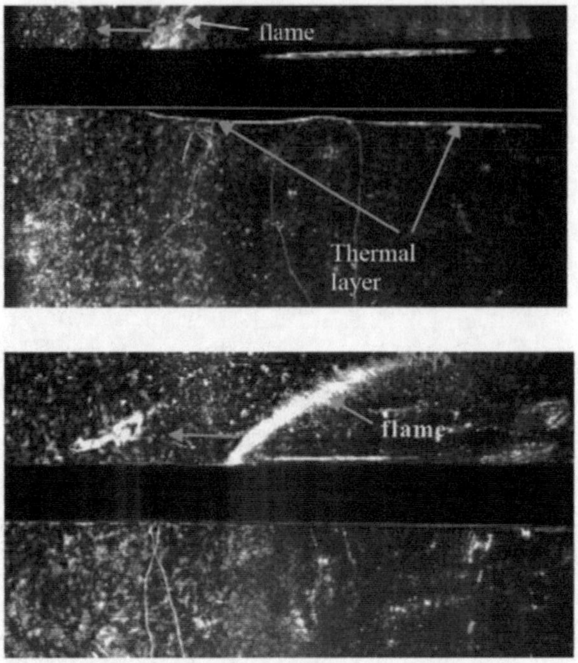

Figure 2. Photos of flame propagation front.

For the existence of a combustion wave, it is important that a combustible mixture be formed before the front. The combustible mixture is formed at mixing the vapors with an oxidant, if vapor concentration is higher than the limiting value. The authors of [2], in the mechanism of flame propagation proposed by them, have determined the role of thermo-capillary phenomena. Due to the temperature gradient formed along the surface and temperature dependence of surface tension coefficient, a force is generated that moves the heated liquid into the pre-flame region. A thermocapillary wave is formed in liquid, and it moves together with the flame along the liquid surface. As a result, vapors of heated liquid, mixing with an oxidant, form a combustible mixture before the flame.

It is obvious that liquid surface bending affects the processes taking place in liquid and gas, and thus, flame propagation. To study the effect of surface bending on the flame velocity, a cylinder or parallelepiped was placed on the channel bottom. On the one hand, the barrier bent the surface of the liquid, on the other hand, the liquid layer thickness decreased. The thickness of the liquid layer begins affecting the velocity of combustion wave, when it is comparable with the characteristic heating thickness (in this case the liquid can move) of the near-surface layer under the flame front. The thickness of the heated layer in turn depends on the flame propagation velocity. The curvature of the surface influences both processes occurring in the liquid and those occurring in the gas phase.

When the flame approaches the barrier: a) it slows down; b) its slope relative to the gravity vector changes; c) its position changes, although insignificantly, relative to the liquid surface; d) its length decreases; e) brightness of its image decreases. The shlieren-image brightness should be related to the flame temperature, so it decreases in the flame front.

The shape of the flame depends on the vapor flow from the liquid surface and oxidant flow. Since the flame velocity in this case does not depend on the oxidant flow rate, it can be assumed that combustion takes place at low gas velocities, i.e., in almost resting gas. In this case, most likely, there is diffusion mixing of vapors and oxidant with formation of a combustible mixture ahead of the combustion front. The greater the surface curvature (determined by the ratio between the thickness of the liquid layer and the height of the obstacle) and the lower the flame velocity, the stronger is the effect of barrier on the flame propagation velocity. During the flame propagation, the barrier should not have a significant effect on the flame velocity until the flame approaches the barrier. When the flame comes close to the barrier, firstly the surface bends, and secondly, the liquid layer thickness changes. These two factors influence the temperature of the near-surface layer.

Diffusion of vapors should also depend on the surface curvature, and the value of this flux should depend on the temperature of the near-surface layer. Dependence of the flame velocity on coordinate and position of the barrier are shown in Fig. 3. The flame spreads from right to left. It can be seen that the velocity decreases, when the flame rises to the barrier. Due to liquid surface curvature, the flame, when moving along the surface, passes the region, where the liquid layer thickness increases first, and then decreases. From the point of view of vapor diffusion and liquid heating ahead of the flame front, the velocity should increase and then decrease, but this is not observed. This can be explained as follows. As mentioned above, the liquid layer thickness starts influencing the flame propagation velocity, when it is comparable to the characteristic thickness of the liquid layer heated under a flame front. If at the distance far from the barrier, characteristic thermal thickness is less than the liquid depth, then the increase in the latter when approaching the obstacle has little effect on the velocity. When the depth decreases and becomes comparable with the characteristic thermal thickness, the flame propagation velocity is reduced that in turn leads to an increase in the characteristic thermal thickness, and this leads to a further slowing of the flame. However, we cannot exclude that the flame velocity is determined by the average characteristics, which are not affected by such a change in thickness.

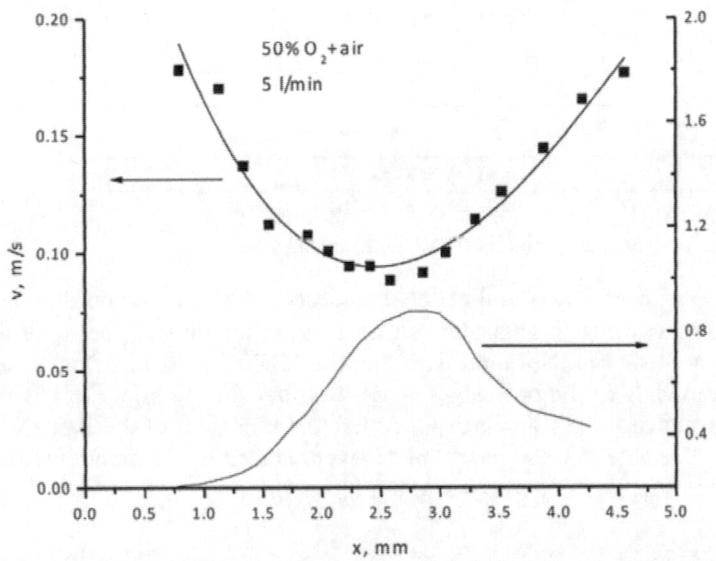

Figure 3. Flame velocity (v) and liquid layer height (h) vs. coordinate (x).

At the place, where the velocity is minimal (Fig. 3), the flame becomes more vertical, it lengthens and its brightness increases. Under some certain conditions (when a parallelepiped is used as a barrier), the flame is stabilized in this place for some time. This behavior can be explained as follows. First, the liquid layer thickness under the flame is small there. Consequently, it is poorly heated due to the heat flux into the solid phase. Second, liquid heating ahead of the front leads to an increase in the vapor flow into the gas phase, but this flow is directed away from the flame by virtue of geometry (fig. 4).

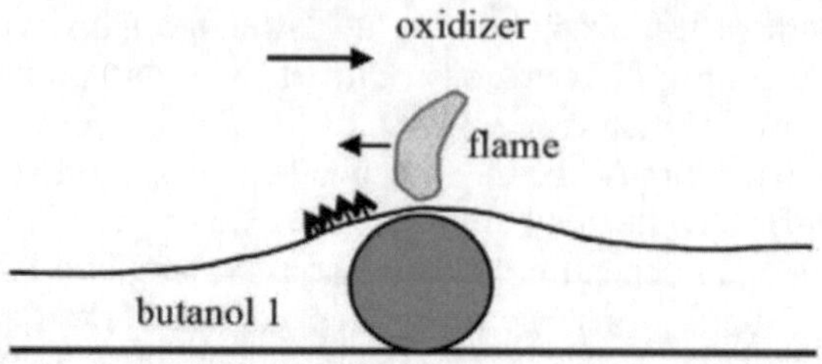

Figure 4. The location of the flame and flow direction.

Therefore, the flame cannot move forward. The following pattern is observed: behind the flame, there are the combustion products and vapors, and before the flame, there is only an oxidant. The vapors and oxidant diffuse towards each other and react where the combustion front is. While liquid is being heated in the pre-flammable region, the vapor flow becomes so large that the combustible mixture is formed already ahead of the combustion front, and the flame moves forward. In favor of this hypothesis is the fact that the flame, when passing through the place where its velocity is minimal, separates slightly from the liquid surface. Perhaps, more precise description of the flame behavior requires taking into account free convection and oxidant flow, although small, but available on the liquid surface. Note that Fig. 4 shows vapor flows in pre-flame region. According to the second mechanism, the main role in the formation of the combustible mixture is played by fuel flows from the area containing combustible products. In this case, the behavior of the flame must most likely be different.

The stabilization of the flame is of practical importance. On the barrier the flame is stabilized for a short time. In the present work, flame stabilization was obtained for a significant change in the oxidant flow rate (Fig. 5).

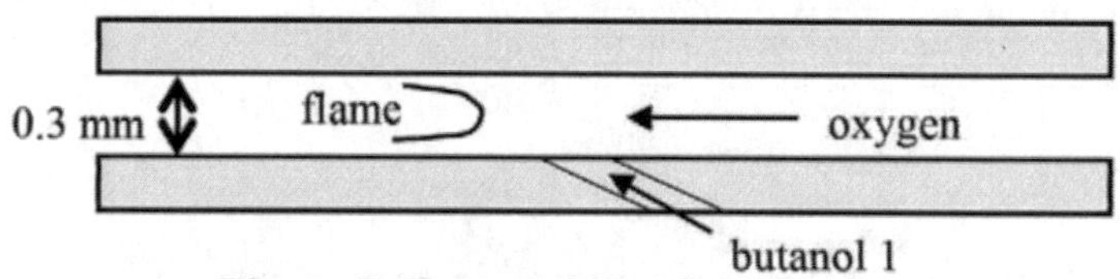

Figure 5. Flame stabilization diagram.

The oxygen was used as the oxidizer. The stabilization was observed when oxygen flow rate changed up to 10 liters/min. However, in order to obtain the stoichiometric relationship between the fuel and the oxidizer, it was necessary to supply the liquid with the low flow rate. At low flow rate, the liquid flowed only along the lateral walls of the channel (Fig. 5). In order for the liquid to cover the whole channel, it was necessary to increase the flow rate, which led to enrichment of the flame with fuel. As a result, not all fuel burned. We note that the quartz plates were heated by the flame, which led to the fact that the main channel was supplied not with liquid but its vapor.

4. Conclusion

Thus, the barrier has an effect on the flame propagation. It affects the processes occurring in the liquid and gas phases. However, the experiments carried out in the presence of a barrier do not give an unambiguous answer about the mechanism of flame propagation.

Acknowledgments

This work was supported by the grant of Russian Science Foundation (Agreement No. 15-19-30038).

References

1. Zamashchikov V.V. 2009 *Combustion Sci. and Technology* **181** 176-189.
2. Ross H.D., Miller F.J. 1996 *Proc. of the Combustion Institute* **26** 1327-1334.
3. Zamashchikov V.V., Korzhavin A.A., Chinnov E.A. 2016 *IJHMT* **102** 470–478.

IOP Conf. Series: Journal of Physics: Conf. Series **925** (2017) 012023 doi:10.1088/1742-6596/925/1/012023

Numerical modelling and experimental study of liquid evaporation during gel formation

B G Pokusaev and D P Khramtsov

Moscow Polytechnic University, Faculty of Chemical and Bioengineering, Department of thermodynamics and nonequilibrium transport processes, B. Semenovskaya, 38, Moscow, 107023, Russia

E-mail: pokusaev2005@yandex.ru

Abstract. Gels are promising materials in biotechnology and medicine as a medium for storing cells for bioprinting applications. Gel is a two-phase system consisting of solid medium and liquid phase. Understanding of a gel structure evolution and gel aging during liquid evaporation is a crucial step in developing new additive bioprinting technologies. A numerical and experimental study of liquid evaporation was performed. In experimental study an evaporation process of an agarose gel layer located on Petri dish was observed and mass difference was detected using electronic scales. Numerical model was based on a smoothed particle hydrodynamics method. Gel in a model was represented as a solid-liquid system and liquid evaporation was modelled due to capillary forces and heat transfer. Comparison of experimental data and numerical results demonstrated that model can adequately represent evaporation process in agarose gel.

1. Introduction

Development of additive technologies, including areas such as medicine and biotechnology requires an understanding of the dynamics of the processes of formation and aging of gels with the aim of developing optimal technological processes for bioprinting technologies [1]. The formation of gels occurs due to the complex hydrodynamic and thermo-physical processes occurring in the internal micro-structure of the gel. The processes of mass transfer in the gel involve the interaction of multiple phases on small linear scales [2]. In order to determine the dynamics of the process of liquid evaporation during gel formation, experimental and numerical study of measuring the rate of evaporation of the liquid phase in the course of cooling of the sample gel was performed.

Currently gels are considered a high-potential structure-forming material for formation of artificial tissues from immobilized cells using the method of additive manufacturing. Due to their rheological properties, gels refer to an intermediate state between liquid and solid. The common definition of gels is a dispersive system with liquid dispersing medium, and the dispersion phase makes up a spatial structured mesh due to intermolecular interaction in the contact areas [3]. Gels are capable of displaying both elasticity and plasticity.

Dynamics of gas liquid flow inside irregular channels and gels are not studied enough, both experimentally and numerically. Gel is a specific two-phase system. One phase represents a relatively sparse spatial network of polymer molecules which are bonded at the intersections of intermolecular bonds. The second phase is a liquid. In the study of additive processes of application of gels in relation to problems of bioprinting, we found the effect of microchannels formation between the layers of gel, filled with gas and fluid.

In the case of the presence of microorganisms within the gel, a carbon dioxide is produced within the gel due to activity of microorganisms. For the normal functioning of microorganisms in the gel it is required a withdrawal of produced products. In this case, the rate of drainage is influenced by the intensity of absorption and the speed of ascent of gas bubbles, which leads to the necessity of conducting experimental and theoretical studies of the process of mass transfer of gas bubbles in the gel. The relevance of such studies is also due to the active development of technologies of three-dimensional printing of living tissues and bodies [1].

2. Numerical model

Model of a gas phase moving inside gel was developed based on the smoothed particle hydrodynamics method. The method is based on representing the fluid as a set of particles, which are located at some distance from each other, the distance called smoothing [4]. The influence of each particle on the properties of nearby particles is calculated on the basis of its density and the distance from the particle for which parameters. Evaporation rate was calculated by procedure proposed in [5]. First, vapour pressure is calculated by applying ideal gas equation, then saturation pressure with a given surface temperature can be calculated using Magnus formula, finally evaporation rate is computed by Smith formula [5].

For the calculation we introduce the notion of kernel functions. As a kernel function, typically uses a Gaussian function. In the case of motion of a gas bubble of carbon dioxide in the channel with obstacles (figure 1). For verification of the developed model was used for calculation of test problem on simulation of freely rising gas bubble in a pure liquid at various angles of inclination of the tube for the purpose of comparison with previously obtained experimental data.

Mathematical model of a liquid flow was based on smoothed particle hydrodynamics method. Method is based on representing liquid as an ensemble of smoothed particles [4]. Influence of each particle on neighborhood particles is calculated based on their density and distance to other particles (1).

$$A(r)=\int A(r')W(r-r',h)dr'. \tag{1}$$

In discrete representation the value of an arbitrary point is presented in the following form (2):

$$A(r)=\sum_i m_i \frac{A_i}{\rho_i} W(|r-r'|,h). \tag{2}$$

For pressure (3) and velocity (4) calculation the following difference scheme is applied:

$$p_i = -\sum_j m_j \frac{p_j - p_i}{\rho_j} W(|r_i - r_j|,h), \tag{3}$$

$$v_i = \mu \sum_j m_j \frac{v_j - v_i}{\rho_j} \nabla W(|r_i - r_j|,h). \tag{4}$$

Evaporation condition is defined as follows (5)

$$evp = \max\left(\frac{\partial m_s}{\partial \tau}\right) \tag{5}$$

Where $\frac{\partial m_s}{\partial \tau} = S\left(a + b\cdot\left\|\vec{v}\right\|\right)\left(p_{sat} - p_c\right)$, p_{sat}- saturation vapor pressure, p_c – vapor pressure, S – surface of liquid-air contact, v – velocity, a,b – arbitrary parameters [5].

The system of equations (3), (4), (5) are solved jointly for each particle over the period of time of the numerical experiment.

3. Experiment

3.1. Experimental setup

In the experimental study an agarose gel with concentration of 0.6 and 1% was used. At the beginning of the experiment, a prepared sample of the gel at a temperature of 60 °C was applied to the surface of the Petri dish. Then Petri dish was mounted on a heat insulating plate and installed at the scale. Further thermopair was injected into the gel. Before starting the experiment a timer was set to measure time of experiment. Differences in values presented by thermopair measurements and electronic scales were recorded in the table with the corresponding time value according to the timer. The experimental setup is shown in figure 1.

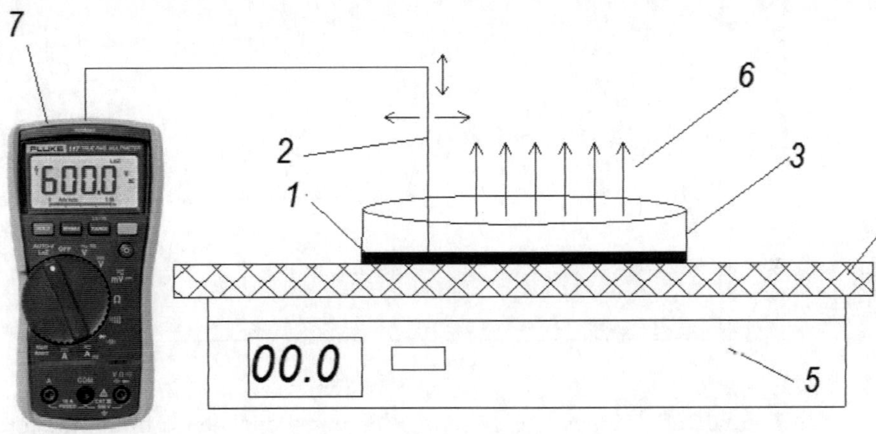

Figure 1. Experimental setup: 1 – layer of agarose gel, 2 – thermopair, 3 – Petri dish, 4 - heat-insulating plate, 5 – electronic scale, 6 – vapour flux, 7 – Multimeter Fluke 117

Thermopair was connected to the multimeter Fluke 117 which supports temperature measurements. Thermopair could be located in arbitrary area of a Petri dish. After series of experiments it was found that temperature distribution among the plate can be considered homogenous and hence thermopair location can be arbitrary. However thermopair is not recommended to locate near the dish wall in order to avoid boundary effects. The thickness of the gel layer was measured by moving thermopair while considering the bottom of the plate as a reference point. The moment when thermopair reached the surface of the gel was fixed by drop in a temperature when thermopair leaved the gel since there is a difference between temperature in gel and atmosphere.

The estimated model of the process of evaporation is based on the idea of considering gel as a system of capillaries. The motion of the liquid within the gel is associated with the movement of the liquid phase in microchannels of complex configuration, where capillary force provides a supply of fluid from the inner layers of the gel to the surface during evaporation [6]. The density of capillaries per unit surface area was determined using qualitative analysis of photographs of the microstructure of the gel obtained by electron microscopy [7].

3.2. Experimental study

Liquid evaporation from gel was observed for 2 hours and 40 minutes. This time interval allows to capture the characteristic of stage of formation and aging of the gel, including the initial stage of structure formation and polymerization. Time step was set to 5 minutes. The data obtained is the dependence of change of weight of gel over the time (figure 2). Data is presented in normalized form where normalized time is current time in experiment divided by end time of an experiment. In this case 1 corresponds to a fully formed gel. Relative mass is mass divided by initial mass of a gal filled with liquid.

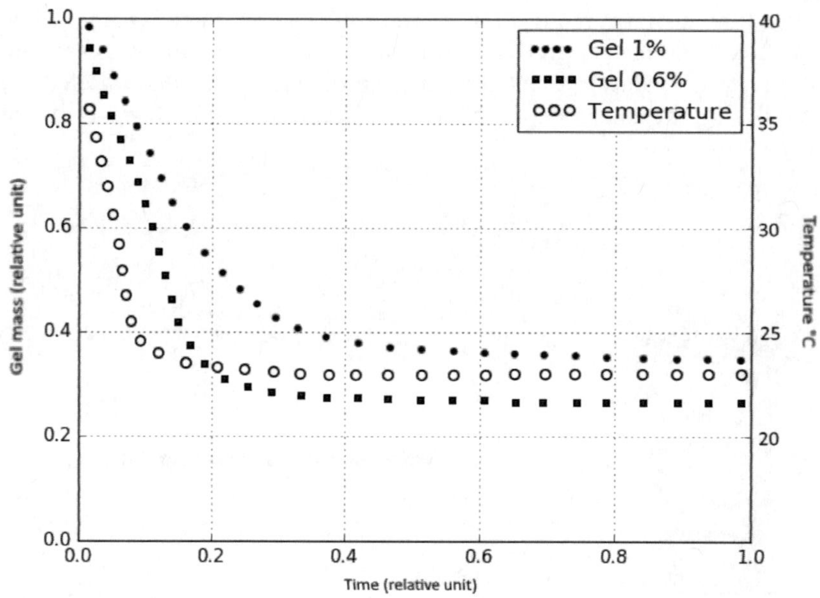

Figure 2. Dynamics of gel evaporation with dependence from time for different gel concentrations

In the first 20 minutes of the experiment there is a linear dependence of change of weight of gel over the time. Dependence may be associated with the fact that in this period liquid evaporation occurs from the surface of the gel, with the supply of fluid from the deep layers of the gel by the action of the capillary forces. In the interval from 20 minutes to 2 hours, dependency becomes nonlinear, which may occur due to the action of additional factors on the evaporation process such as the process of shrinkage of the gel. After 2 hours a stabilization of the evaporation process was observed. The cause of stabilization can be explained by the fact that the remaining liquid is not enough for the action of capillary forces to supply the liquid to the surface of the gel, which leads to the fact that the evaporation process continues within the pores, where the total surface evaporation is less in comparison with the gel surface.

4. Discussion

Temperature measurements in time allowed to identify the mode of establishing a stationary temperature, in which case mathematical model was applied (figure 3).

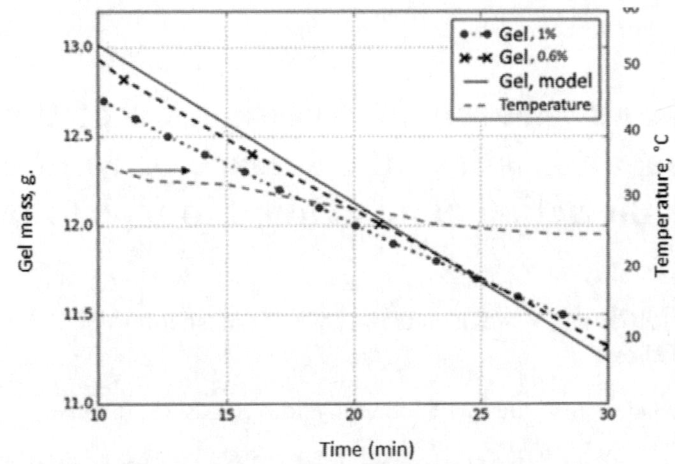

Figure 3. Dynamics of gel evaporation with dependence from time for different gel concentrations

For this mode a numerical calculation of the dynamics of evaporation of the liquid was performed and a comparative analysis with experimental results for agarose gel concentrations of 0.6% and 1% was made. The variation of the mass of the gel is of a linear nature in the area of stationary temperature. Evaporation rate for a given concentrations does not vary significantly. The similarity of the nature of evaporation for these concentrations may be related to the fact that the difference in concentration is insufficient for the occurrence of changes in the internal structure of the gel.

5. Conclusions

Three stages of evaporation of the gel were identified. These stages are: linear which corresponds to the gel formation process, nonlinear, associated with the action of capillary forces and shrinkage of the gel, and the stage of polymerization characterized by plateau. The comparison of results of calculation and experiment shows the effectiveness of the model for the initial stage of gel formation. The character of dependencies of evaporation of the gel over the time at the stage of gel formation demonstrates a linear character, the dynamics of evaporation agarose gel with a concentration of 0.6% differs from the concentration of 1% gel insignificantly, which allows to make a conclusion about the similarity of the internal structure of the gel at these concentrations.

Acknowledgments

This work is financially supported by the Russian Science Foundation (project no. 15-19-00177).

References

[1] Jakab K, Norotte C, Marga F, Murphy K, Vunjak-Novakovic G and Forgacs G 2010 *Biofabrication* **2** 20-34
[2] Pokusaev B, Nekrasov D, Karlov S and Khramtsov D 2016 *Theor Found Chem Eng* **50** 690-696
[3] Arnott S and Fulmer W 1974 *Journal of Molecular Biology* **90** 269-284
[4] Liu G and Liu M 2003 *Smoothed Particle Hydrodynamics: A Meshfree Particle Method* (Singapore: World Scientific) 93
[5] Hochstetter H. and Kolb A 2017 *Evaporation and Condensation of SPH-based Fluids* (Los Angeles: Eurographics Symposium on Computer Animation) 126
[6] Keita E, Faure P, Rodts S, Coussot P and Weitz D 2014 *Evaporation from a capillary tube: Experiment and modelization* (Kona: 5th International Conference on Porous Media and Their Appications in Science, Engineering and Industry) 50-57
[7] Berlin J 2010 *Orthoworld* **9** 120-126

IOP Conf. Series: Journal of Physics: Conf. Series **925** (2017) 012024 doi:10.1088/1742-6596/925/1/012024

Experimental and theoretical investigation of the mechanism of flame propagation above the surface of combustible liquid moving under the action of an oxidant in microchannels

Yu O Kabova[1,2], V V Kuznetsov[3,4], V V Zamashchikov[4,5], E A Chinnov[1,4] and O A Kabov[1,4]

[1]Kutateladze Institute of Thermophysics, SB RAS, 630090, 1 Lavrentiev prosp., Novosibirsk, Russia
[2]Novosibirsk State University of Architecture and Civil Engineering, 630008, 113 Leningradskaya Str., Novosibirsk, Russia
[3]Lavrentyev Institute of Hydrodynamics, SB RAS, 630090, 15 Lavrentiev prosp., Novosibirsk, Russia
[4]Novosibirsk State University,630090, 2 Pirogov str, Novosibirsk, Russia
[5]Voevodsky Institute of Chemical Kinetics and Combustion, SB RAS, 630090, 3 Institutskaya, Novosibirsk, Russia

E-mail: kabova@itp.nsc.ru

Abstract. Flames propagating above the surface of n-butanol in mini and microchannels have been investigated experimentally. It was determined that when oxygen is used as an oxidant, the critical channel size is about 0.3 mm, and according to the estimates this corresponds to the critical size for a premixed homogeneous gas mixture. In order to understand the mechanism of flame propagation above the surface of liquid, in this study, we attempted to simulate the processes, occurring in liquid. The problem of "thermal wave" propagation in a horizontal rectangular channel, when the liquid film moves under the action of gas flow, was studied theoretically. The analysis was carried out taking into account evaporation, specific heat release, and radiation and absorption processes in the gas and on the free surface, as well as the formed zone of strong heating above the liquid film surface for low oxidant velocities in the channel. Based on data obtained, it can be concluded that at high concentration of oxygen and low oxidant velocities, the experimental value of velocity coincides with the calculated value with satisfactory accuracy.

1. Introduction

The tendency to reduce the characteristic sizes of devices in various fields of technology determines the development of fluid mechanics and heat transfer studies in mini and microchannels. Recently, there are many papers on combustion of gaseous fuels in narrow channels of various configurations. This relates to the great practical interest in the miniature power sources, sources of heat and mechanical energy [1], caused by the high energy density of hydrocarbon fuels relative to conventional accumulators of electric energy.

With a decrease in the height of flat channels, the ratio of surface to volume of the channel increases inversely proportional to its minimal transverse dimension, and this causes high intensity of heat transfer in the microsystems. From the point of view of processes taking place in the reacting systems, for instance, combustion of flammable liquids in the small channels, an increase in heat transfer intensity leads, as a rule, to a decrease in the mass burning rate. Finally, at a certain size, the limit of flame propagation comes [2]. The magnitude of the critical size is an important characteristic widely used in practice. On the other hand, high intensity of heat transfer in micro and mini channels allows the creation of the efficient heat exchangers, which can be also useful from the point of view, for example, of thermoelectric converters. In addition, under the certain conditions, high intensity of heat transfer with the walls of microsystem leads to a return of a significant amount of heat released due to the chemical transformation into the initial mixture, which makes it possible to burn even when the characteristic size of the system is less than the critical one. Therefore, investigation of combustion processes in such systems is important for the development of new applications and increasing the energy efficiency of existing engineering solutions.

Most works dealt with investigation of diffusion combustion of liquid, whose saturated vapor pressure at the initial temperature is lower than the limiting one, were performed in large volumes [3]. At diffusion combustion in a limited volume, there is an important feature - impossibility of flame propagation over the liquid surface in the absence of an oxidant flow [4–7]. Experiments [5–7] have shown that the rate of flame propagation in such systems depends on the oxidant flow. In a cylindrical channel, when air is used as an oxidant, the flame moves with pulsations [5]. In a rectangular channel with the height less than 4 mm, flame propagation is possible only when air is enriched with oxygen. Papers [5–7] also show that the flame velocity can vary in a wide range from millimeters per second to several meters per second. The flame propagation velocities obtained under such conditions are comparable with the normal velocities for carbon-oxygen gas mixtures. These facts make us think about the mechanism of flame propagation above the liquid surface. The currently accepted thermocapillary mechanism is reasonably criticized in [7]. The authors point out that the flow of heated liquid from the area with the combustion products to the pre-flame zone due to the capillary forces requires a considerable amount of time. There is no such a process, when the flame propagates in a premixed gaseous mixture; therefore, in order to obtain the velocities close to the normal velocities of carbon-oxygen gas flames, formation of combustible mixture in the pre-flame region should be substantially accelerated. To understand the mechanism of flame propagation above the liquid surface, in this paper we attempted to simulate the processes taking place in the film of a combustible liquid; the model is based on the previously developed mathematical model [8, 9]. To simplify the problem, we consider small oxidant velocities in the channel, whose height was chosen much less than its width and this corresponds to the parameters of flat compact heat exchangers. This was also done in order to create a two-dimensional structure of the flow and flame front, which facilitates process simulation. According to the data available to the authors, there are no theoretical studies on the mechanism of flame propagation above the surface of combustible liquid moving under the action of an oxidant in microchannel.

2. Mathematical model

As a first step, "thermal wave" propagation in a horizontal rectangular channel at the liquid film motion under the action of a gas flow was studied theoretically taking into account evaporation (figure 1). The model is based on the previously developed model of thermocapillary flow of the locally heated liquid film moving under the action of a gas flow in a microchannel. The model takes into account the dynamic effect of gas flow on the liquid as well as diffusion and convective mechanisms of heat and mass transfer in liquid and gas and through the deformable gas-liquid interface [8, 9]. It is assumed that the gas flow consists of oxygen-enriched air and liquid is n-butanol.

The problem is solved in two-dimensional formulation, when the conditions of thin layer approximation are satisfied, taking into account specific heat release and processes of radiation and absorption in gas and at the free interface, as well as the formed zone of strong heating above the liquid film surface. The mathematical model is based on the systems of Navier-Stokes, continuity, energy and diffusion equations (1)–(2).

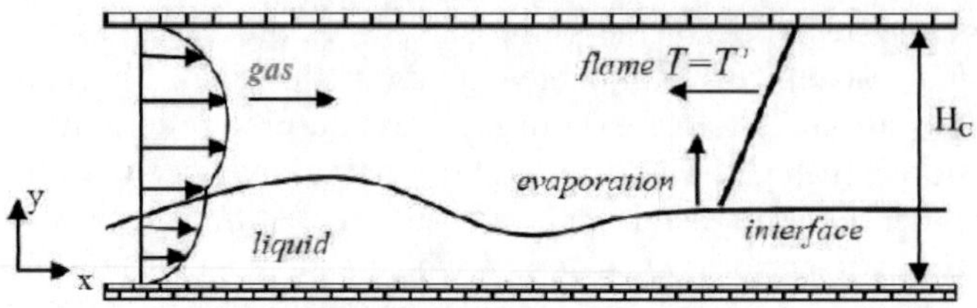

Figure 1. Sketch of the physical model

The unknown functions are $\vec{v}=(u,v),\vec{v}_g=(u_g,v_g),T(t,x,y),T_g(t,x,y),C(t,x,y)$, those functions define correspondingly the velocity and temperature fields in liquid and gas, and distribution of evaporated matter in gas (everywhere, index g indicates that the value belongs to the gas phase). Function $z=H(t,x)$ that specifies the free interface is also unknown. We obtain the system of equations in liquid

$$\rho(dv/dt-g)=-\nabla p+2\mathrm{div}(\mu W),\nabla\cdot v=0,\ \rho c_p dT/dt=\kappa\Delta T, \tag{1}$$

and in gas phase:

$$\begin{gathered}\rho_g(dv_g/dt-g)=-\nabla p_g+2\mathrm{div}(\mu_g W_g),\nabla\cdot v_g=0,\\ \rho_g c_{pg} dT_g/dt=\kappa_g\Delta T+Q\gamma-\omega(T^4-\bar{T}^4),dC/dt=D\Delta C-\gamma/\rho_g,\end{gathered} \tag{2}$$

the components W_{ij} and W_{gij} $(i,j=1,2,3;$ $W_{ij}=W_{ji},\ W_{gij}=W_{gji})$ of rate of strain tensors W and W_g are defined as follows $W_{11}=u_x,W_{g11}=u_{gx},$ $W_{12}=(u_y+v_x)/2,\ W_{g12}=(u_{gy}+v_{gx})/2,$ $W_{13}=(u_z+w_x)/2,\ W_{g13}=(u_{gz}+w_{gx})/2,W_{22}=v_y,\ W_{g22}=v_{gy},$ $W_{g23}=(v_{gz}+w_{gy})/2,$ $W_{33}=w_z,\ W_{g33}=w_{gz}.$

Here p–pressure, $\vec{g}$–vector of gravitational acceleration, D–diffusion coefficient, κ– thermal conductivity, c_p–heat capacity, $\mu=\mu(T)$ –dynamic liquid viscosity, ρ–liquid density, $\sigma=\sigma(T)$ – surface tension. Subscripts in the paper have the following meaning: 0—initial parameters of the flow (at initial temperature $T=T_0$), x, y, t, z—derivatives on x, y, t and z.

With the initial conditions, specifying an undisturbed flow and appeared zone of strong heating (ignition):

$$\vec{v}=\vec{v}_0,\vec{v}_g=\vec{v}_{g0},T=T_0,T_g=T_{g0},C=C_0,H=H_0,T_{g0}=\begin{cases}T_0,x<x_0\\T',x>x_0\end{cases}, \tag{3}$$

boundary conditions on the channel bottom:

$$\vec{v}=0,T=T_0, \tag{4}$$

conditions on the free interface:

$$\begin{gathered}\vec{v}-\vec{n}(\vec{v}\cdot\vec{n})=\vec{v}_g-\vec{n}(\vec{v}_g\cdot\vec{n}),\ \rho(\vec{v}\cdot\vec{n}-V_n)=\rho_g(\vec{v}_g\cdot\vec{n}-V_n),\ (1-C)\rho_g(\vec{v}_g\cdot\vec{n}-V_n)=-D\vec{n}\cdot\nabla(C\rho_g),\\ (P-P_g)\cdot\vec{n}=2\sigma K\vec{n}+\nabla_\Gamma\sigma+\rho_g(\vec{v}_g\cdot\vec{n}-V_n)^2\left(\frac{\rho_g}{\rho}-1\right)\vec{n},\ \kappa\frac{\partial T}{\partial n}-\kappa_g\frac{\partial T_g}{\partial n}=\frac{\rho_g\lambda D}{1-C}\frac{\partial C}{\partial n}+\omega_0 E_n,\\ T=T_g,\ C=C_*(T).\end{gathered} \tag{5}$$

and conditions on the upper wall of the channel:

$$\vec{v}_g=0,T=T_0. \tag{6}$$

Here, T' is the ignition temperature of combustible mixture, $\overline{T}$ is the average temperature of heat release zone in gas, γ is the burning rate, Q is a specific heat release, ω, ω_0 are the radiation and the absorption coefficients.

The problem was transformed to specially developed new variables, so that continuity equations in the liquid and in the gas retain their form in new variables but kinematic condition at the free interface simplifies and become linear. We employ the lubrication theory to derive reduced set of equations, so we assume that the characteristic film thickness and the characteristic film thickness variation are much smaller than the characteristic length scale of the film in streamwise and spanwise directions, thereby the film aspect ratio is taken asymptotically small $\varepsilon = H_0/l << 1$. Considering inertial terms to be negligible, the system of equations with boundary conditions was rewritten in dimensionless form.

Numerical solution of the problem is implemented by the finite difference method. The alternating directions implicit (ADI) method is used to solve the system of grid equations with boundary conditions. To solve the problem on each fractional grid step the Thomas algorithm is used. For more details refer to [8, 9]. The numerical algorithm based on the developed mathematical model allowing theoretical and numerical investigation of the joint nonisothermal liquid and gas motion in the microchannel, as well as heat and mass transfer on the free surface of the liquid film, of the problem was developed. When developing the numerical algorithm, the method of approximation by the mesh functions was used. Herewith the computational domain is divided by the surfaces of constant coordinate's value in equal steps, and the values of calculated functions are determined at all nodal points. The problem was solved numerically using the finite-difference method. The solution algorithm consists of the blocks allowing calculating the temperature, position of the gas-liquid interface, and velocity and pressure fields. The temperature is calculated by the method of alternating directions by introducing the fractional steps, and at each time step, calculations along the longitudinal and transverse directions are performed in turn. Since the temperature is continuous at the interface, calculations in liquid and gas are performed simultaneously, and the heat balance condition plays the role of the missing equation on the layer containing the boundary. This method ensures that the boundary conditions are met exactly. Evolution of the interface is calculated using a two-layer implicit difference scheme for parabolic equations of the fourth order. The velocity and pressure fields are calculated according to the analytical quadrature formulas obtained for the given problem when developing the numerical algorithm. The numerical code is written in Fortran. Each block-procedure in the program code allows independent testing on the check tasks. The corresponding check of the program blocks was carried out, and results give a good match to the tests. The check was made by deriving the solutions with the help of separate blocks of the code for quasilinear parabolic equations of the second and fourth orders for comparison with the known solutions. Totally, the discrepancy does not exceed 1% for all blocks. Stability of calculations was set using the implicit difference schemes with a second order of time approximation and the second and third order of approximation with respect to spatial variables when calculating the temperature and free interface position, respectively. It is known that approximation and stability together guarantee the convergence of numerical solution of the problem. Numerically it was shown that combustion of oxygen-enriched air leads to formation of a "thermal wave" rapidly moving towards the gas flow with the velocity close to constant. In figure 2 one can see evolution of the temperature field in the microchannel at different time moments. Here, all values are given in the dimensionless form: temperature $T=T_0 +[T]\theta$, where the temperature scale is given as $[T] = (T'-T_0)$ and time $\underline{t} = lt/U$. Zone of strong heating moving upstream may be treated as a combustion wave. The velocity of this zone in this case of calculations is equal to 2.7 m/s.

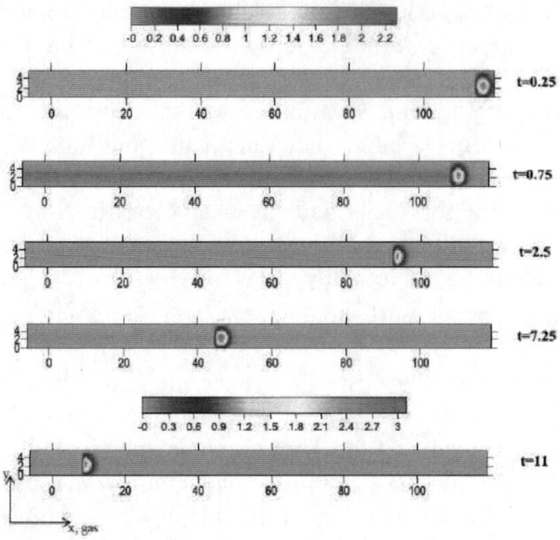

Figure 2. The process of combustion front propagation in time across the channel (x). Temperature in the channel at different moments of time t (all values are in dimensionless form). Channel height is 500 μm, channel length is 30 cm, Reynolds number in gas is Re_g=10, liquid film thickness is 90 μm. Temperature of self-ignition is T'=340 °C, initial temperature is T_0=20 °C.

3. Experimental results

New experimental results on investigation of flames propagating over the surface of n-butanol in mini and microchannels are obtained. The channel height was varied from 0.23 to 0.5 mm. The description of experimental setup is given in [7]. Propagation at a change in oxidant composition was studied experimentally. Experiments showed that the average velocity of flame propagation increases with increasing proportion of oxygen in the oxygen-nitrogen mixture. Therefore, for the channels with low height, pure oxygen was chosen as the oxidizing mixture. Experimental results for oxygen were obtained in the channels 0.3 and 0.5 mm in height. The comparison showed that an increase in the average velocity of oxygen, as in the channel with the height of 0.5 mm, leads to a decrease in the flame velocity. However, if the flame velocity changes significantly with a decrease in the channel height from 1 to 0.5 mm, then with a decrease in the channel height from 0.5 to 0.3 mm, it changes insignificantly at low oxidant velocities with a tendency towards increasing velocity difference with an increase in the oxidant velocity, figure 3.

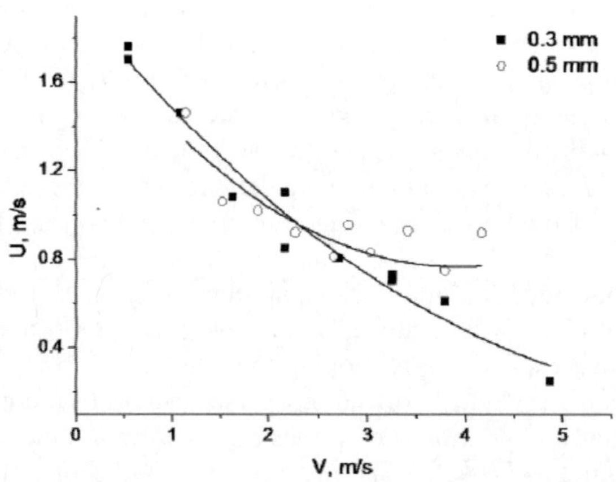

Figure 3. Flame velocities vs. average oxygen velocity for different channel heights

It is determined that in a channel with the height of 0.3 mm, the flame can propagate if pure oxygen is used as an oxidant. The photographs of the flame front moving in this channel are presented in figure 4. Flame extinction in the channel is determined by heat losses to the wall, which depend on the channel height. According to theoretical estimates, the limiting condition is determined by Peclet number $Pe = U_n d/a$ = const. Where U_n is normal velocity (velocity of flame propagation along the normal to its surface relative to the fresh mixture), d is hydraulic diameter of the channel, a is thermal diffusivity coefficient. Experiments of other researchers showed that for hydrocarbon-air mixtures the critical channel height is about 2 mm. Since, according to the critical condition, the size is inversely proportional to the normal velocity and normal velocities for carbon-air and carbon-oxygen mixtures differ approximately by the factor of 10, the critical size for the mixture of n-butanol with oxygen, used in our experiments, should be of the order of 0.2 mm. For the channel of 0.3 mm height, if we take into account that there is a liquid film on the surface, the actual size of the channel will be even smaller, so it is expected that this size is close to the size of extinction. However, experiments showed that this channel is not critical; moreover, the flame velocities with a decrease in the channel height from 0.5 to 0.3 mm vary insignificantly. The result requires further investigation.

a) b)

Figure 4. Flame front moving towards the oxidant flow in the channel with the height of 0.3 mm. a: V(oxygen)=1 m/s, b: V(oxygen)=3.8 m/s.

4. Conclusion

New experimental results on investigation of flames propagating over the surface of n-butanol in mini and microchannels are obtained. The height of the channel was varied from 0.23 to 0.5 mm.

It is found that if oxygen is used as an oxidant, the critical channel size is about 0.3 mm, and according to the estimates, this corresponds to the critical size for a premixed homogeneous gas mixture.

The problem of "thermal wave" propagation in a horizontal rectangular microchannel at the liquid film motion under the action of a gas flow at high oxygen concentration and low oxidant velocities was studied theoretically. When developing the model, evaporation, specific heat generation, and radiation and absorption processes in gas and at the free interface, as well as the zone of strong heating above the liquid film were taken into account. It can be concluded that at high oxygen concentration and low oxidant velocities, the experimental value of velocity coincides with the calculated value satisfactorily.

Acknowledgments
The study was financially supported by the Russian Science Foundation, project no. 15–19–30038.

References

[1] Maruta K 2011 *Proc. of the Combustion Institute* **33** 125–150
[2] Zeldovich B Ya, Barenblatt G I, Librovich V B and Makhviladze G M 1980 *Mathematical theory of combustion* (ed R I Soloukhin , Moscow: Nauka) 478
[3] Ross H D and Miller F J 1996 *Proc. of the Combustion Institute* **26** 1327–34
[4] Bieri J A, Kurdyumov V N and Matalon M 2011 *Proc. of the Combustion Institute* **33** 1227–34
[5] Zamashchikov V V 2009 *Combustion Sci. and Technology* **181** 1 176–189
[6] Zamashchikov V V, Korzhavin A A and Chinnov E A 2014 *Combustion, Explosion and Shock Waves* **50** 4 381–386
[7] Zamashchikov V V, Korzhavin A A and Chinnov E A 2016 *Int. J. Heat and Mass Transf.* **102** 470–478
[8] Kabova Yu, Kuznetsov V V, Kabov O, Gambaryan-Roisman T and Stephan P 2014 *Int. J. Heat and Mass Transf.* **68** 527–541
[9] Kabova Yu, Kuznetsov V V and Kabov O 2014 *Interfacial Phenomena and Heat Transfer* **2** 85–102

IOP Conf. Series: Journal of Physics: Conf. Series **925** (2017) 012025 doi:10.1088/1742-6596/925/1/012025

Foam flows through a local constriction

T Chevalier[1], J Koivisto[1], N Shmakova[2], M J Alava[1], A Puisto[1], C Raufaste[3] and S Santucci [2,4]

[1] COMP Center of Excellence, Department of Applied Physics, Aalto University, Espoo, Finland
[2] Lavrentyev Institute of Hydrodynamics SB RAS, Novosibirsk, Russia
[3] Université Côte d'Azur, CNRS, Institut de Physique de Nice, 06100 Nice, France
[4] Laboratoire de Physique, ENS de Lyon, Univ Lyon, CNRS, UMR 5672, Lyon, France

E-mail: stephane.santucci@ens-lyon.fr

Abstract. We present an experimental study of the flow of a liquid foam, composed of a monolayer of millimetric bubbles, forced to invade an inhomogeneous medium at a constant flow rate. To model the simplest heterogeneous fracture medium, we use a Hele-Shaw cell consisting of two glass plates separated by a millimetric gap, with a local constriction. This single defect localized in the middle of the cell reduces locally its gap thickness, and thus its local permeability. We investigate here the influence of the geometrical property of the defect, specifically its height, on the average steady-state flow of the foam. In the frame of the flowing foam, we can observe a clear recirculation around the obstacle, characterized by a quadrupolar velocity field with a negative wake downstream the obstacle, which intensity evolves systematically with the obstacle height.

1. Introduction

Liquid foams, such as chocolate mousse and shaving foams are soft materials that we all use in our everyday life. Their multi-scale biphasic structure of gas bubbles within a liquid phase leads to a dual mechanical behaviour: they behave as solids at rest, while they can flow like liquids above a given critical yield stress. Such peculiar rheological properties are at the root of their use in diverse applications, ranging from daily-life products developed in food, cosmetics and pharmaceutical industries, to larger scale industrial processes, for instance in oil recovery or soil remediation. For those last applications, understanding and controlling the flow of liquid foams in confined heterogeneous media appears of tremendous importance [1][2][3][4].

Thus, in order to investigate the flow of liquid foams and more specifically, the motion, deformation and topological rearrangements of the bubbles, when forced to invade a heterogeneous porous and fractured medium, we have designed the following simple experimental set-up.

2. Experimental set-up

2.1. A simple model fracture

IOP Conf. Series: Journal of Physics: Conf. Series **925** (2017) 012025 doi:10.1088/1742-6596/925/1/012025

Our set-up allows the direct observation of the motion and deformation of the elementary components of a 2D liquid foam, the bubbles, when forced to invade a confined medium that mimics a simple model fracture. As shown on figure 1, our model fracture consists of an inhomogeneous Hele-Shaw cell, made of 2 glass plates (15 cm wide, 50 cm long and 1 cm thick), separated by a 1 mm gap. Moreover, the cell can present a defect of lateral size 2 x 2 cm2 and height H of typically a few hundred of microns. This bump localized in the centre of the cell reduces locally its gap, and thus decreases locally the permeability of the cell. To ensure identical surface properties of the top and bottom plates of the cell, both surfaces are covered with a thin stretchable hydrophilic plastic film, leading to a trapezoidal shape of the constriction, as shown on the side view of the cell displayed in figure 1, (panel a).

2.2. In-situ generation & forced flow of a liquid foam

We generate in-situ our liquid foam and forced it to flow at a constant velocity in our model fracture, by bubbling filtered pressurized air, at a constant rate of 40 ml/mn (thanks to a mass flow controller) through a needle of inner diameter 0.45 mm, in a vertical chamber connected to the cell (via a 10 mm diameter hole drilled in the bottom plate), filled with a soapy solution. This surfactant solution is obtained by mixing 1% of a commercial dishwashing liquid Dawn© with ultra-pure water.

In these conditions, the bubbles generated are quasi mono-disperse with a diameter larger than the cell gap. Therefore, those bubbles form a 2D confined foam, constituted by a mono-layer of millimetric bubbles. It is also important to notice that those experimental conditions lead to a constant flow rate of the liquid foam, invading the inhomogeneous cell at a velocity around v = 15 mm/s.

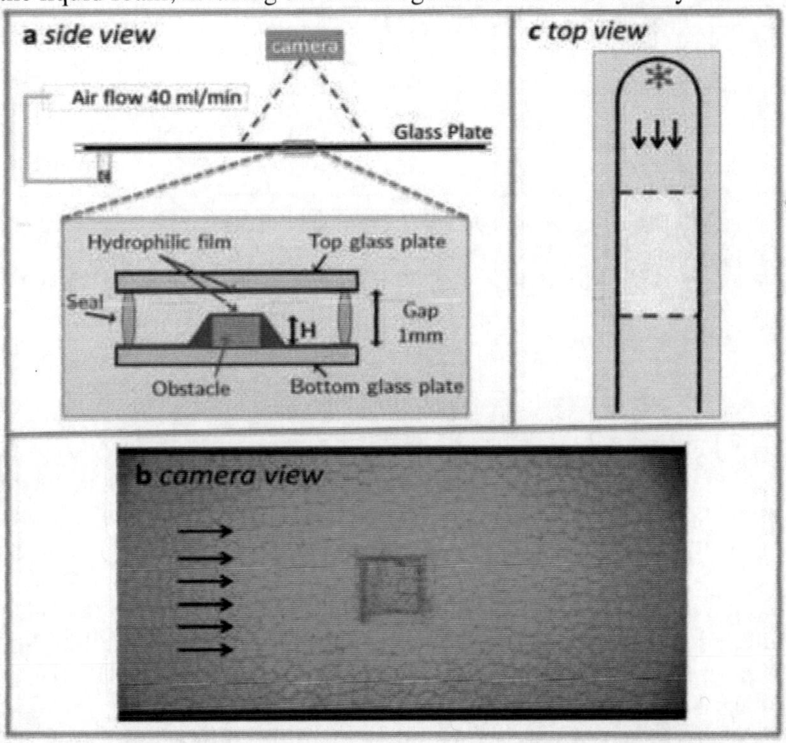

Figure 1: Experimental set-up, allowing the direct observation of a 2D liquid foam flowing in a Hele-Shaw cell with a local constriction, due to an obstacle placed in the middle of the cell (the obstacle height H is smaller than the gap of the cell, G =1 mm). The panels (a) and (c) displays respectively side and top views, while the bottom panel shows a typical recorded image (b), for a flow experiment with an obstacle filling 70% of the gap cell.

2.3. Direct observation
We observe directly the steady-state flow of our 2D liquid foam in the middle of the cell around the obstacle with an imaging area of 20 x 10 cm, using a Ximea XiQ digital camera during 400 seconds at 5 Hz. Thus, we record 2000 frames with a resolution of 2088 x 1048 pixels. A typical image recorded during a flow experiment with an obstacle filling 70% of the gap cell is shown on the bottom panel of figure 1, as well as a zoom in a region around the obstacle on the left panel of figure 2.

3. Image analysis
Using standard image analysis tools, developed within the software ImageJ© and MATLAB©, we can obtain a network of skeletonized bubbles (a zoom around the obstacle of a typical binary image of the bubble network obtained by edge detection is shown on the right panel of figure 2). We can subsequently identify all individual bubbles of the flowing foam, for each recorded image.

Interestingly, one can already observe that the bubbles approaching the constriction get squeezed in the direction of the imposed mean flow, while the bubbles exiting the permeable defect are strongly elongated in this direction. Moreover, the bubbles on the obstacle have an apparent increase of diameter obviously related to the reduction of the gap cell thickness and their incompressible nature.

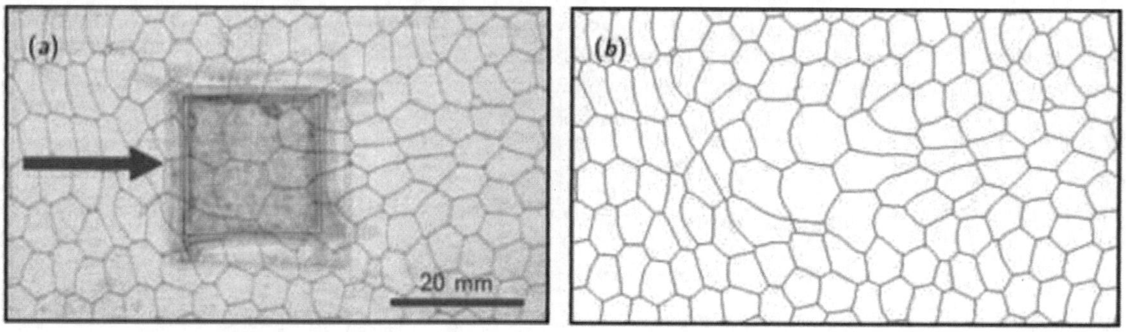

Figure 2: The left panel (a) shows a zoom (on a region around the obstacle) of a raw image recorded by the camera during the foam flow with an obstacle filling 70% of the gap cell. The arrow shows the direction of the imposed flow. The right panel (b) displays the corresponding binary image of the skeletonized bubble network obtained by edge detection, over the same region. The shaded square zone gives the location of the constriction.

4. Averaged velocity field
Tracking those labelled bubbles, we can measure their velocity by computing the displacement vector of the bubble centres, between two subsequent images recorded with a time delay of Δt=0.2 s. The resulting average velocity field for an experiment with an obstacle filling 70% of the gap cell is shown in figure 3.

Furthermore, we have checked that the velocity field of the flowing foam could also be obtained by standard image correlation techniques, using directly the texture provided by the films of the bubbles themselves and their vertices. To perform this task, we took advantage of PIVlab – a Time-Resolved Digital Particle Image Velocimetry Toolbox for MATLAB – developed by W. Thielicke and E. J. Stamhuis to compute the velocity fields shown in figure 4[5]. Furthermore, we could obtain similar velocity fields using solely the vertices of the skeletonized bubble network as tracers of the flow. Such a result appears particularly interesting. Indeed, the same procedure could be extrapolated in 3D to extract easily the velocity field of a 3D liquid foam flowing in a constriction or around an obstacle, such as the one captured recently using ultra fast X-ray tomography [6], without the need of the very tedious and still unclear procedure of the reconstruction of the 3D bubbles shape [7].

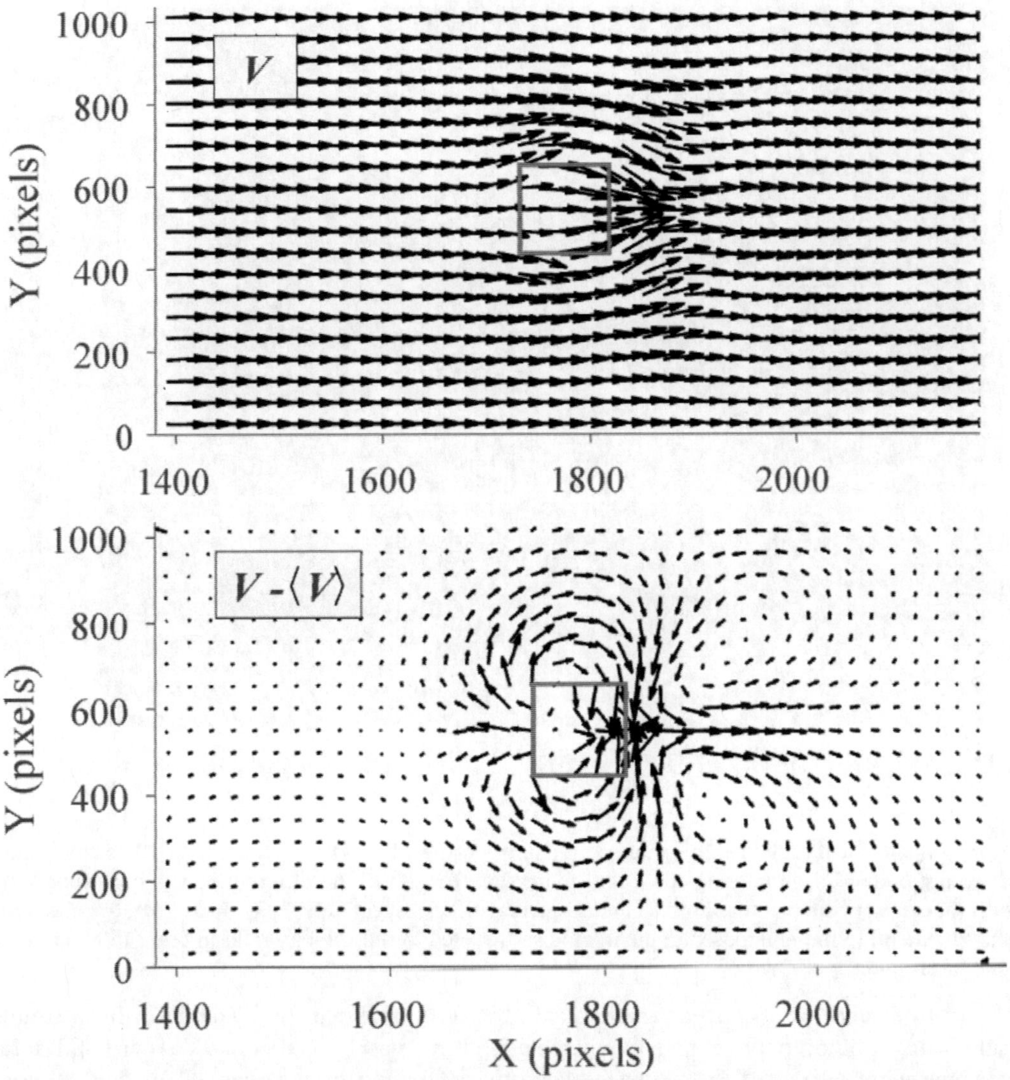

Figure 3: Average velocity field (obtained by tracking bubble centers) for a flow experiment with an obstacle (red square of 2x2 cm^2) filling 70 % of the cell gap. On the bottom panel, the imposed flow rate at the inlet is substracted to the velocity field, revealing the strong disturbance induced by the constriction characterized by a quadrupolar velocity field around the obstacle and an ovsershoot downstream the obstacle.

Far from the constriction, we clearly observe that the velocity field is uniform corresponding to the flow rate imposed by our experimental procedure. Nevertheless, approaching the permeable obstacle, the amplitude of the velocity field decreases, while it increases when leaving the obstacle. Such flow disturbance caused by the constriction is clearly revealed when displaying the velocity field in the frame of the flowing foam (shown in the bottom panel of figure 3, as well as in figure 4). Indeed, a clear recirculation is then observed, characterized by a quadrupolar velocity field around the obstacle. Furthermore, downstream the obstacle, there is a zone where the velocity is larger than the mean velocity (equivalently, in the frame of the flowing foam, the velocity is opposed to the one of the obstacle). Such effect is reminiscent of the observation of a so-called "negative wake" in the flow pattern of rising air bubbles in viscoelastic fluids [8].

IOP Conf. Series: Journal of Physics: Conf. Series **925** (2017) 012025 doi:10.1088/1742-6596/925/1/012025

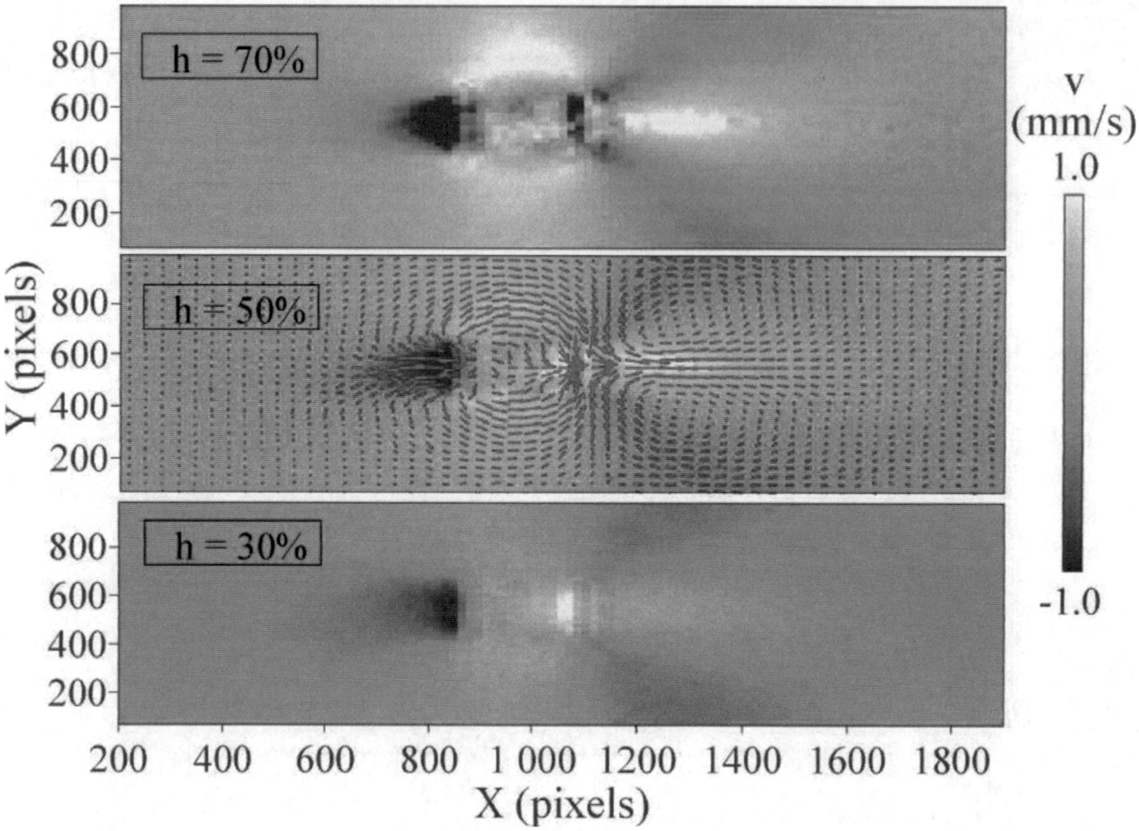

Figure 4: Velocity field (obtained by image correlation – using PIVLAB) averaged over 100 susbequent frames, corresponding to monitoring the liquid foam flow for 20 s. The different panels correspond to flow experiments for different heights of the obstacle, filling respectively 70%, 50% and 30% the gap cell. A flow rate of 15 mm/s imposed at the inlet is substracted to the velocity field in order to reveal the recirculation around the obstacle.

Our observations and results appear analogous to the ones obtained for a 2-D foam flow around a full obstacle (using a liquid pool geometry), characterized in detail by Dollet and Graner [3], Raufaste et al [6], which were attributed to the visco-elasto-plastic nature of the foam [10], that cannot be modelled as a simple visco-plastic fluid.

Interestingly, in our geometry, modifying locally the permeability of the cell (by simply changing the height of the defect/constriction), we can modulate systematically the intensity of the quadrupolar recirculation and specifically the velocity overshoot observed behind the obstacle. Figure 4 shows clearly that while such quadrupolar velocity field is already present for a very mild variation of the gap cell, locally, the amplitude of the disturbance and specifically the intensity of the recirculation increases systematically with the height of the local permeable obstacle.

5. Conclusions and perspectives

We have designed a simple but original experimental set-up to directly observe the motion and deformation of the bubbles of a 2D liquid foam, forced to invade an inhomogeneous confined medium at a constant flow rate. We could show that a constriction (reducing locally the gap of the cell) modifies strongly the average steady–state flow of the foam, leading to a quadrupolar velocity field in the frame of the flowing foam around this permeable obstacle. The intensity of such recirculation (and in particular the velocity overshoot observed downstream the constriction) increases systematically with the constriction height.

Following the procedure and tools developed by Dollet and Graner [3], Raufaste et al [9], Marmottant et al [11], we are currently investigating in detail the influence of the geometrical properties of the localized constriction (shape, lateral extension and height of the permeable defect), as well as other experimental controlling parameters, such as the foam liquid fraction and the imposed flow rate on both the velocity, deformation and the topological rearrangement of the bubbles. On the other hand, we are also implementing the so-called "Bubble model" [12] within our specific geometry in order to check if such simple model is able to reproduce the foam flow observed experimentally.

Acknowledgements
This work is supported by grant no.14.W03.31.0002 of the Russian Government.

References

[1] Weaire D and Hutzler S 1999 *The Physics of Foams* (Oxford University Press)
[2] Cantat I, Cohen-Addad S, Elias F, Graner F, Hohler R, Pitois O, Rouyer F and Saint-Jalmes A 2013 *Foams, Structure and Dynamics* (Oxford University Press)
[3] Dollet B and Graner F 2007 Two-dimensional flow of foam around a circular obstacle: Local measurements of elasticity, plasticity and flow *J. Fluid Mech.* **585** 181
[4] Stevenson P 2012 *Foam Engineering: Fundamentals and Applications* (Wiley)
[5] Thielicke W and Stamhuis E J 2014 PIVlab – Towards User-friendly, Affordable and Accurate Digital Particle Image Velocimetry in MATLAB *Journal of Open Research Software* **2**
[6] Raufaste C, Dollet B, Santucci S and Mokso R 2015 Three-dimensional foam flow resolved by fast X-ray tomographic microscopy *EPL* **111** 38004
[7] Mader K, Mokso R, Raufaste C, Dollet B, Santucci S, Lambert J, Stampanoni M 2012 Quantitative 3D Characterization of Cellular Materials: Segmentation and Morphology of Foam *Colloids Surf. A* **415** 230
[8] Hassager O 1979 Negative wake behind bubbles in non-newtonian fluids *Nature* **279** 402
[9] Raufaste C, Dollet B, Cox S, Jiang Y and Graner F 2007 Yield drag in a two-dimensional foam flow around a circular obstacle: Effect of liquid fraction *Eur. Phys. J. E* **23** 217
[10] Cheddadi I, Saramito P, Dollet B, Raufaste C, Graner F 2011 Understanding and predicting viscous, elastic, plastic flows *Eur. Phys. J. E* **34** 1
[11] Marmottant P, Raufaste C and Graner F 2008 Discrete rearranging disordered patterns Part II: D plasticity, elasticity and flow of a foam *Eur. Phys. J. E* **25** 371
[12] Durian D J 1995 Foam Mechanics at the Bubble Scale *Phys. Rev. Lett.* **75** 4780

IOP Conf. Series: Journal of Physics: Conf. Series **925** (2017) 012026 doi:10.1088/1742-6596/925/1/012026

Experimental investigation of picoliter liquid drops evaporation on a heated solid surface

D P Kirichenko[1,2], D V Zaitsev[1,2], O A Kabov[1,2]

[1]Kutateladze Institute of Thermophysics, Siberian Branch of Russian Academy of Sciences, 630090, Novosibirsk, Russia
[2]Novosibirsk State University, 630090, Novosibirsk, Russia

E-mail: Dmitriy_kirichenko@mail.ru

Abstract. This paper presents a study of the evaporation of sessile picoliter liquid drops on a heated solid surface. It has been shown that during evaporation diameter of the drop is almost constant (the contact line is pinned) and starts to decrease only at the final moment of drop life. It has been shown that the specific evaporation rate of a droplet (droplet weight loss per unit time per unit droplet surface area) is not constant over time: it gradually grows with time up to a maximum, but at the final stage, when the height of the drop becomes on the order of 1 μm, it decreases rapidly.

1. Introduction

The process of liquid drop evaporation, which takes place in a variety of technological systems in power engineering, medicine, agriculture, chemical and other industries, has been actively investigated during the last decade. Effective solutions for removing high-density heat fluxes are systems based on gas-droplet flows or sprays [1, 2] as well as systems using two-phase flow in a microchannel [3, 4], where evaporation of liquid drops in contact with hot wall may occur. Most experimental and theoretical studies deal with evaporation of isothermal water drops. There are many publications studying evaporation of drops with volume from 1 to 1000 μl. In [5,6] it is found that at the final stage of the drop life specific evaporation rate abruptly increases especially for drops with small and moderate contact angle hysteresis. Studies of nonisothermal evaporation of liquid drops of picoliter (pl = 10^{-12} l) volume are almost absent in the literature.

The objective of this work were to study the evaporation of sessile water drops with initial volume of 10-50 pl on a heated solid surface.

2. Experimental setup and methods

The experimental setup is shown schematically in Fig. 1. The working section is made of stainless steel with an embedded copper rod with dimensions of 10x10 mm^2. The surface was heated by a nichrome coil supplied with current from a power source. Substrate surface temperature was measured by thermocouples embedded in the copper rod. The surface temperature of the copper rod was 134 °C. The experiments were performed in the working section open to the atmosphere at an air temperature of 25–27 °C. Liquid microdroplets with a size of the order 10 μm were applied on the working surface

IOP Conf. Series: Journal of Physics: Conf. Series **925** (2017) 012026 doi:10.1088/1742-6596/925/1/012026

by a spray device placed a few centimeters above the heating area. Droplets were examined over the surface of the copper rod. Room-temperature nanofiltered distilled water was used as the working fluid. The droplets were imaged from side using a FASTCAM SA1.1 high-speed camera (5600 fps at a resolution of 1024x1024 pixels) coupled with a microscope objective. The optical resolution of the system was 0.8 μm/pixel. The morphology of the surface was analyzed using scanning electron microscope (JEOL JSM6700F) and atomic force microscope (Solver Pro NT MDT), Fig. 2. The root mean square (RMS) surface roughness was found to be 0.50 μm.

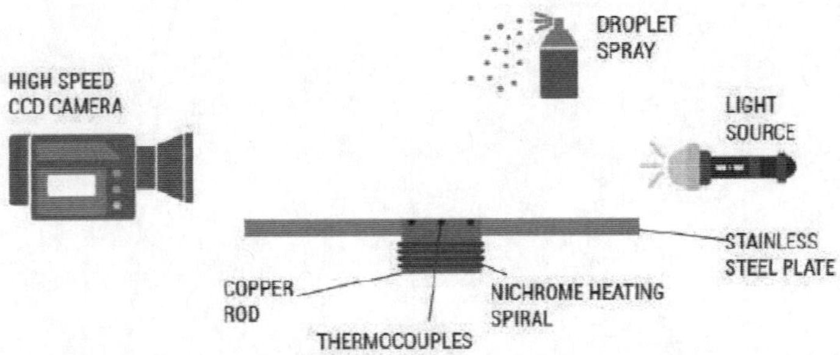

Figure 1. Schematic diagram of the experimental setup

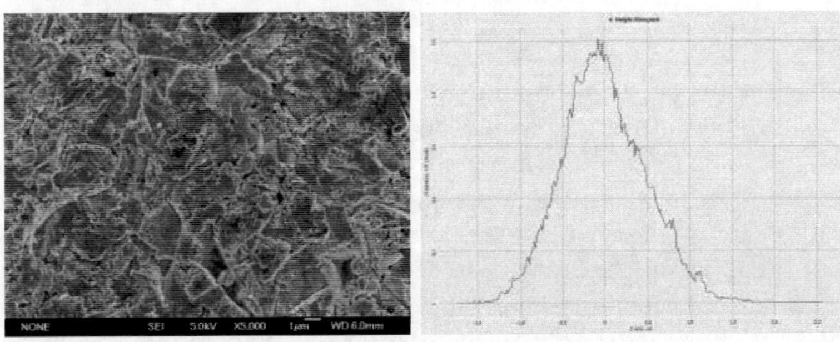

Figure 2. Results of heater surface characterization studies: (left) an image obtained using scanning electron microscope; (right) height histogram obtained using atomic force microscope

3. Results

Figure 3 shows change of the liquid drop profile during evaporation for a liquid drop with initial volume of 51.4 pl. Due to evaporation, the droplet size decreases with time. Depending on the initial size of the drop, life time changes from 0.006 to 0.016 s.

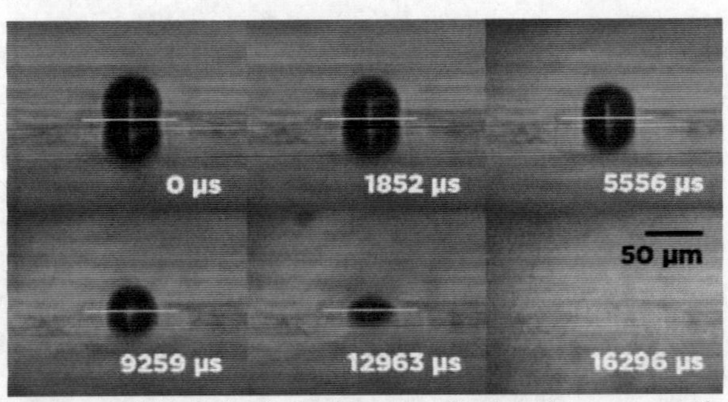

Figure 3. Change of drop shape (side view) during sessile droplet evaporation on a heated substrate, T_w=134 °C

IOP Conf. Series: Journal of Physics: Conf. Series **925** (2017) 012026 doi:10.1088/1742-6596/925/1/012026

It has been found that during evaporation diameter of the drop is almost constant (the contact line is pinned) and starts to decrease only at the final moment of drop life. Figure 4 shows plots of the diameter of an evaporating droplet versus time for different initial droplet sizes.

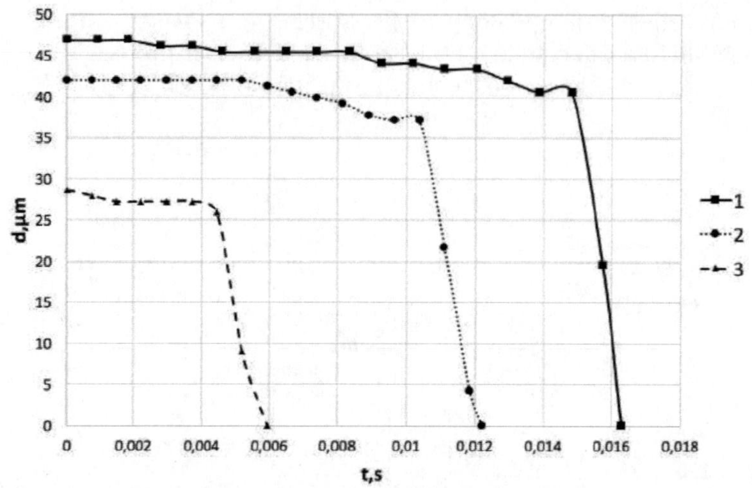

Figure 4. Diameter of wetted area versus time for three different sessile drops (lines 1,2 and 3), T_w=134 °C

Figure 5 shows the volume of an evaporating droplet versus time for different initial droplet sizes. It has been found that volume decreases with time almost linear, but at the final stage of the drop evaporation, the volume starts to decrease much slower. Figure 6 shows the specific evaporation rate of a droplet (droplet weight loss per unit time per unit droplet surface area) versus time. It is seen that the specific evaporation rate is not constant over time: it gradually grows with time up to a maximum, but at the final stage, when the height of the drop becomes on the order of 1 µm, it decreases rapidly. Such decrease of the specific evaporation rate has been observed for the first time. In previous works with evaporation of large drops [5,6] only an increase of the evaporation rate at the final stage was registered.

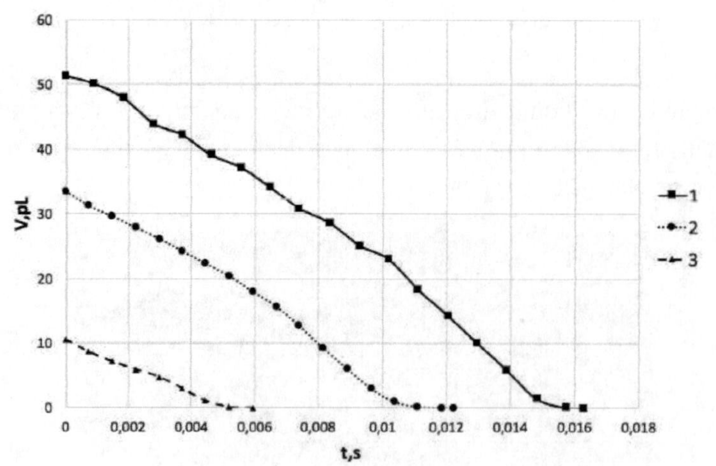

Figure 5: Volume versus time for three different sessile drops (lines 1,2 and 3), T_w=134 °C

IOP Conf. Series: Journal of Physics: Conf. Series **925** (2017) 012026 doi:10.1088/1742-6596/925/1/012026

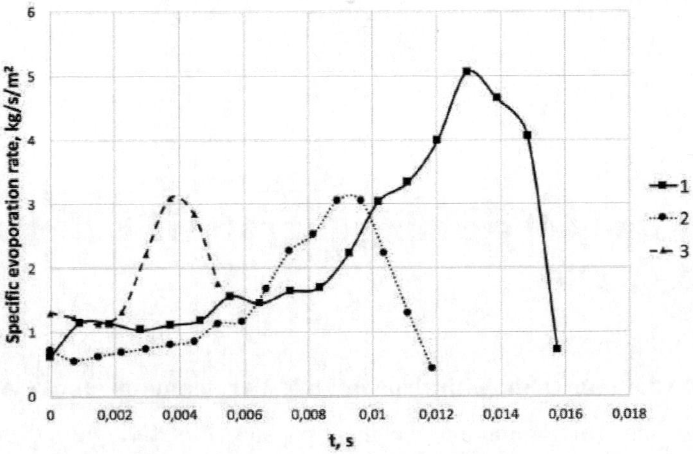

Figure 6: Specific evaporation rate versus time for three different sessile drops (lines 1,2 and 3), T_w=134 °C

4. Conclusions

Evaporation of sessile picoliter liquid drops on a heated solid surface has been studied experimentally and the following results have been obtained.

(1) During the process of evaporation, diameter of the drop is almost constant (the contact line is pinned) and starts to decrease only at the final moment of the drop life.

(2) Specific evaporation rate is not constant over time: it gradually grows with time up to a maximum, but at the final stage, when the height of the drop becomes on the order of 1 μm, it decreases rapidly.

Acknowledgements

This work was supported by the Russian Science Foundation (project 15-19-20049).

References

[1] Jungho Kim 2006 *International Journal of Heat and Fluid Flow* **28** 753

[2] Srikar R, Gambaryan-Roisman T, Steffes C, Stephan P, Tropea C, Yarin A L 2009 *International Journal of Heat and Mass Transfer* **52** 5814

[3] Kabov O A, Lyulin Yu V, Marchuk I V and Zaitsev D V, 2007 *International Journal of Heat and Fluid Flow* **28** 103

[4] Zaitsev D V, Rodionov D A and Kabov O A 2009 *Technical Physics letters* **35** 680

[5] Elizaveta Ya Gatapova, Andrey A Semenov, Dmitry V Zaitsev, Oleg A Kabov 2014 *Colloids and Surfaces A: Physicochemical and Engineering Aspects* **441** 776

[6] A A Semenov, D V Feoktistov, D V Zaitsev, G V Kuznetsov, and O A Kabov 2015 *Thermophysics and Aeromechanics* **22** 771

IOP Conf. Series: Journal of Physics: Conf. Series **925** (2017) 012027 doi:10.1088/1742-6596/925/1/012027

Experimental study of cleaning aircraft GTE fuel injectors using a vortex ejector

O A Evdokimov[1], Sh A Piralishvili[1], S V Veretennikov[1] and A A Elkes[2]

[1] Department of general and technical physics, P.A. Soloviev Rybinsk State Aviation Technical University University, Pushkina st., 173a, Rybinsk, 152934, Russian Federation
[2] D-30KP/KU Department, JSC NPO Saturn, Lenina av., 163, Rybinsk, 152903, Russian Federation

E-mail: yevdokimov_oleg@mail.ru

Abstract. The main ways of cleaning the fuel injectors and the circuits of jet and vortex ejectors used for pumping gas, liquid and two-phase media, as well as for evacuation of enclosed spaces are analyzed. The possibility of organizing the process of pumping the liquid out of the fuel injection manifold secondary circuit using a vortex ejector is shown experimentally. The regimes of manifold evacuation at various inlet liquid pressure values are studied. The technology of carbon cleaning fuel injectors using a washing liquid at various working process parameters is tested.

1. Introduction

While operating the aircraft gas turbine engines, many problems related to the deterioration of the quality of fuel spraying by injectors due to their coking arise. At present, the carbon cleaning of fuel injectors is carried out mechanically, which lies in their disassembling during the aircraft engine removal, which is quite costly and inefficient.

Hydraulic study of cleaning D-30KU/KP/KU-154 engines duplex fuel injectors (Fig. 1), conducted by JSC NPO Saturn earlier [1], has shown that the washing liquid supply through the primary circuit manifold and pumping out of the secondary circuit manifold is quite promising. This, on the one hand, allows to clean the injectors directly “over the wing” without disassembling them. On the other hand, it requires the hydraulic circulation circuit to be developed and implemented, that would provide the required pressure drops within the fuel injection manifold circuits. The scheme of the circulation process is shown in Fig. 2.

The problem above can be solved if the necessary pressure drop is provided through the creation of a rarefaction area at the fuel injection manifold secondary circuit inlet. Pumping the contaminated liquid out according to the circuit above is usually additionally accompanied by air masses sucking and requires the separation and filtration processes to be implemented for the purpose of its reuse organization.

IOP Conf. Series: Journal of Physics: Conf. Series **925** (2017) 012027 doi:10.1088/1742-6596/925/1/012027

2. Problem Statement

Analytical and numerical calculations of the liquid flow in the flow section of D-30KU/KP/KU-154 engines fuel manifold and injectors show that in order to organize their evacuation the absolute pressure within the range of 0.02–0.05 MPa is to be provided. Jet ejectors, despite their reasonable study and successful application in many industries, as a rule, cannot provide a sufficient evacuation level, and moreover, have large mass-dimensional characteristics [2-4].

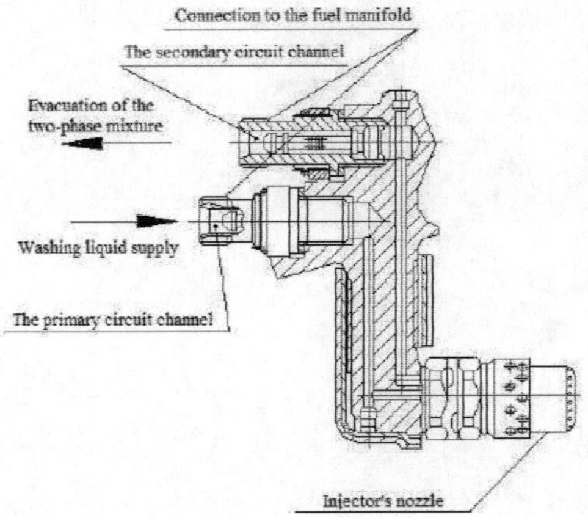

Figure 1. Design of the fuel injector

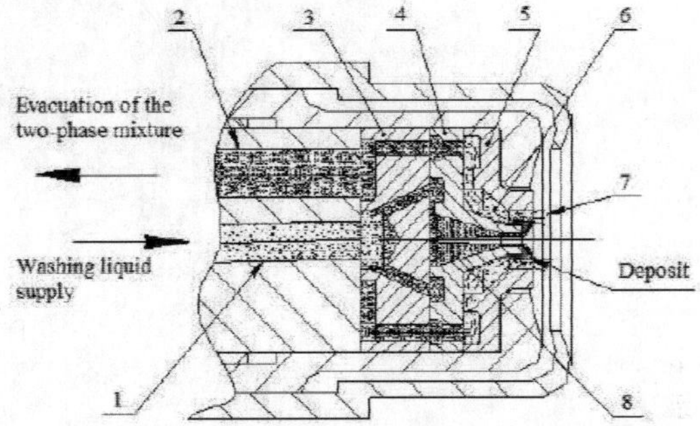

Figure 2. Scheme of the washing liquid circulation: 1) the primary circuit channel; 2) the secondary circuit channel; 3) adapter; 4) spray nozzle of the primary circuit; 5) spray nozzle of the secondary circuit; 6) two-phase mixture; 7) ingestion of the washing liquid and air; 8) washing liquid

The main advantage of vortex ejectors in comparison with jet ejectors in terms of evacuation and transportation is the possibility of organizing the gas-liquid media separation. Studies of vortex ejectors (M. G. Dubinskiy, V.I. Metenin et al. [3]) show a much broader possibility of achieving higher evacuation pressure values (up to 0.001 MPa).

One of the possible solutions to the problem of cleaning the duplex fuel injectors within the manifold is the organization of a circulation cleaning, which includes the following: a heated washing liquid is supplied to the primary circuit channel, which is connected to the central jet nozzle of the injector.

Based on the parameters required for the liquid pumping out, a vortex direct-flow ejector with a 15 mm diameter mixing chamber diameter was designed, the exterior of which is shown in Fig. 3.

Figure 3. Vortex direct-flow ejector

To experimentally study the evacuation of the fuel manifold using a vortex ejector, a test stand has been designed and installed, the functional diagram of which is shown in Fig. 4.

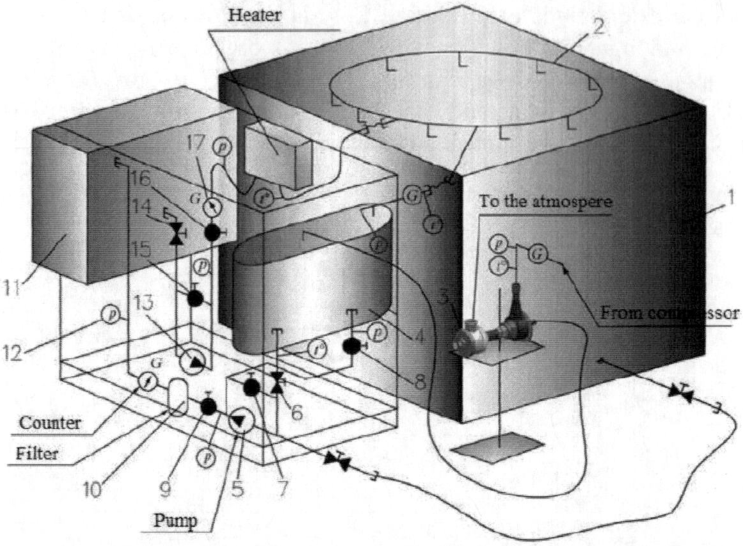

Figure 4. Test stand functional diagram: 1 – spraying tank; 2 – fuel injection manifold; 3 – vortex ejector; 4 – receiving tank; 5 – intake circulating pump 25/40; 6 – shutoff valve; 7 – bypass pipeline flow valve; 8 – return pipeline flow valve; 9 – flow valve regulating the washing liquid supply to the filter; 10 – mechanical cleaning filter; 11 – supply tank; 12 – liquid counter; 13 – feed circulating pump 25/60; 14 – feed circulating pump shutoff valve; 15 – feed circulating pump bypass pipeline flow valve; 16 – flow valve supplying the washing liquid to clean the injectors; 17 – liquid counter

The test stand is operated as follows. The liquid is supplied to clean the fuel injection manifold 2, installed in the spraying tank 1. The required liquid supply parameters are provided by the feed circulating pump 13, transferring the liquid from the supply tank 11. Bypass pipeline flow valve 15 provides a stable pump operation. The fluid supply to the fuel injection manifold is controlled by flow valve 16.

The stand includes a receiving tank 4, the upper part of which consists of an inlet for receiving the mixture from the fuel injection manifold, a tap for measuring the tank pressure, and an outlet for intaking the separated gas phase using a vortex ejector 3. The washing liquid separated from the gas phase is taken through two lower openings by the pump 5.

The pump 13 is used to pump the washing liquid into the fuel injection manifold primary circuit, sucked at the outlet through the secondary circuit into the receiving tank 4 due to organizing the tank evacuation by a vortex ejector 3. The gas and liquid components are separated inside the receiving tank, and the gas component is sucked from the tank by a vortex ejector 3 mounted on its lid, thus creating a vacuum inside the tank. The liquid phase, which is a contaminated washing liquid, is pumped by the pump 5 through the lower tank openings and supplied to the filter for cleaning. In addition, the receiving tank is also used as a surge tank, allowing to ensure a uniform vacuum inside the fuel injector secondary circuit.

The compressed air from the compressor main line is supplied through the vortex ejector nozzle, performing the function of an active ejecting flow. The compressed air temperature, flow rate and pressure are monitored at the inlet of the ejector discharge nozzle.

The required flow rate ratio is regulated by the flow characteristics of the intake circulating pump 5, the control valve 8 installed on the return pipe and the shutoff valve 6. Bypass pipeline flow control valve 7 is installed for smooth control of the flow characteristics of the intake circulating pump 5. Part of the liquid pumped by the pump 5 is supplied to a mechanical cleaning filter 10, which provides the washing decontamination, removing particles greater than 5 μm. The amount of liquid supplied to the filter is regulated by valve 9 and is measured by a counter 12, which allows to determine the amount of washing liquid to be returned to the system. The return washing liquid is supplied to the supply tank 11, which ensures the feed circulating pump 13 stable operation and allows to compensate for the washing liquid loss the during the system operation. The amount of washing liquid lost during the washing period is determined by the washing liquid level changing in the supply tank before and after washing.

3. Experimental Results

Preliminary tests of the vortex ejector not connected to the fuel manifold of the gas turbine engine have shown the possibility of ensuring the secondary circuit evacuation along with the organization of the two-phase flow return pumping.

The results of studies on ejecting of liquid supplied through the fuel injection manifold primary circuit showed that pumping out by a vortex ejector is a nonstationary pulsating process. This is expressed by pumping the liquid by an ejector in small portions and is illustrated in Fig. 5 as the time dependent evacuation pressure function.

The nonstationarity is characterized by a clear-cut periodicity with an average frequency of 1.15 Hz, which causes only 52 % of the total flow rate of fluid supplied to the fuel injection manifold primary circuit return to the test stand hydraulic network. It is necessary to note that the time dependence in Fig. 5 illustrates the periodic evacuation process generally, more detailed study of the pressure pulsations requires to increase a sample rate of the measurements but this was not the purpose of the current experiment.

The physics of the process under consideration is as follows. When the vortex ejector in the operating regime, during the fuel injection manifold evacuating to $0.82 \cdot 10^5$ Pa and further liquid supply at $p_l = 0.5 \cdot 10^5$ Pa pressure the latter is ejected and pumped through the main line connecting the manifold outlet to the ejector. After some time, during which the first portion of the liquid moves from the fuel injection manifold to the ejector inlet, the passive nozzle cross section is completely filled with water, which leads to the pumped up flow "blocking" and the vacuum drop in the main line. After the first portion of the liquid slipping into the ejector, the main line gets reevacuated and the subsequent portions of water are sucked, further repeating the process.

The main line vacuum pressure periodic change determines the pulsating operating regime of the fuel manifold injectors, characterized by successive liquid spraying and sucking stages, which ensures only 52 % of the water mass fraction returned to the system.

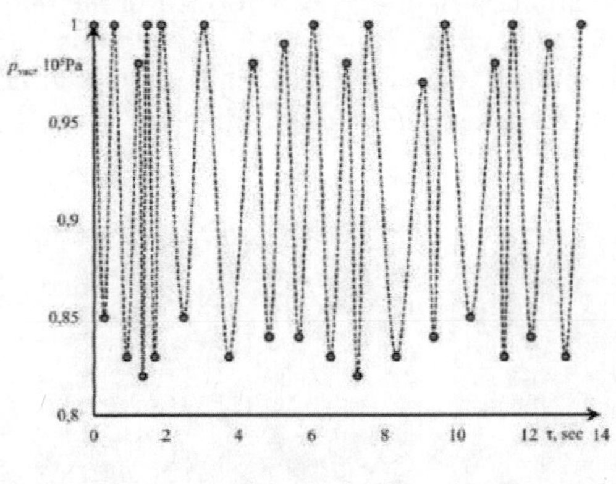

Figure 5. Time dependence of the absolute vacuum pressure at absolute air pressure at the ejector inlet $p_a = 2.5 \cdot 10^5$ Pa, and liquid supply pressure $p_l = 0.5 \cdot 10^5$ Pa

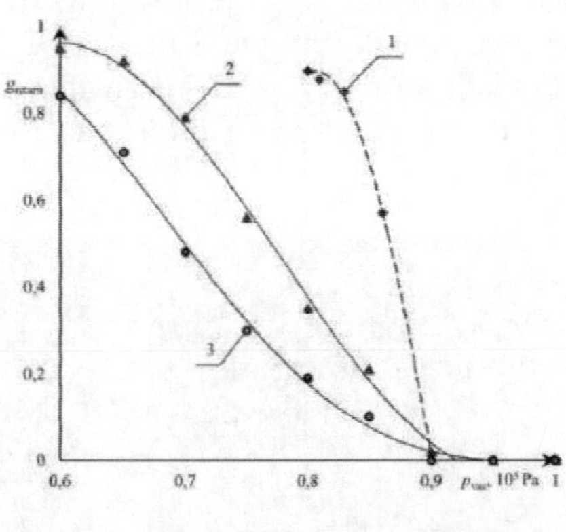

Figure 6. Absolute vacuum pressure dependence of the liquid mass fraction returned to the system: *1* – $p_l = 0{,}55 \cdot 10^5$ Pa; 2 – $p_l = 1{,}0 \cdot 10^5$ Pa; 3 – $p_l = 1{,}9 \cdot 10^5$ Pa

To eliminate the noted negative moments, the scheme of experimental studies organization was changed by installing the receiving tank 4 between the fuel manifold 2 and the vortex ejector 3. This led to the experimental evacuation of not only the fuel manifold secondary circuit, but of the receiving tank 4 as well, which allowed to additionally organize the two-phase mixture separation by mass forces within the receiving tank volume and practically completely eliminate the washing solution slipping into the vortex ejector flow section. In case of applying the fuel manifold evacuation and the two-phase flow separation schemes above at the injectors industrial cleaning stage, the washing liquid slipping into the evacuating device flow section and potential loss shall be regulated by using the liquid pump to control the level of liquid inside the receiving tank and organizing the counterflow motion scheme for the air inside the receiving tank. When organizing the operating conditions corresponding to the temperature of the washing liquid close to the evaporation temperature, with the possible occurrence of cavitation along the test stand return line and fine dispersion within the receiving container, a simple condensing-type heat exchanger installation is reasonable.

The stated system optimization also allowed to completely eliminate the pulsating regimes of vortex ejector operation. The experimental results of the fuel injection manifold evacuation at various liquid supple pressure values are shown in Fig. 6.

It can be seen that the manifold evacuation to a pressure of $0.8\cdot10^5$ Pa at a water supply pressure of $p_l = 0.55\cdot10^5$ Pa allows to pumping out more than 96 % of the liquid from the total mass supplied to the manifold. At evacuation pressure $p_{vac} = 0.6\cdot10^5$ Pa for both fluid flow regimes above, the fraction of the liquid returned to the system exceeds 0.84. On the basis of the above, it can be concluded that, within the framework of the experimental scheme proposed for organizing the evacuation and separation of a two-phase flow, the only practical criterion for selecting an evacuating device (ejector, vacuum pump, etc.) is the possibility of providing the required evacuation level values.

Fig. 7 shows the photos of the fluid pumping by a vortex ejector in various operating regimes. It can be seen that with the absolute vacuum pressure $p_{vac} = 0.95\cdot10^5$ Pa (Fig. 7a) some spray jets begin to shift to the periphery, contributing to an increase in the opening angle, but no liquid ejected into main line of the fuel injection manifold secondary circuit.

An increase in the evacuation pressure to $p_{vac} = 0.9\cdot10^5$ Pa (Fig. 7b) facilitates the initiation of liquid pumping by the ejector, and the mass fraction of the water returned is $g_{return} = 0.04$. The subsequent very insignificant increase in the evacuation pressure leads to a significant increase of g_{return}.

The observed behavior of the function $g_{return} = f(p_{vac})$ is due not only to an increase in the evacuation pressure. Fig. 7 clearly shows that a quasi-stationary liquid layer is formed in the injector nozzle area, bounded by the walls on the sides, and by the counter-ejected air from the bottom. It is obvious that a sufficiently large influence on the formation and retention of such a layer is exerted by the air-side surface tension forces, as well as by the adhesion forces of the injector nozzle side surfaces.

a) **b)** **c)**

Figure 7. Visualizing the injector nozzle evacuation at $p_l = 0.55\cdot10^5$ Pa: a) $p_{vac} = 0.95\cdot10^5$ Pa; b) $p_{vac} = 0.9\cdot10^5$ Pa; c) $p_{vac} = 0.8\cdot10^5$ Pa

The conducted fuel injectors washing tests, both individually and as part of the manifold, demonstrated the high quality of spray nozzles exterior cleaning from varnish and carbon deposits (Figure 8). It is shown in Fig. 8 for two different injectors.

The development of various washing regimes in terms of liquid flow and pressure, vacuum pressure with kerosene spillage tests allowed to conclude that the nozzles operation requirements are complied with a wash time of at least 90 minutes. This is associated with the need for long-term softening and removal of surface deposits from the channels of the primary and secondary circuits, filters and spray nozzles.

a) **b)** **c)** **d)**

Figure 8. Photos of injector nozzles: a) injector 1 before cleaning; b) injector 1 after cleaning without circulating washing liquid; c) injector 2 before cleaning; d) injector 2 after cleaning with circulating washing liquid

4. Conclusion

The conducted fuel manifold injectors evacuation and washing tests determined the possibility of up to 99 % of the liquid mass fraction returned to the system and demonstrated the high quality of spray nozzles exterior cleaning from varnish and carbon deposits.

Acknowledgments

This work was financially supported by RFBR (Russian Foundation for Basic Research) grant No. 16-08-00974 "Development of a model description of heat and mass exchange processes within swirling limited flows and its application in the development of advanced technology devices."

References

[1] Lanin V 1987 Method of hydraulic circulation injectors cleaning *USSR Patent 1690439*
[2] Biruk V, Veretennikov S, Guryanov A and Piralishvili Sh 2014 Vortex effect. Technical applications
[3] Suslov A, Ivanov S, Murashkin A and Chizhikov Yu 1985 Vortex devices
[4] Piralishvili Sh and Ivanov R 2012 *Russian Aeronautics* **55** 179-183

IOP Conf. Series: Journal of Physics: Conf. Series **925** (2017) 012028 doi:10.1088/1742-6596/925/1/012028

The thermocapillary deformations in the locally heated shear-driven liquid film flowing in minichannel

V V Cheverda [1,2]

[1] Kutateladze Institute of Thermiphysics SB RAS, Novosibirsk 630090, Russia
[2] Novosibirsk National Research State University, Novosibirsk 630090, Russia

E-mail: slava.cheverda@gmail.com

Abstract. Using phase schlieren system we have measured the shape of thermocapillary bump in the gas-driven (nitrogen) locally heated water film flowing in the minichannel. The measurements were conducted for liquid and gas flow rates corresponding to Re_l=5.6 and Re_g=73.5, respectively, and heat flux of 2 W/cm^2. The paper shows a qualitative coincidence of the experimental results obtained for the locally heated flowing liquid film and the numerical results on the deformation in a locally heated shear-driven water film flowing in the minichannel.

1. Introduction.

Film flow is widely used in the food industry and when cooling of the heating elements of a nuclear reactor, as well as in other devices with a high heat generation. Cooling of microelectronic equipment in space flight conditions is especially acute problem. Two-phase flows are a promising cooling solution. The review on the regimes of two-phase flows in the channels of different geometries under different conditions is presented in [1]. The experimental investigations of two-phase flows in rectangular channels are given in [2, 3]. A stratified flow with shear-driven liquid film is a sub regime of two-phase flow. In [4, 5] it is revealed that the critical heat flux in a shear-driven liquid film is 10 times greater than that in the gravitationally flowing liquid film at the same liquid flow rate.

At a local heating of the flowing liquid film, a thermocapillary bump arises due to thermocapillary effects [6]. In [7], thermocapillary bump was measured in flowing film of 25% solution of ethyl alcohol in water, using schlieren method. Similar experiments were carried out on measuring the thermocapillary bump in a flowing film of a 10% solution of ethyl alcohol in water using an optical fiber sensor [8].

Numerical simulation of thermocapillary bump formation in the falling liquid film of 25% ethyl alcohol in water was carried out in [9], as well as compared with experimental data. The authors' studies in [10, 11] were focused on thermocapillary structures formation in underheated thin fluid films flowing down through a locally heated plate. Numerical simulation of thermocapillary bump formation in the gas driven liquid film flowing in the minichannel was carried out in [12, 13]. The analysis of deformation calculated numerically under local heating of a liquid film of 25% ethanol solution in water, flowing down (the liquid Reynolds was equal to 1) a vertical plate with local heating (heat flux of 0.57 W/cm^2) has shown that relative deformation (absolute value of deformation divided by the initial thickness of the liquid film) was about 1.02 [14]. The analysis of numerically calculated deformation of locally heated FC-72 film flowing under a constant shear stress that simulates the gas driven flow of a liquid film, revealed that shear stress influences the magnitude of deformation [15]. For example, at FC-72 liquid flow rate corresponding to Re=1, the tangential tension of 0.012 kg/s^2m, heat transfer coefficient of 20 W/m K, and heat flux of 0.1 W/cm^2, the relative deformation (absolute deformation divided by the initial thickness of the liquid film) was about 1.2, which is comparable with the deformation in a locally heated falling down liquid film [14]. The difference can be explained by the fact that liquids vary in physical properties, as well as due to difference in heat fluxes. The authors have also shown that with increasing shear stress, the magnitude of the deformation increases. This is due to the fact that with increasing shear stress, the film thickness is reduced, and consequently the shear stress on the film surface caused by thermocapillary forces is growing, and thus, the magnitude of the deformation increases.

The calculation of heat transfer and thermocapillary deformations in a nitrogen gas driven water film in the microchannel (height of 250 micron) was carried out in [16, 17]. The authors of [16] investigated the effect of acceleration on heat transfer and deformation of the water film, and showed that with increasing acceleration from 0 to 20g deformation decreases. At a fixed heat flux on the heating element, with increasing gas flow rate, the flow rate of the evaporated liquid is reduced, and at a fixed temperature of the heating element, increase of gas flow rate leads to the growth of the amount of evaporated liquid. In [17] it is shown that the thermocapillary bump is formed near the upper edge of the heater, while the minimum thickness of the liquid film is achieved near the lower edge of the heater (along the flow). Besides, the deformation in the form of jets occurs on both sides of the channel due to evaporation and thermocapillary forces. Described deformations occur because of evaporation and thermocapillary forces.

In the course of preparation for experiments aboard the International Space Station (ISS) it is scheduled to carry out an analysis and select the optical system to measure thermocapillary deformations at local heating of a gas driven liquid film flowing in the minichannel under zero gravity conditions. The obtained results will be used for verification of existing numerical models.

2. Experimental setup.

Installation for carrying out laboratory experiments is open loop (Fig. 1). In the experiments, the liquid and gas are transferred from the tanks. Set gas flow rate and pressure in the test section are maintained automatically through Bronkhorst regulators. The degassed liquid is fed into the test section by means of the syringe pump (Cole-Parmer EW-74905-54) through the thermal stabilization system. Thermal stabilizing systems based on Peltier elements are used to maintain the desired temperature of the liquid and gas at the inlet of the test section. These systems are cooled by means of a water system including pump, radiator, and safety valves.

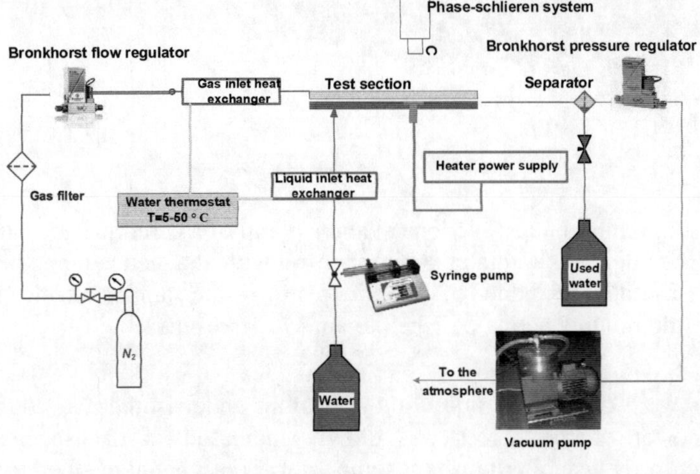

Figure 1. Schematic diagram of the experimental setup

IOP Conf. Series: Journal of Physics: Conf. Series **925** (2017) 012028 doi:10.1088/1742-6596/925/1/012028

In the course of experiment, a two-phase mixture is evacuated into the atmosphere through the separation system employing a membrane pump (KNF N0150 AN.12E VP). The separation system is used to separate used water from the gas phase. All experimental parameters are monitored and recorded using software developed in LabView.

The test section (Fig. 2) consists of a textolite plate with mounted stainless steel plate serving a substrate. A frame, which is covered on top by optical glass, thus forming a minichannel with a height of 5 mm and a width of 30 mm, is attached to the textolite plate. Liquid film flow is created by means of the liquid nozzle with a height of 150 μm and a width of 30 mm. Gas flow enters the minichannel and flows over the liquid knife (42 mm wide), after which affects the fluid. The stainless steel substrate with copper bar (10 x 10 mm^2) pressed into the substrate at a distance of 29 mm from the liquid nozzle is polished to a mirror-quality surface. The temperature of the substrate is measured by 10 K-type thermocouples. Two-phase mixture is evacuated from the test section through the orifice.

3. Experimental results.

The liquid film profile was measured using a phase-schlieren system (NIMO). The operating principle of this system is based on a combination of the schlieren method and the principles of interferometry. Using a special transparent liquid-crystal filter with variable transmittance, a schlieren image with a specific phase shift first is created horizontally (Fig. 3a) and then vertically (Fig. 3b). A series of obtained images are processed using the principles of interferometry, and then a three-dimensional profile of the resulting deformation is constructed. The measurement area is a circle with a diameter of 30 mm. The system's sensitivity is 50 nm, while the error is 1 μm. A detailed description and operating principle is given in [18].

Based on obtained experimental results we have measured deformation profiles of the locally heated shear-driven water film. The heat flux was supplied to the copper rod by means of a ceramic electric heater. Part of this heat is removed by the film, while the other part is distributed over a substrate of stainless steel, and the remaining small portion is lost in heating the air surrounding the copper rod with attached ceramic heater.

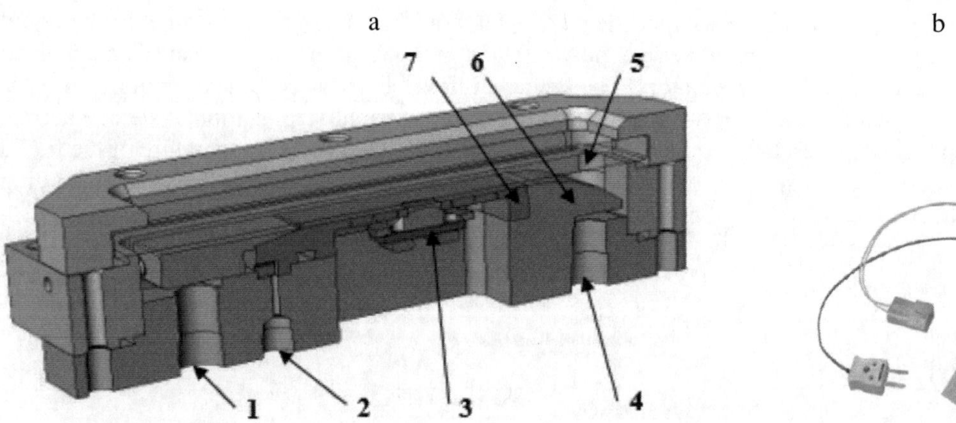

Figure 2. Schematic diagram of the test section (a) and a photo of a ceramic heater with a built-in thermocouple. (b): 1 - gas inlet, 2 - fluid inlet, 3 - copper rod with attached ceramic heater, 4 – two-phase mixture outlet, 5 - optical window, 6 - textolite plate, 7 - the stainless steel substrate with aluminum coating polished to a mirror-like surface.

Profile measurements were carried out first without heating under isothermal conditions of water film flow. Then the power of the heater was slowly increased to the specified value. The measurements were taken after reaching constant temperature not changing over five minutes. We have obtained three-dimensional measurements of deformation profile (Fig. 3). Figure 3c presents the deformation of the thermocapillary bump across the flow (along line A passing through the center of the heater). Figure 3d presents the deformation profile of the film along the flow (along the line B passing through the middle of the heater).

Figure 3c shows considerable thinning of the film near the side walls of the heater. This thinning is caused by two factors: 1) the shear stress from the gas is not high enough and as a result the film is slightly convex in the center, and 2) there is an additional convexity due to thermocapillary bump in the center of the liquid film. At that, thermocapillary deformations are formed near the edges of the heater [16, 17]. However, they are not visible in Fig. 3, because they are located outside the field of view of the optical system.

It is seen that deformation in this region reaches 200 μm and even more. According to the research presented in [19], the film located at the edges of the heating element is the most likely to rupture and form dry spots, that is confirmed by conducted measurements of surface deformation.

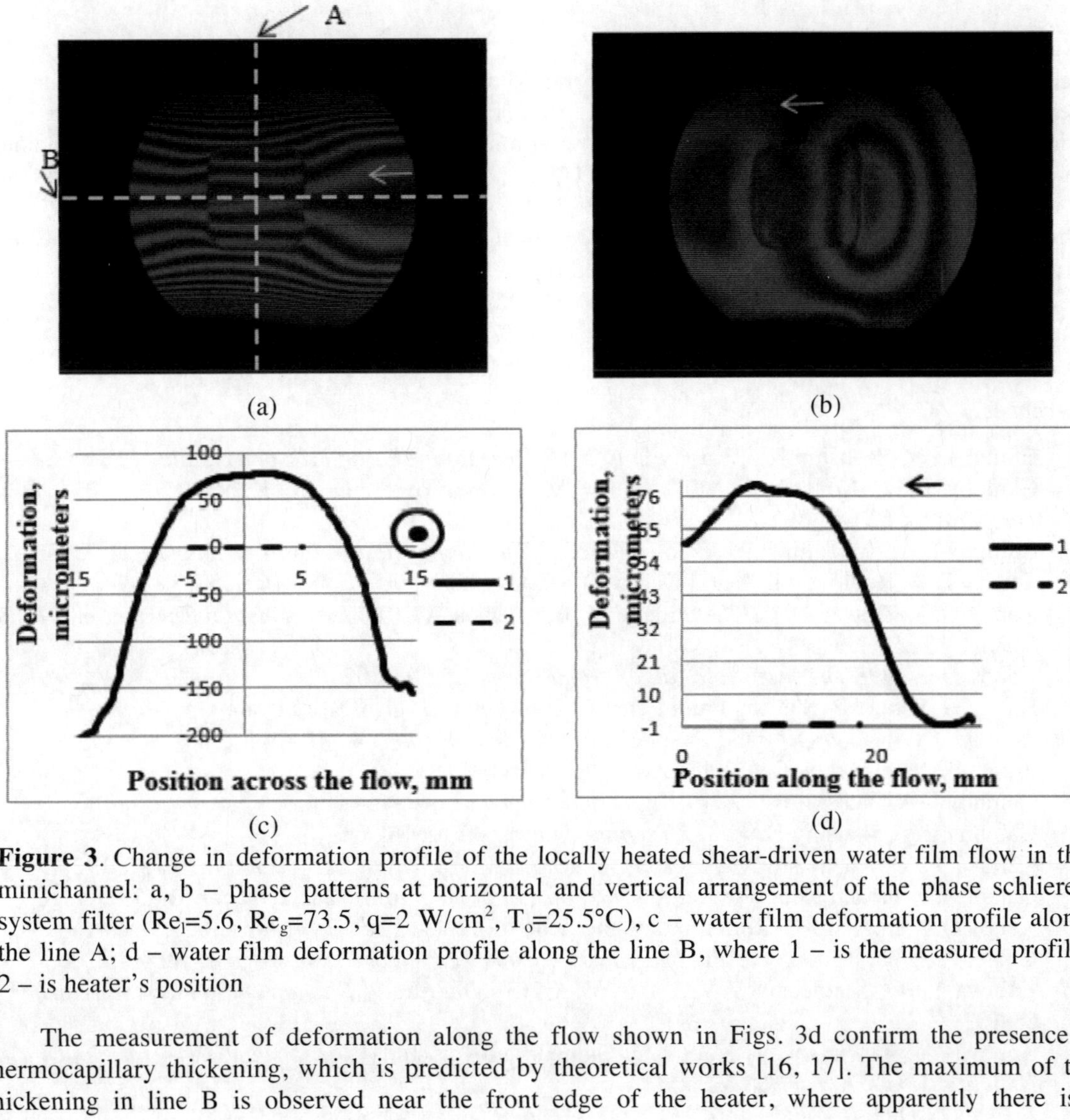

Figure 3. Change in deformation profile of the locally heated shear-driven water film flow in the minichannel: a, b – phase patterns at horizontal and vertical arrangement of the phase schlieren system filter (Re_l=5.6, Re_g=73.5, q=2 W/cm^2, T_o=25.5°C), c – water film deformation profile along the line A; d – water film deformation profile along the line B, where 1 – is the measured profile, 2 – is heater's position

The measurement of deformation along the flow shown in Figs. 3d confirm the presence of thermocapillary thickening, which is predicted by theoretical works [16, 17]. The maximum of this thickening in line B is observed near the front edge of the heater, where apparently there is a maximum temperature gradient. For the above parameters, the maximum deformation along the line B was 77 μm.

It should be noted that the lateral deformation profile qualitatively agrees well with the calculation presented in [17]. At the same time, the longitudinal deformation profile along the line B is somewhat different from that obtained in [17]. In the calculations [17], the maximum in the region of the heater's front edge is much more pronounced. Perhaps this is due to different boundary conditions. In [17] calculations were made for the conditions q=const, while in this paper, the condition T = const is satisfied at the heater.

The conducted measurements confirm an important conclusion that has been derived theoretically in [17]. These calculations have shown for the first time that for locally heated shear-driven liquid film, transverse deformations significantly exceed the longitudinal ones. In [17], this difference was almost two times. In the present experimental work, the difference is 2.5 times. It is known that in locally heated flowing liquid film, the longitudinal and transverse deformations are of the same order [17].

4. Conclusions

Finely, we can conclude that the measurements of deformations in a locally heated nitrogen gas driven water film flow in minichannel were carried out for the first time and the data obtained well agree with the numerical simulation of such liquid film flow in minichannel, conducted in [9, 12, 14]. Obtained data partially coincide with the results of works [15-17]. The difference in the results is due to the fact that numerical simulations were carried out for film flow in microchannel rather than minichannel.

In conclusion, we recommend using this optical technique when designing experiments aboard the ISS to study the heat transfer in the locally heated liquid film driven by gas in a microchannel.

Acknowledgments
The work was funded by RFBR and Novosibirsk region (research project № 17-48-540124).

References:

[1] Chinnov E A, Ronshin F V, Kabov O A 2015 *Thermophysics and Aeromechanics* **22** 265-284
[2] Chinnov E A, Ronshin F V, Guzanov V V, Markovich D M, Kabov O A 2014 *High Temperature* **52** 681-687
[3] Chinnov E A, Ron'shin F V, Kabov O A 2014 *Thermophysics and Aeromechanics* **21** 759-762
[4] Zaitsev D V, Rodionov D A and Kabov O A 2009 *Technical Physics Letters* **35** 88-94
[5] Kabov O A, Zaitsev D V, Cheverda V V, Bar-Cohen A 2011 *Experimental Thermal and Fluid Science* **35** 825
[6] Kabov O A 1998 *Thermophysics and Aeromechanics* **5** 597-602
[7] Kabov O, Scheid B, Sharina I and Legros J 2002 *Int. J. Therm. Sci.* **41** 664
[8] Zaitsev D V, Kabov O A, and Evseev A R 2003 *Exp. Fluids* **34** 748–754
[9] Frank A M and Kabov O A 2006 *Phys. Fluids* **18** 032107
[10] Chinnov E A, Shatskii E N 2014 *High temperature* **52** 461-464
[11] Chinnov E A, Shatskiy E N 2014 *Technical physics letter* **40** 7-9
[12] Kabova Yu O, Kuznetsov V V, Kabov O A 2008 *Microgravity Sci. Technol.* **20** 187–192
[13] Gatapova E Ya and Kabov O A 2008 *Int. Journal of Heat and Mass Transfer* **51** 4797–4810
[14] Kabov O A, Legros J-C, Marchuk I V and Scheid B 2001 *Fluid Dynamics* **36** 521–528
[15] Gatapova E Ya, Kabov O A and Marchuk I V 2004 *Technical Physics Letters* **30** 46-52
[16] Kabova Yu O, Kuznetsov V V, Kabov O A 2014 *Interfacial Phenomena and Heat Transfer* **2** 85-102
[17] Kabova Yu, Kuznetsov V, Kabov O, Gambaryan-Roisman T, Stephan P 2014 *Int. Journal of Heat and Mass Transfer* **68** 527–541
[18] Joannes L, Dubois F and Legros J C 2003 *Applied Optics* **42** 5046-5053

IOP Conf. Series: Journal of Physics: Conf. Series **925** (2017) 012029 doi:10.1088/1742-6596/925/1/012029

Study of pool boiling of distilled water on SiO_2 nanoparticle-coated wire

M I Pryazhnikov[1,2], A V Minakov[1,2]

1 Siberian Federal University, 79 Svobodny pr., Krasnoyarsk, 660041, Russia
2 Institute of Thermophysics SB RAS, 1 Acad. Lavrentiev pr., Novosibirsk, 630090, Russia

E-mail: arrivent@yandex.ru

Abstract. Experimental results of pool boiling of distilled water on bare nichrome wire and SiO_2 nanoparticle-coated wire are presented. Nano-coated wires were obtained by first boiling them in SiO_2-water nanofluid. The nanofluid was prepared based on distilled water and SiO_2 nanoparticles. The average size of the nanoparticles was 100 nm. The volume concentration of particles was equal 2%. Boiling curves of distilled water were obtained on bare nichrome wire and SiO_2 nanoparticles coated wire. The coating increased the critical heat flux during boiling of water.

1. Introduction

The use of different methods for modifying the heat exchange surface to increase critical heat flux (CHF) is an actual task of mitigating the limitations of the critical heat flux density. In recent years, profound interest has been shown in the study of heat and mass transfer during boiling on surfaces coated with nanoparticles [1-4]. In particular, this is due to the appearance of new types of coolants, which have a significant effect on the properties of the heating surface during boiling (change of wettability and roughness). Nanofluids can be attributed to such coolants. Nanofluids are two-phase systems consisting of a based fluid and nanoparticles. The size of nanoparticles is less than 100 nm.

A significant (up to two or more times) increase of the critical heat flux density at low (0.001 – 0.1% vol.) concentrations of nanoparticles is one of the remarkable properties of nanofluids [1-4]. The results of experimental studies and observations [1-9] of the heating surface show that the main reason for this phenomenon is the appearance of a layer (coating) of nanoparticles on heater surface, which is formed directly during boiling. The coating of nanoparticles affects the boiling crisis in several ways. First, the coating has a specific surface morphology, which leads to a significant increase in roughness and wettability of the surface. Secondly, an additional suction of liquid to the surface of the heater is organized due to the formation of a porous structure, which is due to capillary forces.

The problem of describing the formation of surface coating during boiling and its influence on the boiling process and the boiling crisis is connected with the study of the physicochemical properties of the surface and the effect of surface phenomena on the layer formation and morphology.

In the present work, experimental results of pool boiling of distilled water on SiO_2 nanoparticle-coated wire are presented.

2. Experimental apparatus and procedure

The scheme of the used experimental setup is presented in figure 1. The investigated fluid in amount of 300 ml was placed in a high sealed glass flask (4) 8 cm in diameter. A nickel-chromium wire heater with the length L of 40 mm and the diameter of 0.1 mm was emerged into the flask filled with fluid. The wire was fixed by copper bus leads with the cross section of 10×2 mm to supply voltage. The heater was energized by means of AKIP-1122 power source. The applied voltage and load amperage in the circuit was recorded using a GDM-78261 voltmeter ("V" and "A" in figure 1, respectively). The flask with the test fluid was sealed, so that the condensate formed in the upper part of the flask dripped back into the flask, maintaining saturation conditions in the working chamber. The fluid temperature in the flask was controlled by means of a chromel-copel thermocouple, which was connected to the TRM-200 temperature meter (5). The tail end of the thermocouple was located at the same level with the nickel-chromium heater at a distance of 2 cm from it. The absolute pressure in the flask was measured using digital precision manometer DM5002M (6). In the experiments the pressure was atmospheric and constant.

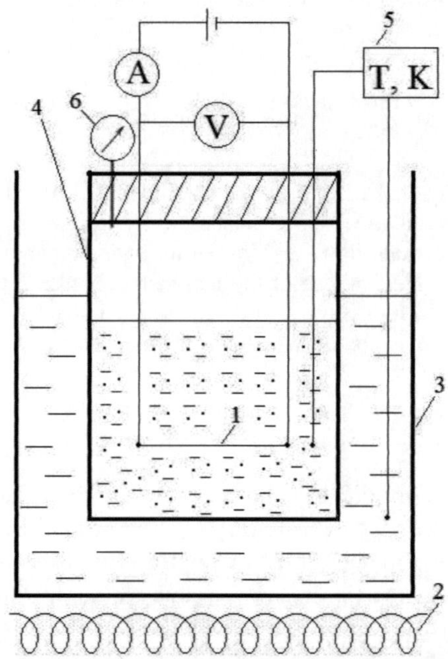

Figure 1. The scheme of the experimental setup.

The flask with the test fluid was placed in a water bath (3) with the constant temperature, which was about 0.5°C below the boiling point, maintained by means of electric heater (2). Thus, in this paper we investigated the boiling close to saturation conditions. The temperature control in a water bath was conducted using another chromel-copel thermocouple connected to the TRM-200 meter. Some times after placing the flask into a water bath, the temperature in the flask and water bath equalized. After the nickel-chromium heater was energized, the heat flux density was measured. The programmable current power supply allowed increasing the heater voltage with a fixed predetermined step. Thus, it was possible to control the heat flux growth rate and, thereby, to fix the onset of the boiling crisis.

The boiling heat flux density on the heater was determined by the ratio: $q = Q/S = IU/S$, where $S = \pi dL$, S – is the lateral surface area of the heater between the current-carrying wires, Q – is the heat flux released by the heater, I – is the electric current in the heater circuit, U – is the voltage drop in the heater. Heat generated at the lead wires was negligible. The total error in determining the heat flux density was about 2%.

Studying the boiling processes requires also the knowledge of the heater surface temperature. Since for the nickel-chromium the dependence of resistance R on temperature within the investigated temperature range is close to linear $R = R_0[1 + A(T - T_0)]$, then by measuring the resistance of the heater and knowing the temperature coefficient of resistance A, we can find the average temperature of heater: $T = T_0 + (R/R_0 - 1)/A$. Here R_0 – is the heater resistance at the temperature T_0. To determine the temperature coefficient of nickel-chromium electrical resistance, we have performed special series of resistance measurements of the used nickel-chromium wire. It was found that A=0.00012 1/K. The total error in determining the temperature of the heater was about 3 %.

3. Experimental results

Nano-coated wires were obtained by first boiling them in SiO_2-water nanofluid. The nanofluid was prepared based on distilled water and nanoparticles of silicon. The volume concentration of particles was equal 2 %. For preparation of nanofluid we used the standard two-step process. After adding the necessary quantity of nanopowder to water, obtained nanofluid was first thoroughly mixed mechanically and then it was placed in an ultrasonic disperser for half an hour to destruct the particles conglomerates. No surfactants were added to the nanofluid. The average size of the nanoparticles was 100 nm. The first three kinds of nanopowder were acquired from the "Plazmoterm" company. The SiO_2-water nanofluid was preheated to the boiling point. After the nickel-chromium heater was energized. The heat flux density was 1 MW/m^2. The SiO_2-water nanofluid was boiling on wire for 25 minutes under these conditions. The deposits of SiO_2 nanoparticles were formed on the wire surface.

The series of water boiling experiments was conducted on SiO_2 nanoparticles coated wire. Boiling curves of distilled water were obtained on bare wire and SiO_2 nanoparticles coated wire (see figure 2). The dependence of the heat transfer coefficient on heat flux density was also obtained (see figure 3). According to our information, distilled water has critical heat load q_{CHF} in the range 1.0–1.3 MW/m^2, which agree with many works in which the boiling of water on thin wires of similar size was studied, for example, in work [10] q_{CHF}=1.15 MW/m^2, in work [11] q_{CHF}=1.15 MW/m^2, in work [12] q_{CHF}=1.2 MW/m^2. Figures 2-3 show that the coating of SiO_2 nanoparticles increased the critical heat flux during boiling of water. The relative critical heat flux density is 2.7 (the ratio of critical heat flux density on coated wire to the critical heat flux on bare wire).

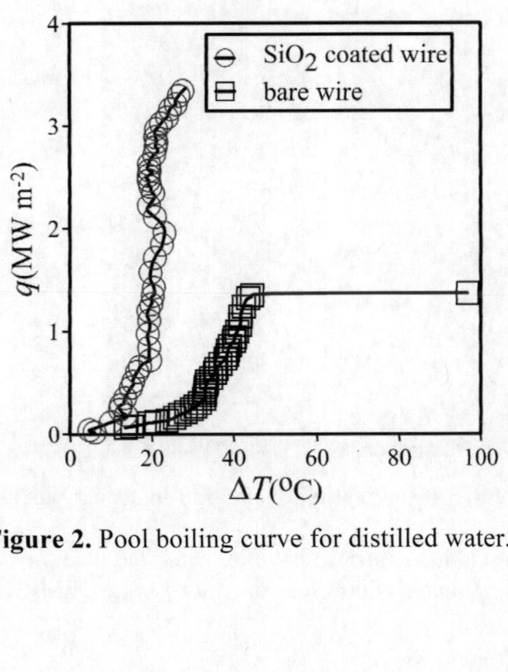

Figure 2. Pool boiling curve for distilled water.

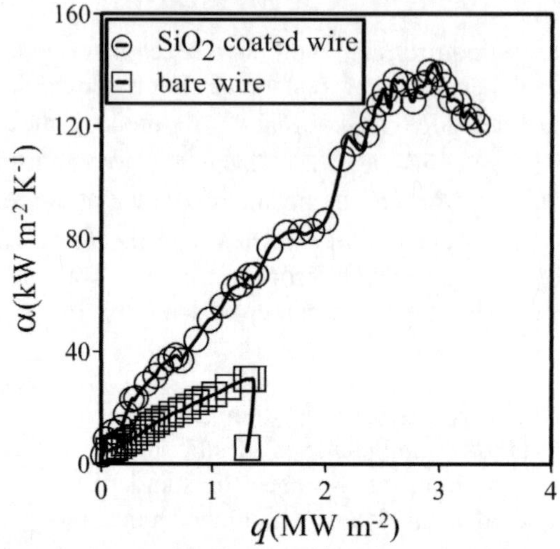

Figure 3. Dependence of heat transfer coefficient on heat flux density.

Figure 2 shows that the boiling curve of the coated wire is shifted to the left (in the region of smaller surface superheating). This suggests that the heat transfer coefficient at boiling of distilled water is higher than for bare wire (see figure 3). This agrees with the results of other experimental studies, for example [4].

The video of the boiling process was recorded using the Sony Cyber-shot DSC-RX100. The Sony RX100 IV allows to provide for high frame-rate video at up to a staggering 1000 frames per second. Figure 4 shows photographs of pool boiling at different heat flux densities on bare wire and on SiO_2 nanoparticles coated wire. It can be seen that the flow visualization of the boiling process is different for the same heat flux density.

Heat flux	≈ 220 kW/m^2	≈ 1 200 kW/m^2	≈ 1 500 kW/m^2
Bare wire	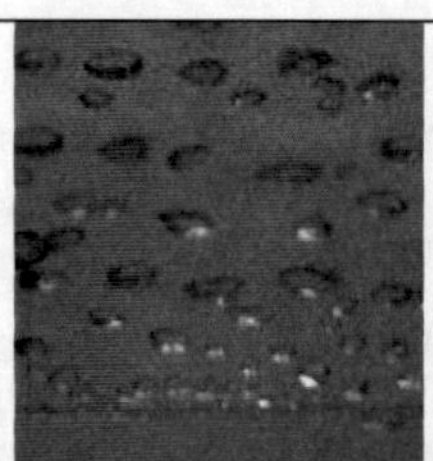	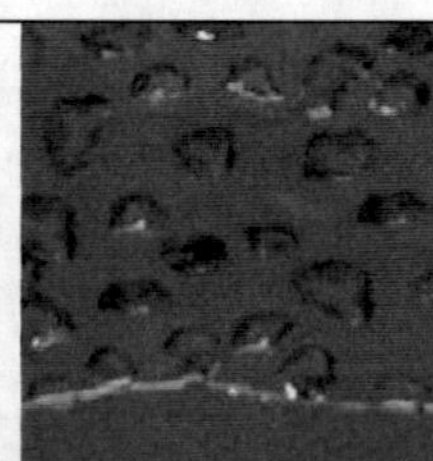	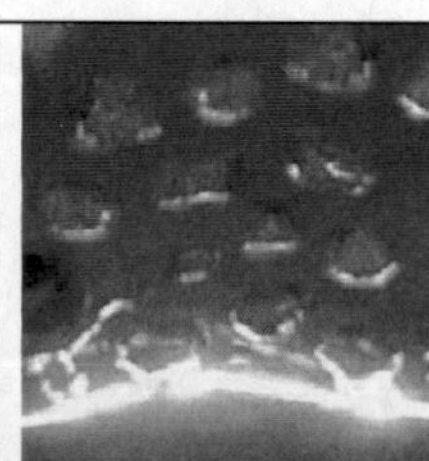
SiO_2 coated wire	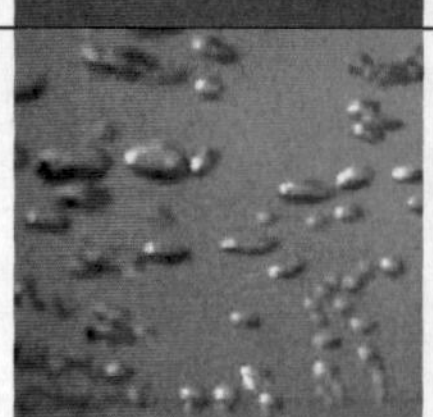	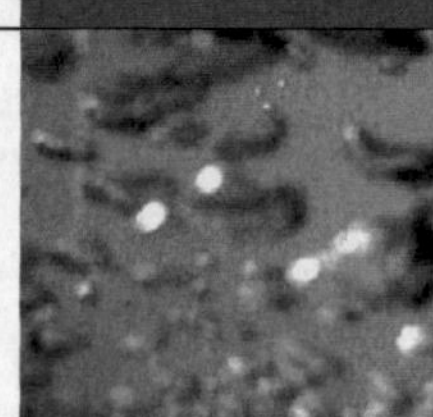	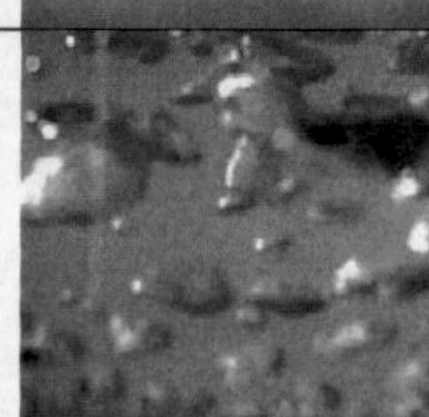

Figure 4. Flow visualization of the boiling at various heat fluxes.

Besides measuring the critical heat flux we have investigated also the heaters surface structure after boiling. This was done using transmission electron microscope JEM-2100 (JEOL, Japan) equipped with an energy dispersive spectrometer Oxford Inca x-sight. Figures 5 and 6 show SEM images of bare wire and SiO_2 nanoparticles coated wire after boiling of distilled water.

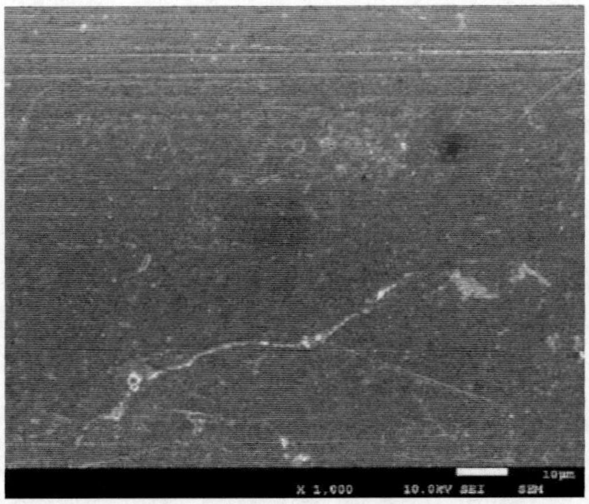

Figure 5. SEM images of nickel-chromium wire after boiling of distilled water.

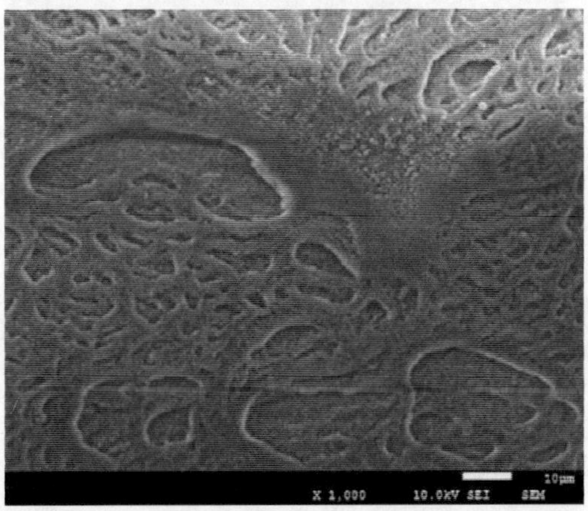

Figure 6. SEM images of SiO_2 nanoparticles coated wire after boiling of distilled water.

4. Conclusion

In this work experimental study of pool boiling of pure distilled water (without nanoparticles) on nanoparticle-coated wire are conducted. Nano-coated wires were obtained by first boiling them in SiO_2-water nanofluid. Boiling curves of distilled water were obtained on bare wire and SiO_2 nanoparticles coated wire (see figure 2). The dependence of the heat transfer coefficient on heat flux density was also obtained (see figure 3). Figures 2-3 show that the coating of SiO_2 nanoparticles increased the critical heat flux during boiling of water. The relative critical heat flux density is 2.7 (the ratio of critical heat flux density on coated wire to the critical heat flux on bare wire).

It can be seen that the heater surface is covered by fairly dense deposits (see figure 6). A microscopically rough surface is formed by depositing particles on the surface of the heater. Particle deposition on the heater surface plays the key role in the enhancement of CHF. Thus, the critical heat flux is directly related to deposits of nanoparticles on the heater surface. The electron microscopy has shown that the thickness of deposits reaches about 5-10 µm. Additional work is needed to study characterization of layer growth to understand the whole mechanism.

IOP Conf. Series: Journal of Physics: Conf. Series **925** (2017) 012029 doi:10.1088/1742-6596/925/1/012029

Acknowledgments
The reported study was funded by Russian Foundation for Basic Research, Government of Krasnoyarsk Territory, Krasnoyarsk Region Science and Technology Support Fund to the research project №16-48-243042.

References
[1] Kim H, Kim J and Kim M 2006 *Nucl. Eng. Technol.* **38** (1) 61-68.
[2] You S M, Kim J H and Kim K H 2003 *Appl. Phys. Lett.* **83**:3374.
[3] Vassallo P, Kumar R and D'Amico S 2004 *Int. J. Heat Mass Transfer* **47** 407-411.
[4] Chopkar M, Das A K, Manna I and Das P K 2008 *Heat Mass Transfer* **44** 999-1004.
[5] Das S K, Putra N and Roetzel W 2003 *Int. J. Heat Mass Transfer* **46** 851–862.
[6] Golubovic M N, Madhawa Hettiarachchi H D, Worek W M and Minkowycz W J 2009 Appl. Therm. Eng. **29**:1281-1288.
[7] Kim H 2011 *Nanoscale Res. Lett.* **6**:415.
[8] Hedge R N, Rao S S and Reddy R 2012 *Therm. Sci.* 16 (2) 445-453.
[9] Okawa T, Takamura M and Kamiya T 2012 *Int. J. Heat Mass Transfer* **55** 2719–2725.
[10] Anokhina E V 2010 *Tech. Phys. Lett.* **55** (5) 645-653.
[11] Ahn H S, Kim H M 2011 *ASME. J. Heat Transfer.* **134** (2) 024001.
[12] Matre G A, Karwande R L 2015 *Int. J. Res. Eng. Technol.* **4** (5) 177-185.

IOP Conf. Series: Journal of Physics: Conf. Series **925** (2017) 012030 doi:10.1088/1742-6596/925/1/012030

The effect of dry spots on heat transfer in a locally heated liquid film moving under the action of gas flow in a channel

D V Zaitsev[1,2], E M Tkachenko[1,2] and E F Bykovskaya[1]

[1]Kutateladze Institute of Thermophysics, Siberian Branch of Russian Academy of Sciences, 630090, Novosibirsk, Russia
[2]Novosibirsk State University, 630090, Novosibirsk, Russia

E-mail: egor_tkachenko@mail.ru

Abstract. Intensive evaporation of a thin liquid film, moving in a flat micro-/minichannel under the action of gas flow is very promising for the use in cooling systems of modern semiconductor devices with localized heat sources of high intensity. In this work, using the high-speed visualization, the effect of the formation of dry spots on heat transfer in a locally heated liquid film shear-driven in a channel was investigated. It was found that the maximum intensity of heat removal from the heater is achieved in the mode, when the film flow continuity is broken. During the experiment the total area of dry spots increases with increasing heat flux and heater temperature, but when the heater reaches a certain temperature (≈100°C), the total area begins to decrease. However, the length of contact line increases with increasing heat flux and reaches a maximum in the pre-crisis regime. Intensive evaporation in the region of the contact line may explain the achievement of high heat fluxes in the shear-driven liquid film.

1. Introduction

One of the most important problems of thermal physics today is cooling of microelectronic equipment. Currently, the production of processors changes over from the 14 nm manufacturing process technology to the 7 nm technology. New IBM processors use 7 nm transistors. This has become possible through the use of a silicon-germanium alloy in the transistor design instead of pure silicon, the traditional material for the manufacture of microchips. A normal-size processor can accommodate more than 20 billion transistors having a size of 7 nm. In addition to reducing the size of the microprocessor unit, the use of the silicon-germanium alloy has improved the switching speed of transistors and lowered the power consumption. The average heat flux density on the chips of commercially available computers and other electronic devices is up to 200-300 W/cm^2. In local areas from 100 μm^2 to a few square millimeter, the heat flux density reaches values of 1 kW/cm^2 or even higher (IBM). Devices of the next generation may have even higher heat fluxes and pulsed loads [1], which requires a new thermal management level.

There are three "classical" methods of heat removal from localized heat sources of high intensity: 1) flow boiling in microchannels, 2) spray cooling, and 3) micro-jet cooling. The authors of works [2-4] proposed the fourth method of effective cooling, in which heat removal is due to intensive evaporation of a thin liquid film, moving in a flat micro-/minichannel under the action of gas flow.

Recent experimental investigations conducted in [5] proved possible removal of heat fluxes of up to 870 W/cm^2 from the heating area of 1x1 cm^2 using this method. In [6] heat transfer in a liquid film shear-driven in a channel with an extended heater has been studied (under comparatively low heat fluxes). Our present paper deals with systematic experimental studies of the flow and destruction of a water film, shear-driven in the channel, under heating from the local heat source of 1x1 cm^2. The influence of the film rupture on heat transfer and CHF is being investigated.

2. Experimental setup

Figure 1 shows the design of the test section. Its main part is a stainless steel plate with a pressed-in copper rod. The rod surface has a shape of a square head 1x1 cm^2, imitating the surface of a computer chip. The rod is heated using a nichrome coil, wound around its lower part. This design of the heater provides constant temperature on the surface of the rod, T_w= const (as confirmed by thermocouple measurements). The working area is covered with a transparent cover of optical glass (not shown in Fig. 1), thus forming a flat channel. The channel height is 1.5 mm, and the width is 40 mm.

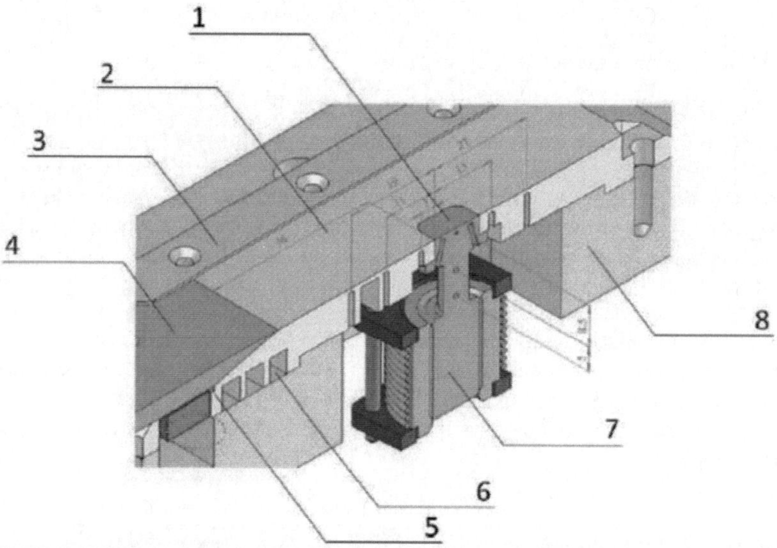

Figure 1. The scheme of the test section: 1 — source of local heating; 2 — stainless steel plate; 3 — channel height regulator; 4 — knife; 5 — liquid nozzle; 6 — thermal stabilizer; 7 — copper rod; 8 — textolite plate. Locations and distances for thermocouples are shown

Gas is introduced into the working area from a compressor. The fluid is supplied from the thermostat, enters the channel through the liquid nozzle and moves over the stainless steel plate as a film under the influence of gas friction. After passing the working area the gas is vented to the atmosphere, and the fluid flows back to the thermostat. Distilled water with initial temperature of 24°C is used as the working fluid. Air with temperature of 24-27°C and relative humidity of 15-30% is used as the working gas.

3. Experimental results

To carry out high-speed visualization of small-scale dry spots that occur on the heater under intense heating, the FASTCAM SA1.1 CCD camera is used (5400 frames per second at a resolution of 1024x1024 pixels and up to 675000 frames per second at lower resolutions). The CCD camera is equipped with an optical system of high spatial resolution (up to 500 nm per 1 pixel of the camera sensor).

Figure 2. The flow of a water film moving under the action of gas in a 1.5 mm high channel with a heater of 1x1 cm^2. Reynolds number of liquid $Re_l = 14$, superficial gas velocity $U_{Sg} = 34$ m/s, heat flux q=240 W/cm^2, temperature of the heater surface is T_w=126°C. The flow is directed from top to bottom

With the help of high-speed imaging, it was found that the maximum intensity of heat removal from the heater is achieved in the mode, when the film flow continuity is broken. The heater is covered with small (of about 100 microns in size) dry spots with the lifetime of about 1/100 – 1/1000 s. At that, the number of spots that exist simultaneously on one square centimeter of the surface can reach several hundred (Figure 2). Typical speed of the contact line in these dry spots is about 10 cm/s, which is about 100 times higher than the speed during rupture of the spot-heated liquid film without gas flow [7].

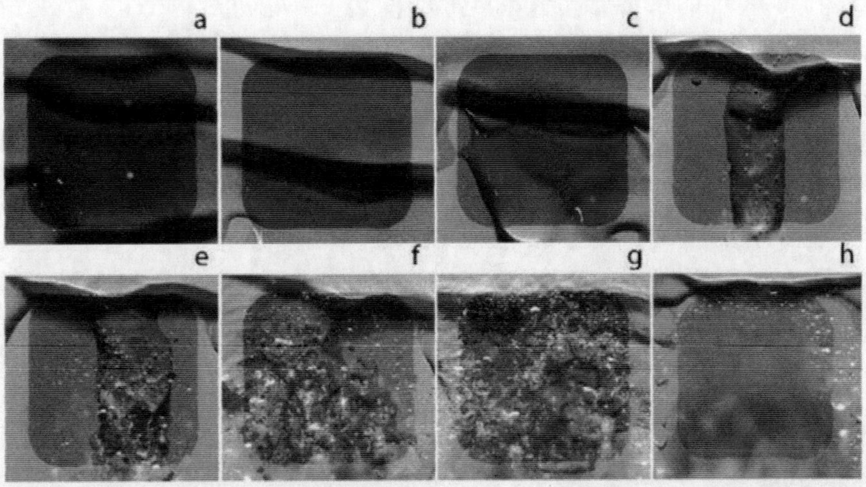

Figure 3. The dynamics of liquid film at $Re_l = 45$ and $U_{Sg} = 7.4$ m/s. a – q = 45.1 W/cm^2; b – q = 61.1 W/cm^2; c – q = 73.6 W/cm^2; d – q = 87.5 W/cm^2; e – q = 109 W/cm^2; f – q = 142.9 W/cm^2; g – q = 185.5 W/cm^2; h – q = 186 W/cm^2, crisis

Also, there is another interesting phenomenon connected with the total area of dry spots on the heater surface. During the experiment, the total area of dry spots increases with increasing heat flux and heater temperature, but when the heater reaches a certain temperature (≈100°C), the total area begins to decrease. This process is shown in Fig. 3.

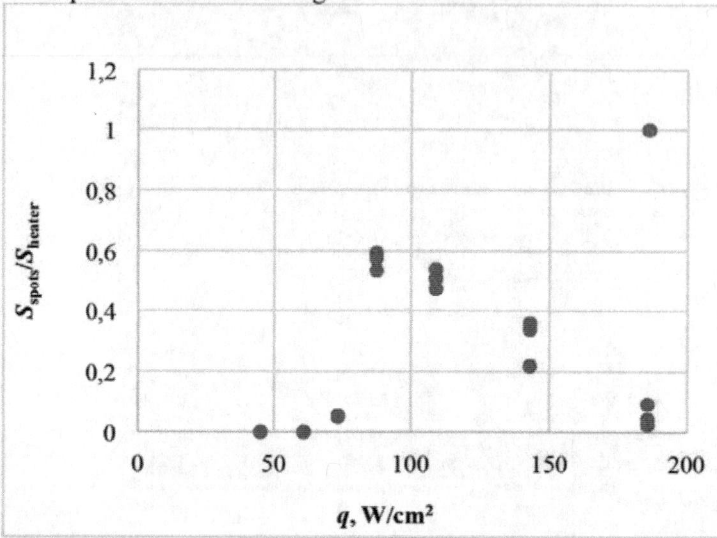

Figure 4. The dependence of the total area of dry spots, divided by the area of heater, on the heat flux at $Re_l = 45$ and $U_{Sg} = 7.4$ m/s

Figure 4 shows the photo processing results (photos were processed using ImageJ software) in the form of the total area of dry spots on the heater, divided by the area of the heater, vs. heat flux for $Re_l = 45$ and $U_{Sg} = 7.4$ m/s.

The dependence of the total length of the contact line, divided by the length of the heater (1 cm), vs. heat flux at $Re_l = 45$ and $U_{Sg} = 7.4$ m/s is shown in Figure 5. In contrast to the area of dry spots, the length of contact line increases with increasing heat flux and reaches a maximum in the pre-crisis regime. Intensive evaporation in the region of the contact line [8] may explain the achievement of high critical heat fluxes using thin shear-driven liquid films [2-5].

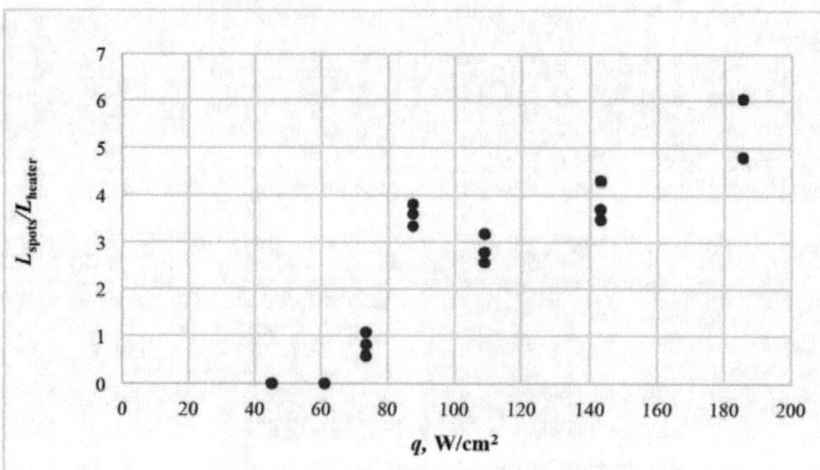

Figure 5. The dependence of the total contact line length, divided by the length of the heater, on the heat flux at $Re_l = 45$ and $U_{Sg} = 7.4$ m/s

IOP Conf. Series: Journal of Physics: Conf. Series **925** (2017) 012030 doi:10.1088/1742-6596/925/1/012030

4. Conclusions

It was found that the maximum intensity of heat removal from the heater is achieved in the mode, when the film flow continuity is broken. The heater is covered with small (of about 100 microns in size) dry spots with the lifetime of about 1/100 – 1/1000 s. At that, the number of spots that exist simultaneously on one square centimeter of the surface can reach several hundred. During the experiment, the total area of dry spots increases with increasing heat flux and heater temperature, but when the heater reaches a certain temperature (≈100°C), the total area starts to decrease. In contrast to the area of dry spots, the length of contact line increases with increasing heat flux and reaches a maximum in the pre-crisis regime. Intensive evaporation in the region of the contact line may explain high heat fluxes achieved in the experiment.

Acknowledgements

This work was supported by the Ministry of Education and Science of Russia (Agreement No 14.604.21.0053, project identifier RFMEFI60414X0053).

References

[1] Bar-Cohen A, Holloway A 2016 *Journal of Physics: Conference Series* **745**
[2] Kabov O A, Lyulin Yu V, Marchuk I V and Zaitsev D V 2007 *International Journal of Heat and Fluid Flow* **28** 103
[3] Zaitsev D V, Rodionov D A and Kabov O A 2009 *Technical Physics letters* **35** 680
[4] Kabov O A, Zaitsev D V, Cheverda V V, Bar-Cohen A 2011 *Experimental Thermal and Fluid Science* **35** 825
[5] Zaitsev D, Tkachenko E, Orlik E, Kabov O 2016 *MATEC Web of Conferences* **92**
[6] Hirokawa T, Ohta H and Kabov O 2015 *Interfacial Phenomena and Heat Transfer* **3** 303
[7] Lyulin Yu V, Spesivtsev S E, Marchuk I V, Kabov O.A 2015 *Technical Physics Letters* **41** 1034
[8] Kabov O A, Zaitsev D V, Kirichenko D P, Ajaev V S 2017 *Nanoscale and Microscale Thermophysical Engineering* **21** 60

IOP Conf. Series: Journal of Physics: Conf. Series **925** (2017) 012031 doi:10.1088/1742-6596/925/1/012031

Numerical investigation of flow dynamics and scalar transport in a wall-bounded turbulent jet

M Hrebtov[1], A Bazhenov[1] and K Borynyak[1]

[1] Institute of Thermophysics of SB RAS, 630090, Ac. Lavrentiev ave. 1, Novosibirsk, Russia

E-mail: weexov@ya.ru

Abstract. We present the results of Large Eddy Simulation of a turbulent jet discharging into a confined slot with the wall spacing to jet inlet width ratio = 0.1 and $Re = 10^4$. The jet exhibits a meandering motion accompanied by formation of checkerboard pattern of large vortices in the mixing layer. A passive scalar transport was simulated with uniform inlet distribution of the scalar. It was found that the fluctuations of spanwise velocity component grow downstream and their maximum location is migrating from the lateral boundaries of the jet (free-shear layers) to the wall-boundary layers. Some evidence of counter-gradient turbulent scalar transport was found in the mixing layers of the jet, which may be attributed to the influence of observed large-scale checkerboard-type vortices.

1. Introduction

The plane turbulent wall-bounded jet is a flow with a lot of important practical applications in industries dealing with heat and mass transfer. It also retains of special importance in theoretical studies of turbulence, as a plane wall jet is a prototypical near-wall flow that is significantly more complex than a simple boundary layer. Its dynamics is an interplay of two wall boundary layers and two free shear layers in orthogonal planes.

The instabilities generated at the wall interact with the main Kelvin-Helmholtz type vortices in the shear layer. For different aspect ratios between the jet width and the distance between the walls the flow develops large-scale meandering oscillations and quasi-two-dimensional checkerboard pattern of vortices resembling von Karman's vortex street. For large (~10) aspect ratios these features become more pronounced, especially in a turbulent regime for large enough Reynolds numbers ($\sim 10^4$).

The processes inside the jet are affected to a large degree by inflow conditions (the shape of the nozzle, harmonic forcing, etc.) and by the geometry of the flow. Such a big influence is caused by the high sensitivity of the shear layer to small flow perturbations that may lead to the change of dominant instability mode. Thus it is necessary to set the inlet conditions that are closest to physical ones.

This meandering behaviour of the flow must affect the turbulent diffusion of passive scalars enhancing the transport of the scalar in spanwise direction [1]. The present work was aimed to investigate this process.

For present numerical simulation we consider a wall-bounded shallow jet flow in a rectangular domain of size $H \times L \times D$, where $L = 267H$ and $D = 200H$. L and D are much larger than H (see Fig. 1). This dimensions of the domain was taken to mimic the experimental setup [2].

The inflow channel is a short duct of the length $2\pi H$ (with the cross-section $B \times H$ where $B = 9.6H$). The inlet velocity distribution was taken from a precursor DNS simulation of a fully developed turbulent duct flow. The Reynolds number of the flow was 10^4 based on the bulk velocity and the distance between the walls (H).

We used LES approach with the dynamic Smagorinsky model and the following equations for the momentum and scalar transport:

$$\frac{\partial \bar{u}_i}{\partial t}+\frac{\partial}{\partial x_j}\left(\bar{u}_i\bar{u}_j\right)=-\frac{1}{\rho}\frac{\partial \bar{p}}{\partial x_i}+\frac{\partial}{\partial x_i}\left(\left(\nu+\nu_{sgs}\right)\bar{S}_{ij}\right), \quad (1)$$

$$\frac{\partial \bar{C}}{\partial t}+\frac{\partial}{\partial x_j}\left(\bar{C}\bar{u}_j\right)=\frac{\partial}{\partial x_i}\left(\left(\frac{\nu}{Sc}+\frac{\nu_{sgs}}{Sc_t}\right)\frac{\partial \bar{C}}{\partial x_j}\right). \quad (2)$$

The Schmidt number in the equation (2) was set to be 0.9, and its turbulent analogue Sc_t was also set to this value.

We used a classical form of subgrid viscosity

$$\nu_{sgs}=\left(C_s\Delta\right)^2\left|S_{ij}\right|, \quad (3)$$

where Δ is a grid spacing and C_S is Smagorinsky coefficient, which should be evaluated dynamically.

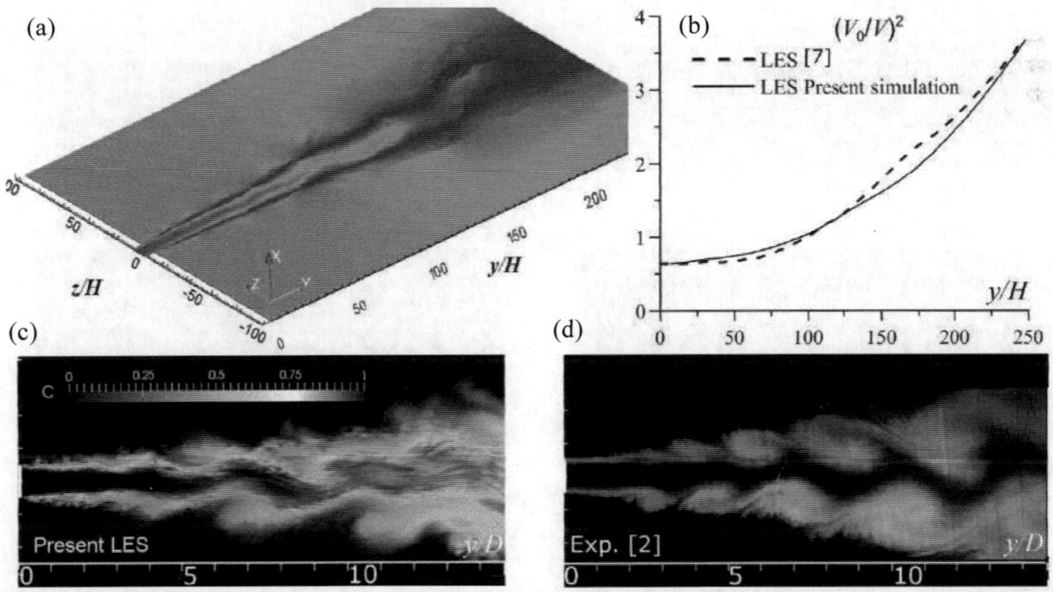

Figure 1. The computational domain for numerical simulation (a); inverse mean streamwise velocity profile through the jet axis (b); instantaneous concentration distribution (c) (note the false coloring) compared to the experiment [2] (d).

The classical dynamic Smagorinsky [3] model computes C_S after applying explicit spatial filtering to the velocity field and then minimizing the difference between the subgrid and residual stress tensors. This procedure may lead to negative turbulent viscosity (i.e., energy backscatter from the subgrid to resolved scales). This in turn was shown to cause a blow up of the solution because the amount of energy which may be transferred from the small scales is unlimited. It is common to use a clamping of Smagorinsky constant, leaving only positive values. But such procedure is an *ad hoc* and non-physical in nature. It was shown that averaging of the Smagorinsky constant over homogeneous directions resolves that issue, but only a few types of flow have suitable homogeneous directions for that averaging. Another way which was adopted in the present study was introduced in [4] and consists in averaging of the residual tensor magnitude over the lagrangian trajectories. The model has shown a good agreement with direct numerical simulations and can be adopted for almost every flow.

2. Computational domain and numerical scheme

The numerical grid consisted of 16×10^6 hexahedral cells with the wall-adjacent grid-cell height ($x+$) of about 1.5 in wall units. The grid was refined in the near-wall regions and also in the mixing layer. The grid resolution was chosen based on preliminary simulations with a set of grids with different number of cells.

The momentum and scalar transport equations (1), (2) were discretized using finite volume scheme with second order accuracy in time and space. The modified solver from OpenFoam [5] package was used.

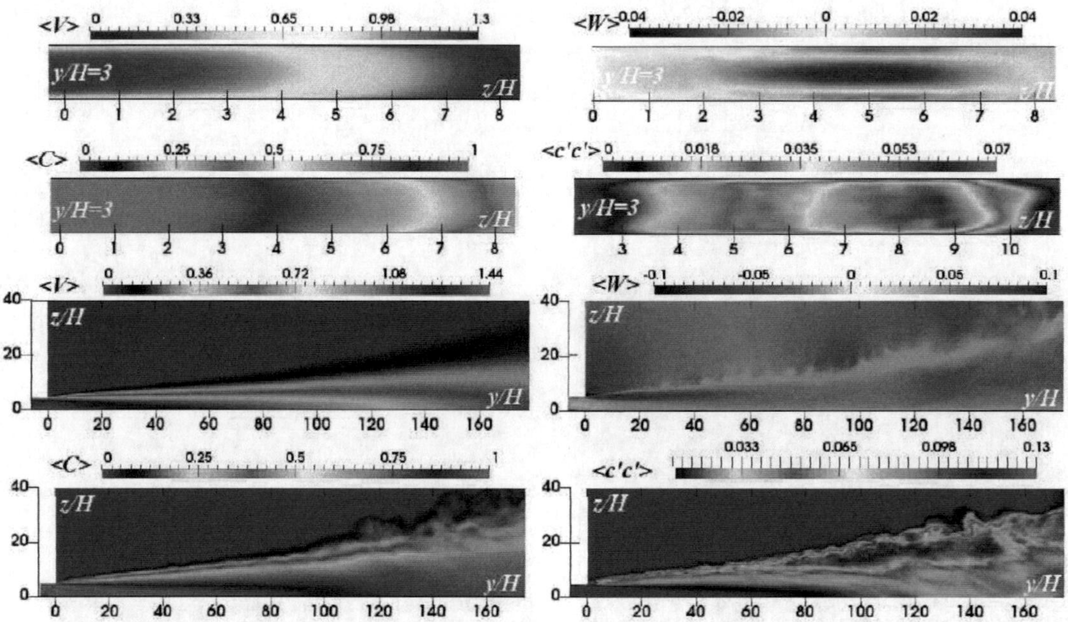

Figure 2. Mean flow fields in spanwise (top) and streamwise (bottom) cross-sections.

3. Inlet boundary condition

Proper boundary conditions are necessary for complex flow simulations. Currently, there are several ways to generate turbulent initial conditions at the inlet.

First and the simplest way is a random fluctuations generation. This method is based on a white noise generation. Fluctuations are generated without temporal and spatial correlation, so the spectra of the signal will be governed by the grid spacing and the time step of the simulation. Also, continuity constraint might be violated, that will lead to the appearance of numerical artifacts in pressure fields.

The next way is precursor simulations. The main idea of the method is to let the fluctuations to evolve naturally in a separate simulation and then copy them to the inflow. This method provides the best quality inlet conditions but form many practical applications its computational costs may be too high.

To avoid costly preliminary computations we use the third way: the anisotropic divergence-free synthetic turbulence generation at inflow bound, introduced in [6].

The method utilizes the idea of obtaining the velocity from a vector potential field with the prescribed properties. Using the vector potential ensures the divergence-free condition of the inlet field. The second idea of the method is to represent the vector potential field in a form of "blobs" - constructed random objects of the given scales and intensities. The blobs's characteristics are defined from known components of Reynolds stress tensor at the inlet and the two-point velocity correlations that represent the fluctuations scale. The method was shown to represent the turbulent statistics of the shear flows quite well.

IOP Conf. Series: Journal of Physics: Conf. Series **925** (2017) 012031 doi:10.1088/1742-6596/925/1/012031

In the present case for the inlet fields generation we used previously obtained distributions of Reynolds stress components and the two-point velocity correlations of the flow in a rectangular duct with the same geometry as the inlet. The mean inlet velocity was set to be unity and the inlet scalar concentration was set uniform with the value of 1.

4. Simulation results

The instantaneous velocity and concentration fields of the wall-bounded jet obtained in the simulation show the expected meandering behavior with a Strouhal number of the order of 0.1 in a good agreement with experimental observations. Largescale checkerboard-type vortex structures appear after *y/H*=50. The meandering behavior of the jet was shown [7] to be connected with instabilities caused by longitudinal vortices originating from the wall boundary layer interacting with a Kelvin-Helmholtz instabilities in the mixing layer.

The instantaneous scalar concentration distribution obtained in the simulation (Fig. 1) shows the appearance of large billow-like structures associated with Kelvin-Helmholtz instabilities. These structures start to form at *y/D*=2-3. The vortex structure lose lateral symmetry and form checkerboard pattern at *y/H=50* where the two mixing layers collide. The comparison of the scalar distribution with experiment [2] shows a good agreement, with a similar type of structures observed in both the simulation and experiment.

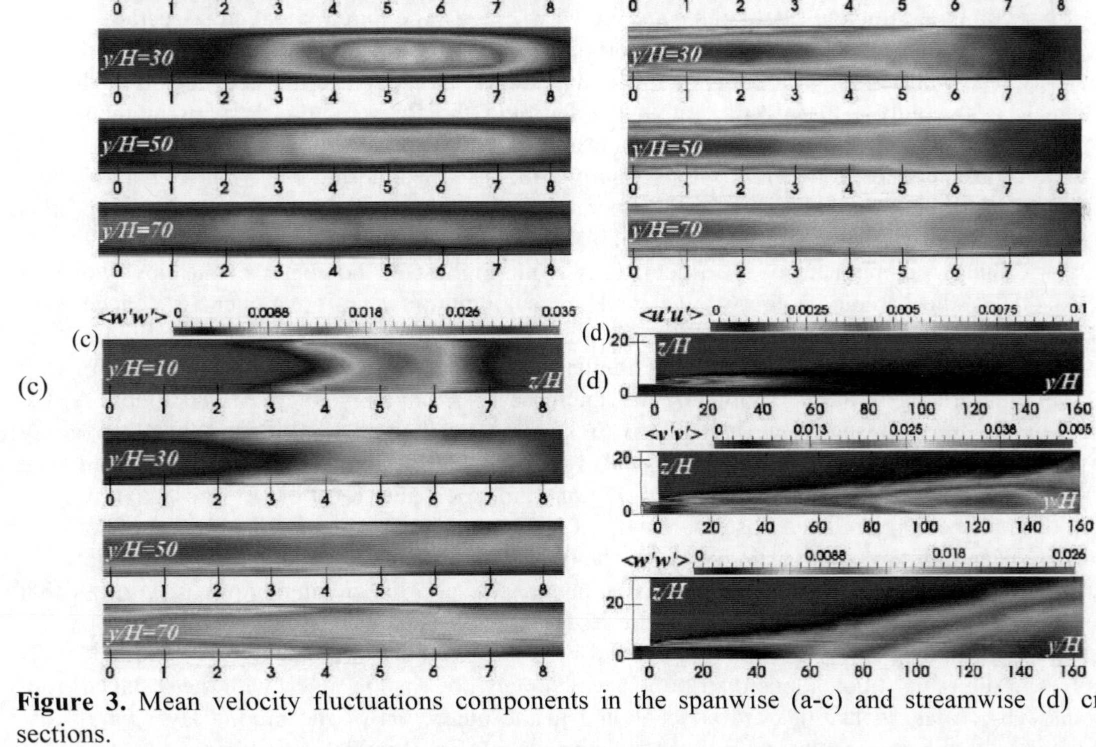

Figure 3. Mean velocity fluctuations components in the spanwise (a-c) and streamwise (d) cross-sections.

The longitudinal profile of mean velocity was compared to the high resolution LES of [7] and shown a satisfactory agreement. The difference may be due to the too-coarse mesh in the far region of the jet and an insufficient statistics to average over all the low frequency modes of the flow. In present case the averaging time to gain statistics was equal to twice the travelling time of the fluid particle over the computational domain. The expected behavior of the mean longitudinal velocity along the center line for plane jets is $V \sim y^{-1/2}$ in the self-similar region, however this regime was not exactly achieved in our simulations, which may be because the domain size was not large enough in longitudinal direction.

4.1 The mean fields

The mean velocity distribution along the streamwise and spanwise cross-sections are shown in Fig. 2. It can be seen that spanwise velocity component has a maximum in the mixing layer region close to the nozzle exit. As shown by [7], this is due to intense lateral vortex stretching of the "corner" vortices when they are exiting the nozzle.

The longitudinally directed vortices shed from the corners of the nozzle starts to interact by inducing the velocity fields that lead to the spreading of the jet in the mid-plane. However the three-dimensional instabilities that start to grow due to this motion lead to the loss of coherence of the motion of these vortices and eventually to their dissipation. This all is reflected by a maximum of the spanwise velocity close to the nozzle exit. The action of the "corner" vortices is reflected in the mean spanwise velocity in lateral cross-section by the appearance of zones with negative (i.e. directed toward the jet axis) velocity. The zones of negative mean spanwise velocity appear in a shape of stripes close to the wall with the width of several H.

The mean scalar concentration distribution has in general the behavior similar to the longitudinal velocity but the lateral spreading of the concentration is faster, which may be due to a significant role of turbulent scalar flux generated by a large-scale quasi-two-dimensional vortices.

4.2 Turbulent fluctuation intensities

The cross-sections of turbulent fluctuations are shown in Fig. 3. The longitudinal cross-sections (Fig.3 d) are in accord with [7], showing strong anisotropy between the different components of turbulent fluctuations. The wall-normal component of the fluctuations has a peak in the near region of the jet, and then decays rapidly. This means that in the far field the flow should show two-dimensional features. However, the effect of the walls should lead to the appearance of longitudinal vortices so the three-dimensional features of the flow should also present. The meandering behaviour of the jet's core leads to growth of spanwise and streamwise fluctuations amplitudes downstream. We extended this observation by looking at the lateral cross-sections of the fluctuations components (Fig.3 a-c).

The wall-normal (x) component of the fluctuations is the highest in the center of the flow which is expected due to the damping effect of the walls. The longitudinal (y) component is concentrated mostly in the near-wall region where the shear is maximal. While going downstream the longitudinal near-wall fluctuations become weaker while simultaneously growing in the inner part of the mixing layer. Quite surprising is the behaviour of the spanwise (z) component of the fluctuations. At the nozzle exit the most of spanwise fluctuations are concentrated in the mixing layer. When going downstream some fluctuations appear in the near-wall region of the jet. The magnitude of span-wise velocity fluctuations grow in the near-wall region and soon the location of the maximum of fluctuations moves to a thin layer close to the wall. This means that the meandering motion of the flow affects the near-wall longitudinal velocity fluctuations leading to their tilting, thus some of this fluctuations are transferred to a transversal component. The non-linear interactions of vortices that accompany this processes is the subject of future research.

The scalar fluctuations (Fig 2) are mostly affected by a large-scale two-dimensional vortices. This is evident from the scalar fluctuations distribution which does not change much with the distance from the walls. The scalar fluctuations are concentrated in the outer part of the mixing layer, and grow downstream with the growing intensity of the meandering motion of the jet.

4.3 The turbulent fluxes

The large-scale checkerboard-type vortices should affect the mass transfer in spanwise direction quite significantly. To investigate this we calculated the z component of turbulent scalar flux ($\langle c'w' \rangle$) and yz component of Reynolds stress (Fig. 4). The streamwise distribution of the Reynolds stress shows two maxima, one close to the nozzle where the "corner" vortices play the major role, and the other further

IOP Conf. Series: Journal of Physics: Conf. Series **925** (2017) 012031 doi:10.1088/1742-6596/925/1/012031

downstream (*y*/*H*=120) where the meandering motion of the core become the most pronounced. The spanwise cross-section in the near field (*y*/*H*=30) shows the appearance of negative Reynolds stress in the near-wall region. This may be connected to the effect of "corner" vortices inducing the negative values of span-wise velocity in the same region. However the appearance of zone with negative stress means that turbulence act pushing the longitudinal fluctuations laterally toward the jet axis in the near wall region.

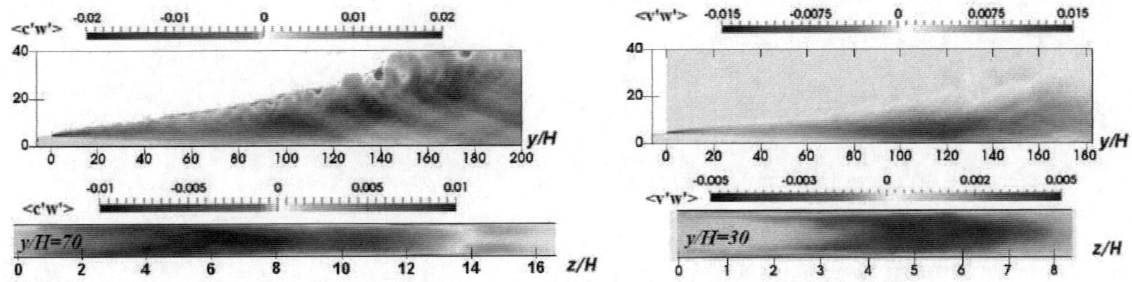

Figure 4. Spanwise component of turbulent scalar flux (left); and *yz* component of the Reynolds stress (right) in the lateral and longitudinal cross-sections.

The longitudinal distribution of turbulent scalar flux shows some negative values at the outer part of mixing layer. This is the evidence of counter-gradient transport of the scalar by the turbulent vortices. The distribution of scalar flux in a spanwise cross-section shows that negative values have two symmetric peaks close to both walls and are caused by the effect of longitudinal vortices that recapture some of the scalar concentration at the mid-plane section and advect it toward the walls and back in the direction of jet axis.

5. Conclusion

The confined jet is a complex flow with an interplay of two- and three-dimensional features. Large quasi-two-dimensional checkerboard vortices observed in the flow contain many smaller vortices with wall-parallel directions. The effect of the wall is very significant for that flow. The present study shows that spanwise velocity oscillations grow in the near-wall region as the flow progresses. This is caused by the interaction of wall-induced longitudinal vortices with mixing layer instabilities. The longitudinal vortices act to produce a counter-gradient transport of the passive scalar at the outer boundaries of the mixing layer. Also these vortices counteract lateral diffusion in the near-wall region of the flow. Further research is needed to investigate the nonlinear interaction between the vortices and the mean flow in the near-wall region.

Acknowledgements

This work was supported by Russian Science Foundation (grant №14-19-01685).

References

[1] Rowland J C, Stacey M T, and Dietrich W E 2009 *J. Fluid Mech.* **627** pp. 423-449
[2] Shestakov M V, Dulin V M, Tokarev M P, Sikovsky D P, Markovich D M 2015 *Int. J. Heat Fluid Flow* **51**, 335–352
[3] Germano M, Piomelli U, Moin P, Cabot W 1991 *Phys. Fluids A*. **3** (7): 1760–1765
[4] Meneveau C, Lund T and Cabot W1996 *J. Fluid Mech.* **319**, pp. 353-385
[5] Weller H G, Tabor G, Jasak H, Fureby C 1998 *Comp. in Phys.*, **12**, 6
[6] Kroger H, and Kornev N 2015 *ICHMT digital library online*. Begel House Inc.
[7] Mullyadzhanov R, Ilyushin B and Hanjalić K 2015 *Int. J. Heat Fluid Flow* **56**: 284-289

IOP Conf. Series: Journal of Physics: Conf. Series **925** (2017) 012032 doi:10.1088/1742-6596/925/1/012032

Research on droplets growth of Marangoni condensation during the time-series process

Y Li, J Wang, K Xia, N Chen and J Yan

State Key Laboratory of Multiphase Flow in Power Engineering, Xi'an Jiaotong University, Xi'an, 710049, P.R. China

E-mail: wangjinshi@mail.xjtu.edu.cn

Abstract. In this paper, the condensate droplets growth of Marangoni condensation during the time-series process was experimentally studied. The photographs taken in the experiments were handled by image processing method. The droplets coverage ratio was introduced to analyse the growth process. As the time-series process went on, the condensate droplets coverage ratio increased fast at first, and then increased slowly until reaching a stable state. The droplets coverage ratio increased with the increase in the vapor-to-surface temperature difference and decreased with the increase in the ethanol vapor concentration.

1. Introduction

Marangoni condensation is caused by the surface tension instability on the condensate film surface of binary mixtures in a positive system, where the surface tension of the mixtures has a negative gradient with the mass fraction of the more volatile component, such as ethanol-water mixtures. The Marangoni force, induced by the surface tension gradient, pulled the condensate with lower surface tension towards the higher part along the liquid surface, which tore the condensate film and resulted in the dropwise mode. It is a new way to achieve dropwise condensation and could maintain for a long time, which depends merely on the physical proprieties of the mixtures.

The heat transfer coefficient of Marangoni condensation was found to vary non-monotonously with the increase in the vapor-to-surface temperature difference and the ethanol vapor concentration, increase with the increase in the vapor velocity and pressure [1, 2]. Several condensation modes were observed during the experiments and variations of droplet size distribution and maximum diameter were analysed [3].

Previous studies on the Marangoni condensation were mainly on the vertical surfaces, horizontal tubes and vertical tubes, where the gravity effect acted on the condensate with the same direction of the surface tension gradient. However, the essential reason for dropwise modes in the Marangoni condensation was the surface tension gradient. When the gravity effect operated with the same direction of the surface tension gradient, the condensation modes were diversified under the effects of both surface tension gradient and the gravity effect. Thus, the condensation modes under the interference of the gravity effect were unrepresentative. Moreover, the interference of the gravity effect could make the condensate droplets leave the condensing surface early and the growth of the droplets was not complete. When the condensing surface was horizontal, the gravity effect was perpendicular with the direction of the surface tension gradient on the condensate. Then the interference of the gravity effect could be neglected and the condensate droplets could grow freely.

IOP Conf. Series: Journal of Physics: Conf. Series **925** (2017) 012032 doi:10.1088/1742-6596/925/1/012032

Then present Marangoni condensation modes were caused merely by the surface tension gradient, which could be regarded as the essential ones. Experimental study was conducted on the horizontal surface for Marangoni condensation of ethanol-water mixtures and droplets growth was analysed during the time-series process in this paper.

2. Experimental system

The experimental system consists of four closed loops, which are named as the main vapor loop, the impinging cooling loop, the condenser loop and the auxiliary condenser loop, respectively. The details could be found in the work by Li et al. [4]. As the experiments were conducted horizontally, the condensate film couldn't be removed automatically. For the continuity of the condensation process, a wiper was designed and employed to remove the condensate film on the condensing surface, as shown in Fig. 1. Moreover, two pieces of magnets were adopted to transmit the driving power from a micro-motor for keeping the excellent vacuum condition. The test copper with a condensing surface area of 25.0 mm × 40.0 mm was horizontally installed and fixed with epoxy plates to ensure the heat flux inside transferred vertically.

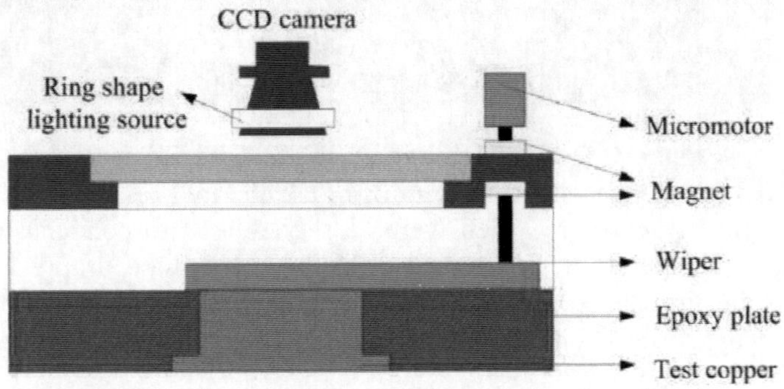

Figure 1. Sketch diagram of the condensing chamber

3. Image processing method

The image processing method was also one focal point in this study. As the number of condensate droplets formed on the condensing surface was huge, it was difficult to figure out the numbers and sizes of the condensate droplets manually. So the image processing program was written by MATLAB herein and the flow chart was shown in Fig. 2. The ring-shape lighting source was equipped for improving the accuracy of the image processing. The result of the image processing method was also shown in Fig. 2. The white parts in the photo were the identified droplets and their numbers and sizes were obtained by the program automatically.

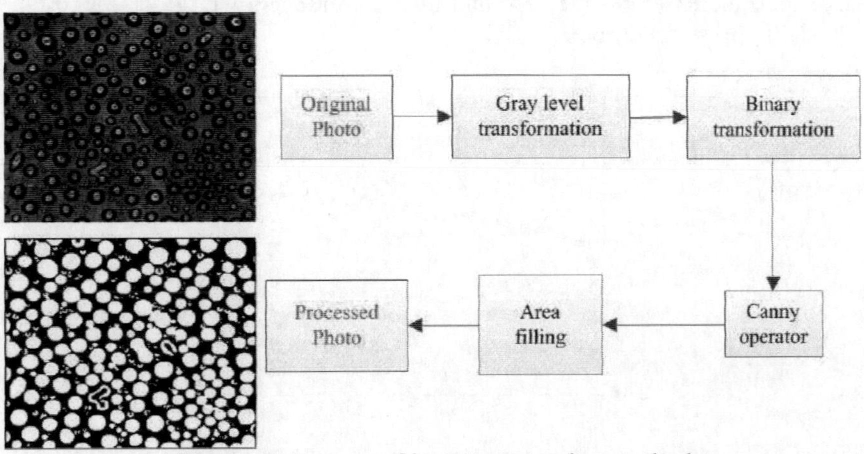

Figure 2. Process of image processing method

IOP Conf. Series: Journal of Physics: Conf. Series **925** (2017) 012032 doi:10.1088/1742-6596/925/1/012032

4. Results and discussions

The time-series process expressed the subsequent process after the wiper just brushed over the center of the condensing surface and the condensation process developed freely. The wiper didn't work during this process. Figure 3 showed the visual figures during one time-series process. As the time-series process went on, huge number of small droplets appeared at the beginning and grew up fast mainly by direct condensation. As the time-series process went on, the droplets grew slowly mainly by coalescence.

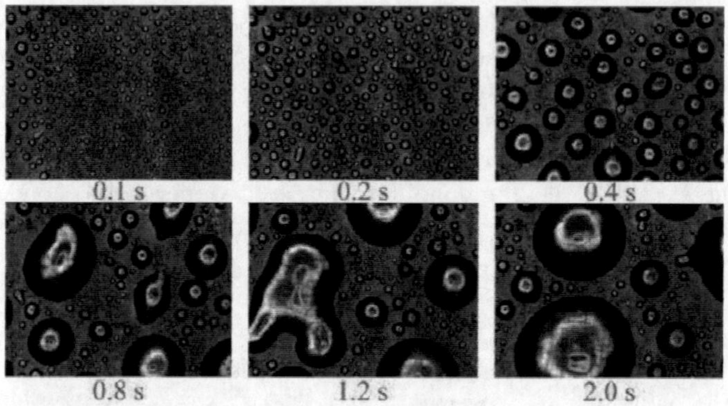

Figure 3. Droplets growth during time-series process

As the primary heat transfer resistances came from the thermal resistances from the condensate droplets and the vapor diffusion layer, which were closely related to condensate droplets coverage ratio and ethanol vapor concentration, respectively. So the condensate droplets coverage ratio was introduced to evaluate the growth of the droplets, which expressed the areas ratio between all the droplets and the condensing surface. The definition equation of condensate droplets coverage ratio was shown in Eq. (1).

$$C(t) = \frac{\sum_{1}^{n} A_i(t)}{A_s} \tag{1}$$

Where $C(t)$ expressed the condensate droplets coverage ratio during the time-series process, $A_i(t)$ expressed the occupied area on the condensing surface of one condensate droplet and A_s expressed the whole area of the condensing surface.

Figure 4 showed the variation of condensate droplets coverage ratio during the time-series process corresponding to the photos in Fig. 3. As the time-series process went on, the condensate droplets coverage ratio increased fast at first, and then increased slowly until reaching a stable state, with the growth of the condensate droplets.

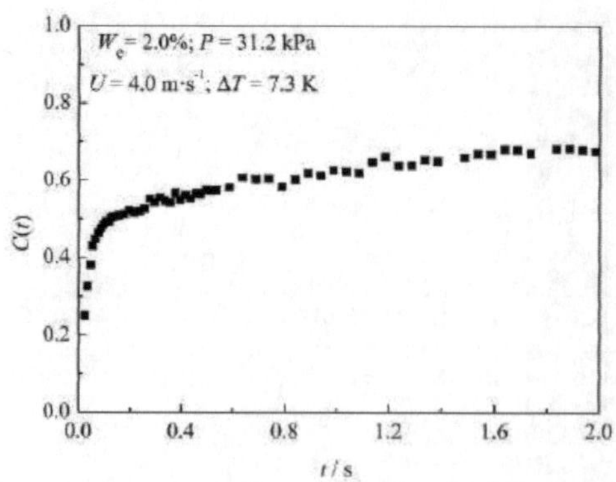

Figure 4. Coverage ratio during the time-series process

Figure 5 showed the effects of vapor-to-surface temperature difference on the condensate droplets coverage ratio at the ethanol vapor concentrations of 2.0% and 5.0% respectively. The droplets coverage ratio increased obviously with the increase in the vapor-to-surface temperature difference, especially when the time-series reached the stable state. As the vapor-to-surface temperature was the dominant driving force for the Marangoni condensation, when the vapor-to-surface temperature difference increased, the condensate droplets grew up faster and the condensate droplets coverage ratio increased correspondingly.

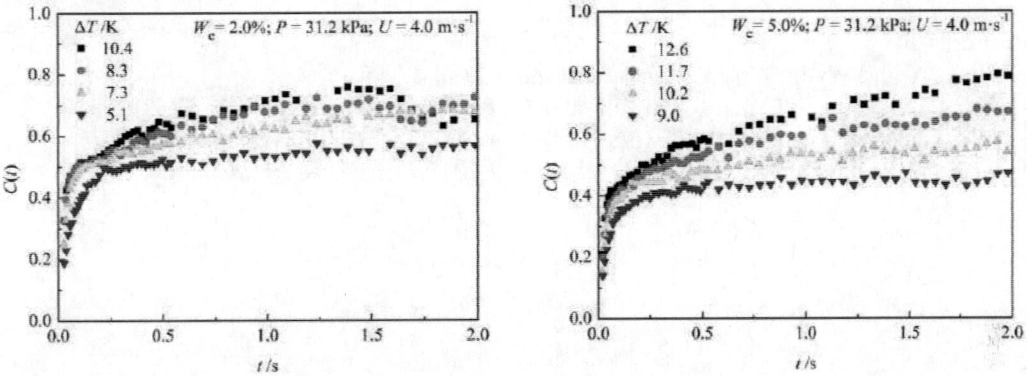

(a) (b)

Figure 5. Effects of vapor-to-surface temperature difference

Figure 6 showed the effects of ethanol vapor concentration on the condensate droplets coverage ratio at the impinging cooling water temperatures of 50 °C and 55 °C respectively. The condensate droplets coverage ratio decreased obviously with the increase in the ethanol vapor concentration. The effect of the ethanol vapor concentration mainly came from the thermal resistance in the vapor diffusion layer. As the ethanol vapor concentration increased, the thermal resistance from the vapor diffusion layer increased accordingly, which weakened the heat transfer process and slowed down the growth of the condensate droplets with the smaller droplets coverage ratio. The fluctuation in the data was mainly caused by the movement of condensate droplets.

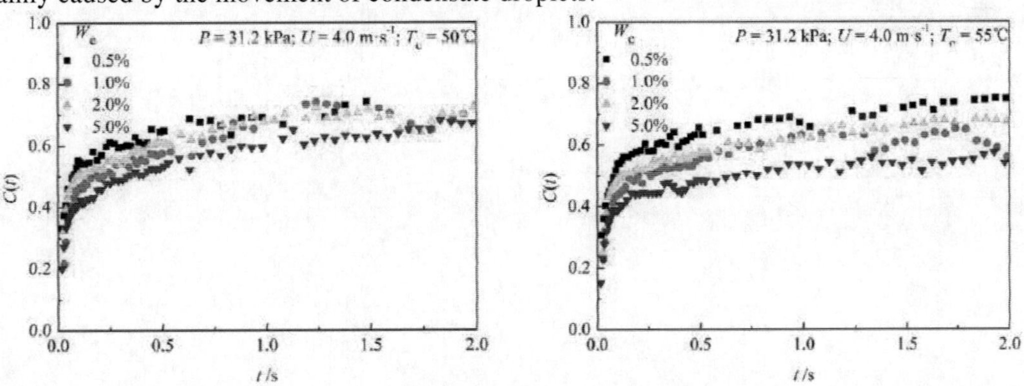

(a) (b)

Figure 6. Effects of ethanol vapor concentration

5. Conclusions

The condensate droplets growth process was experimentally studied on the horizontal surface during the Marangoni condensation for ethanol-water mixtures. The image processing program was written by MATLAB to identify the droplets in the photographs taken in the experiments. The condensate droplets coverage ratio, which expressed the areas ratio between all the condensate droplets and the condensing surface, was introduced to analyse the growth process. Results indicated that the condensate droplets coverage ratio increased fast at first, and then increased slowly until reaching a

IOP Conf. Series: Journal of Physics: Conf. Series **925** (2017) 012032 doi:10.1088/1742-6596/925/1/012032

stable state during the time-series process. And the condensate droplets coverage ratio increased with the increase in the vapor-to-surface temperature difference while decreased with the increase in the ethanol vapor concentration.

Acknowledgments
This project has been supported by the National Natural Science Foundation of China (No. 51436006) and the National Basic Research Program of China (No.2015CB251504).

References
[1] Utaka Y and Wang S X 2004 *Int J Heat Mass Tran* **47** 4507
[2] Yang Y S, Zhen K J, Yan J J, Liu J P and Hu S H 2006 *J Chem Ind Eng* **57** 2816
[3] Yan J J, Wang J S, Yang Y S, Hu S H and Liu J P 2009 *Microgravity Sci Tec* **21** 77
[4] Li Y, Wang J S, Xia K, Chen N N and Yan J J 2017 *Exp Therm Fluid Sci* **83** 141

IOP Conf. Series: Journal of Physics: Conf. Series **925** (2017) 012033 doi:10.1088/1742-6596/925/1/012033

Enhancement of heat transfer at pool boiling on surfaces with silicon oxide nanowires

E A Chinnov[1,2], E N Shatskiy[1,2], S Ya Khmel[1], E A Baranov[1], A O Zamchiy[1], V V Semionov[1] and O A Kabov[1,2]

[1] Kutateladze Institute of Thermophysics SB RAS,630090 Novosibirsk Ac. Lavrentiev Ave. 1, Russia
[2] Novosibirsk State University, 630090, Pirogov str. 2, Novosibirsk, Russia

E-mail: shataskiy.itp@gmail.com

Abstract. The boiling heat transfer on the local heaters with microstructured and nanomodified surfaces was studied. As nanomodified surfaces we used copper ones where the microropes of silicon oxide nanowires were grown. The aging of the nanomodified surface was observed after first series of experiments. It was shown that both finning and nanostructuring of the surface result in increase of heat transfer. The heat flux density of 1400 W/cm^2 was reached.

1. Introduction

Cooling the heated surfaces in various technologies is an important task, whose solution requires application of such processes as boiling. Today much attention is paid to development of the methods of heat transfer enhancement at boiling and increasing the critical heat flux [1]. One of directions is the study of the influence of various nanocoatings on heat transfer intensity. The review of recent studies on the influence of micro/nanomodifiedsurfaces and coatings on heat transfer enhancement and critical heat flux increase at boiling is presented in [2]. There are another techniques for heated surfaces such as spray cooling, systems based on evaporation of a thin liquid layer or rivulet[3, 4], thermoelectric coolers. Dynamics of evaporation depends on the conditions in the layer [5]. For instance, the rupture of thin liquid layer can terminate the cooling process [6].

The aim of the present work is studying of pool boiling of water on the surfaces with different structure.

2. Experimental setup

In this work, we created an experimental setup for studying heat transfer from a local heat source. The test section is a PTFE cylinder with diameter 28 mm. The working liquid (distilled water) is supplied to the test section using a peristaltic pump, where a horizontal layer of liquid is formed. The thickness of the liquid layer is controlled by the amount of liquid (the geometry of test section is known) and varies depending on the operating parameters of the experiment (20-50 mm). In the center of the test section there is a local heating performed by a heater. The heating element is a cylindrical core made of copper. The upper part of the heater facing the liquid has diametrical dimension D = 5 and 1 mm. The source of heat is a nichrome wire wound around the lower part of the core. In order to minimize heat losses, the heating element was carefully insulated. The fiberglass was used as a heat insulator, which was wound in several layers around the core of the heating element. It was estimated that the whole incoming power transferred to the liquid and heat losses did not exceed 20%.

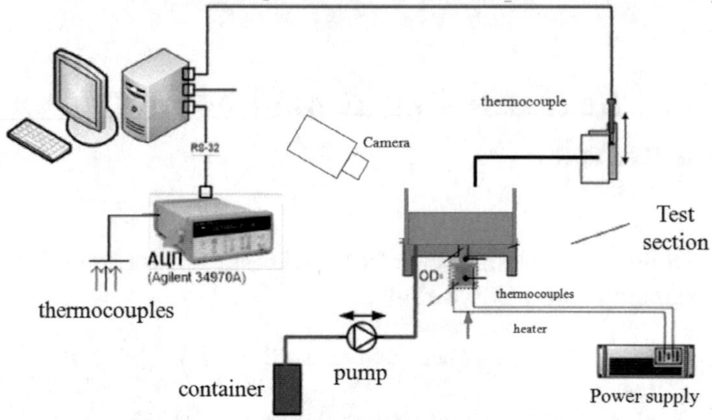

Figure 1. Scheme of experimental setup

Boiling heat transfer on the local heaters with nanomodified surfaces has been investigated. In this study, we carried out the experiments under the conditions of a large volume of liquid and maintaining the temperature of water at the level of saturation. The copper surfaces coated with a layer of silica, molybdenum or tungsten, where the microropes of silicon oxide nanowires were grown, were used as the nanomodified surfaces. The technology of their production is described in [7]. Unfortunately, a layer of silica on copper surfaces was destroyed after heating.

For the 1 mm diameter heater, special headers (heat sinks) have been developed, which is shown in Figure 2. These headers were heated and placed on a 1-mm-diameter heater cooled in liquid nitrogen, which provided good thermal contact.

Figure 2. Photo of the heat sink with finned surface for 1 mm heater

3. Results

The results obtained on 1mm heater with different headers are shown in Figure 3. On finned surfaces, the superheat with respect to the saturation temperature decreases up to four times. More than three times the heat transfer coefficient increases in finned surface compared with the smooth. It should be noted that the type of finning did not have a significant effect on the boiling curve. Application of nanocoatings leads to an intensification of heat transfer in comparison with a smooth surface that is comparable with the use of radial finning. As a result of the experimental work, the heat flux density of 1400 W/cm^{2}was reached.

IOP Conf. Series: Journal of Physics: Conf. Series **925** (2017) 012033 doi:10.1088/1742-6596/925/1/012033

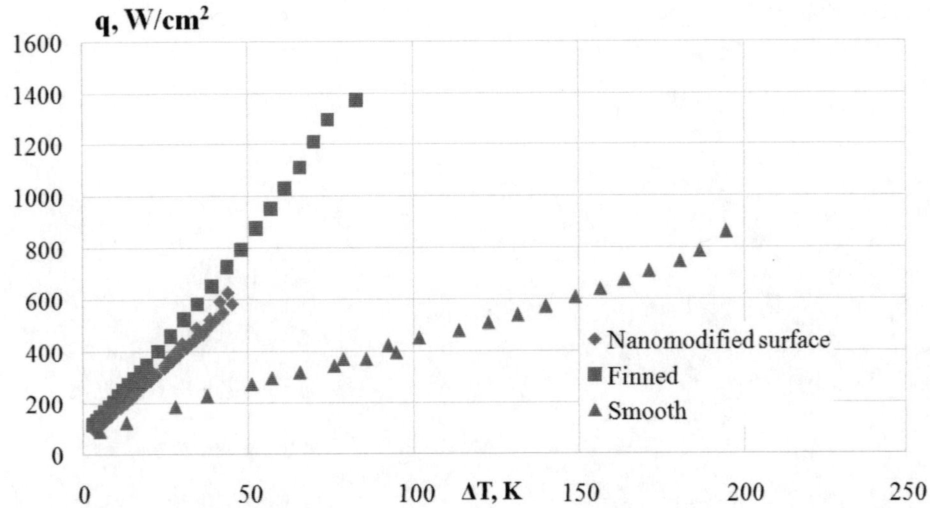

Figure 3. Dependence of heat flux density on overheating of heater surface relative to saturation temperature for heater D = 1 mm with heat sink

Experimental data on the 5 mm heaters with the smooth and nanomodified surfaces are presented in Figure 4. On the surface with a molybdenum coating, we failed to grow the silicon oxide nanowires. The contact angle with a water drop on the resulting coating is 54° and it almost coincides with the wetting angle on the polished copper surface.

The best results were obtained on the copper surface with a sublayer of tungsten. On the intermediate layer of tungsten was obtained arrays of microropes and "cocoons", consisting of particles of tin coated with silicon oxide nanowires Figure 4a. The size of the cocoon and the distance between them was up to 1 μm. The contact angle with a water drop on the resulting coating, measured before the experiment, is 18°, and this is significantly less than that on the copper surface or molybdenum, tungsten coating.

According to Figigure 5, the heat transfer coefficient for this coating is substantially higher than for a smooth molybdenum surface. Two series of experiments were performed with limitation of heating. After the first series of experiments, microropes were gone (Figure 4b), cocoons were left. During the second series of experiments, the area of nanomodified surface began decreasing, which was reflected in the boiling curve behavior. The coefficient of heat transfer decreased.

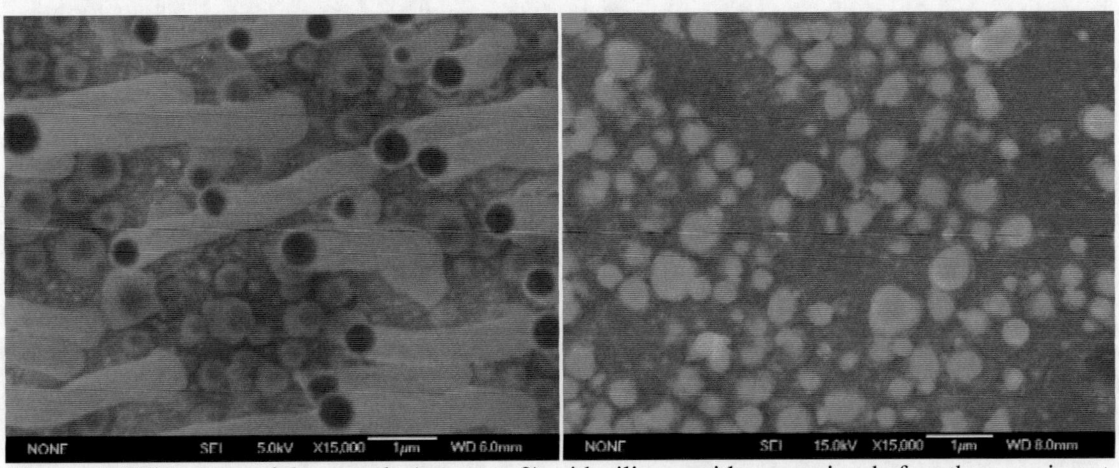

Figure 4. SEM photo of the sample (tungsten 2) with silicon oxide nanowires before the experiment (a) and after the experiment (b), tungsten 2 (1series)

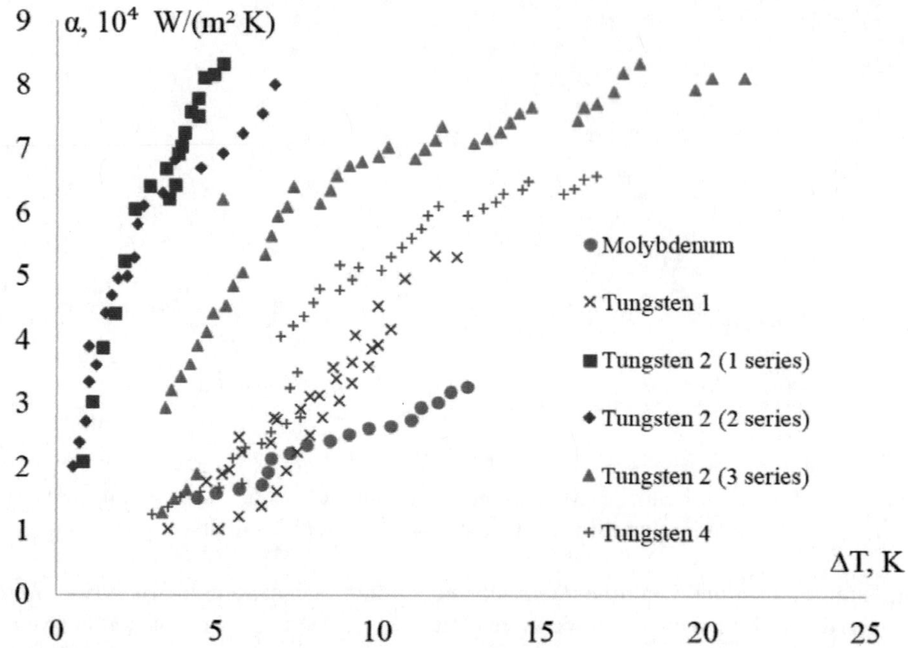

Figure 5. Heat transfer coefficient against overheating of heater surface relative to saturation temperature for heater D = 5 mm

4. Conclusion

In our experiments the layers of silica on copper surfaces were destroyed after heating. Arrays of microropes on heated surfaces with silica, molybdenum and tungsten sublayers disappear after boiling. The best results were obtained on the copper surface with a sublayer of tungsten. Surfaces with "cocoons" on tungsten sublayers were stable at boiling.It can be concluded that a new type of microfinning and nanocoatings provides heat transfer enhancement during boiling. Heat transfer coefficient of 80000 W/(m^2K) was reached on nanomodified surfaces. The heat flux density of 1400 W/$cm^2$was reached on microfinned surface.

Acknowledgements

The study was carried out by the grant of Russian Science Foundation (agreement no. 15-19-30038).

References

[1] D. Attinger, et al. 2014 *MRS Energy & Sustainability* **1** E4

[2] A. Surtaev, V. Serdyukov, A. Pavlenko 2016 *Nanotechnologies Russ.* **11** 696

[3] V. V. Cheverda, I. V. Marchuk, A.L. Karchevsky, E.V. Orlik, O.A. Kabov 2016 *Thermophysics and Aeromechanics* **23** 415-420

[4] V. V. Cheverda, A. Glushchuk, P. Queeckers, S. B. Chikov, O. A. Kabov 2013 *Microgravity Sci. Technol.* **25** 73–81.

[5] Yu.V. Lyulin, O. A. Kabov 2013 *Technical Physics Letters* **39** 795-797

[6] Yu.V. Lyulin, S.E. Spesivtsev, I.V. Marchuk, O.A. Kabov 2015 *Technical Physics Letters* **41** 11

[7] S. Khmel, E. Baranov, A. Zamchiy, E. Maximovskiy, D. Gulyaev, K. Zhuravlev 2016 *Phys. Status Solidi. (a)* **213** 1790

IOP Conf. Series: Journal of Physics: Conf. Series **925** (2017) 012034 doi:10.1088/1742-6596/925/1/012034

Two-phase flow regimes in a horizontal microchannel with the height of 50 μm and width of 10 mm

V P Fina[1,2] and F V Ronshin[2,3]

[1] École Nationale d'Arts et Métiers de Bordeaux, Esplanade des Arts et Métiers, 33405 Talence CEDEX, France
[2] Kutateladze Institute of Thermophysics, Siberian Branch, Russian Academy of Sciences, 1 Lavrentiev ave., Novosibirsk, 630090, Russia
[3] Novosibirsk State University, 1 Pirogova str., Novosibirsk, 630090, Russia

E-mail: valentin.fina@ensam.eu

Abstract. Two-phase flows of distilled deionized nanofiltered water and nitrogen gas in a microchannel with a height of 50 μm and a width of 10 mm have been investigated experimentally. The schlieren method has been used to determine main features of the two-phase flow in the microchannel. This method allows detecting the liquid film on the lower and upper walls of the microchannel as well as droplets of various shapes and sizes or vertical liquid bridges. Two-phase flow regimes have been observed, and their boundaries precisely determined using post-processing of the recordings. The following flow regimes have been distinguished: bubble, churn, jet, stratified and annular. Comparison of regime maps for channels of different widths has been carried out, and this parameter showed to have a significant impact on the boundaries between the regimes in microchannels of a height of less than 100 μm.

Introduction

The last decades have seen the miniaturization of devices in various fields of engineering, for example: in the space and chemical industries, in hydrogen and nuclear power, in treating materials, and in medicine. It is well-known that the existing cooling systems do not meet the modern requirements for heat removal from these highly heat-intense sources, and it seems probable that the next generation of heat exchangers will be based on flows in mini and micro channels. The studies show that such systems are more energy efficient than macrosystem with free-flow of liquid [1,2]. Indeed, as the thickness of the channel decreases, the ratio of the channel surface to its volume increases inversely proportional to its minimal cross dimension, which leads to heat exchanges of dramatic intensity.

The logical consequence of this need of a new technological solution is the recent interest on two-phase flows in these kind of conditions, as many studies have been published on this matter over the last years. Overviews of the publications on two-phase flows in micro channels of various configurations are given in [3-5]. Even if the results of investigations on two-phase flows in mini- and micro channels can be sometimes ambiguous, contradictory and open to different interpretations, some breakthrough have been made on the subject. It is now established that the two-phase flow structure is mainly influenced by the following parameters: the channel geometry and the dimensions as well as the parameters of the input section [6,7] and the properties of the liquid, such as viscosity and surface tension.

Research originally gave priority to circular flows, to get insights on these new phenomena's characteristics, but as rectangular shapes tend to be more convenient to design heat exchangers, studies have recently focused on two-phase flows within rectangular channels. However, only one study has investigated flows' behaviours in wide rectangular channels with a height of less than 100 μm and was carried by Ronshin et al. [8] on a 50 μm × 20 mm channel. The present study comes in its wake, as flows in a 50 μm × 10 mm channel have been investigated using the same methods to better understand the behaviours previously witnessed.

Experimental setup

As stated above, the purpose of this study is the experimental investigation of two phase flows of distilled deionized water and nitrogen gas in a microchannel with the height of 50 μm and a width of 10 mm. This paper focuses mainly on two-phase flows regimes and on the boundaries between them.

The test section, figure 1, consists of two parallel plates with a length of 160 mm and a width of 55 mm (the top plate being made of antireflecting glass and the bottom one of stainless steel); the distance between them is set by two constant spacers with a thickness of 50 μm. The liquid nozzle is placed in the lower plate at an angle of 11°. The microchannel dimensions are: length: 160 mm, width: 10 mm, height: 50 microns. Nitrogen of high purity is supplied into the central part of the micro channel from the tank. Its flow rate is adjusted from 20 to 1000 ml/min and kept constant with the aid of the flow controller with a precision of 0.5 %. It enters the gas chamber before going in the micro channel through a gas nozzle. The liquid flow rate varies from 0.5 to 50 ml/min using a high-precision (with an accuracy of 0.355 %) syringe pump. The liquid, extra-pure deionized nanofiltered water, is supplied into the microchannel through the liquid inlet after being pre-cleared. Its measured electrical conductivity after cleaning is 0.05 μS/cm at 25°C. The distance between the gas and liquid nozzles is about 70 mm. The pressure in the gas chamber is measured by a pressure sensor. The data it provides and the current gas flow rate are recorded on a file on a personal computer.

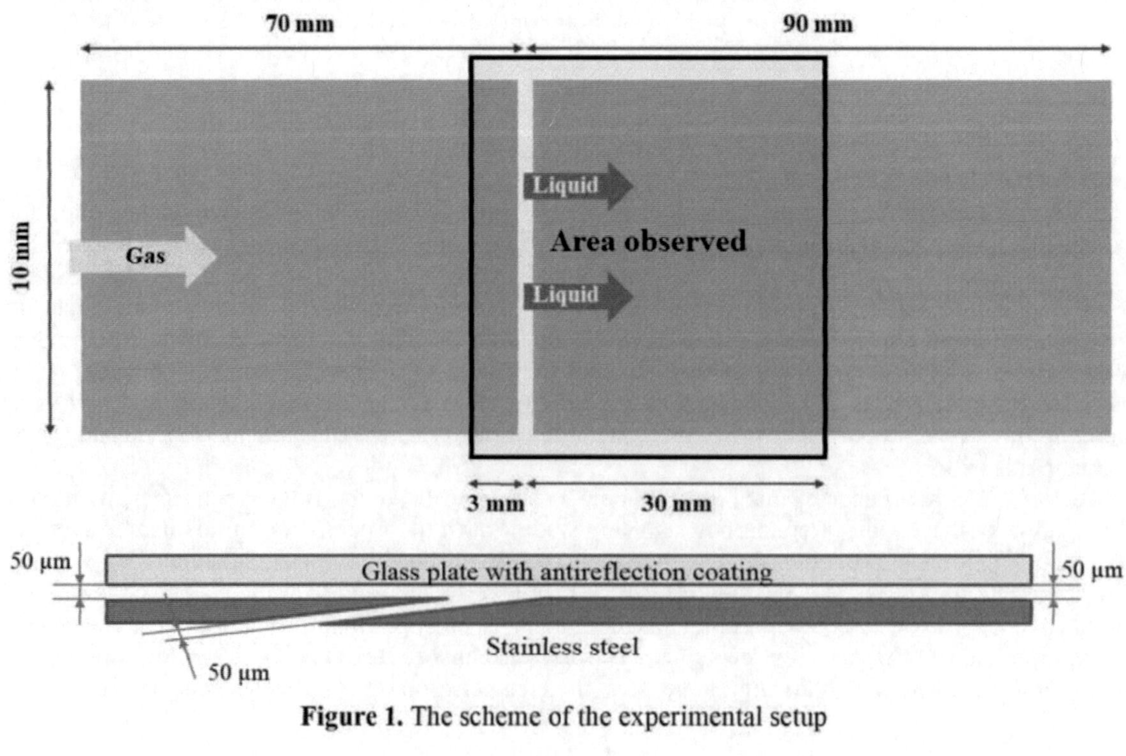

Figure 1. The scheme of the experimental setup

Interactions of gas and liquid in the microchannel are visualized in the observation area using digital video cameras according to the requirements of the schlieren method [9], figure 2, an experimental process used for registration and visualization of surface deformations of a thin liquid film which works as follows: the light enters the micro channel hosting the gas-liquid flow through a diffuser (1), lens (2), a beam splitter (3) and an optical glass (4). The reflection of the gas-liquid interface passes through a beam splitter (3), a lens (5) and a camera lens filter (6). The schlieren knife-edge (7), shifted by a micro-screw, highlights the central part of the light flux. As a result, the camera captures a grayscale image, where each grey level corresponds to a certain angle of inclination of the liquid-gas interface.

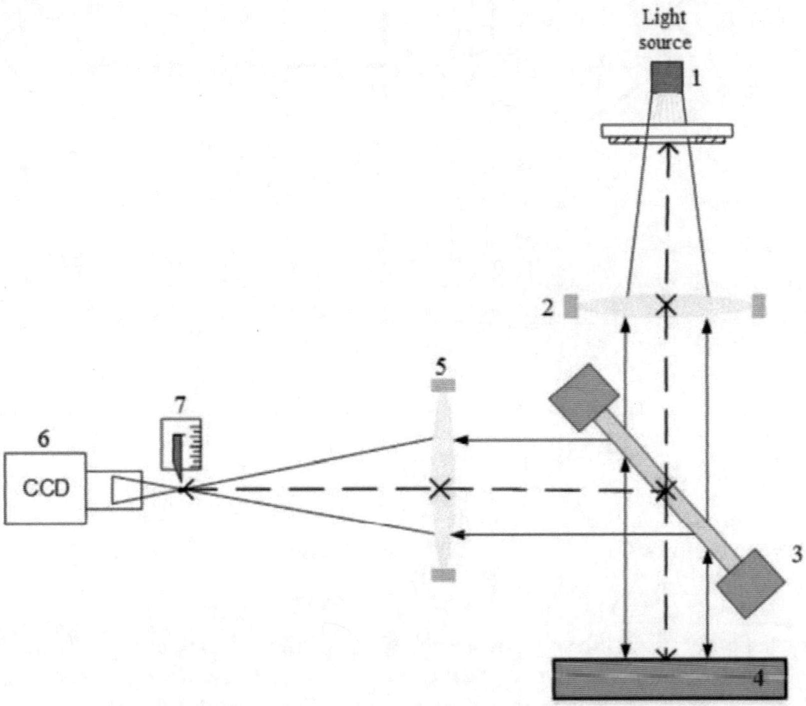

Figure 2. The scheme of the schlieren method

To allow us to perform our experiments in the best conditions possible, the glass has been pre-deposited with an antireflective coating and the stainless-steel plate has been pre-treated by polishing with an abrasive P400 with a grain size of 28 to 40 μm. After assembling, the microchannel height has been measured at several points using a confocal method. The wetting contact angles of the stainless steel and glass plates have been measured using the sessile drop method using the KRUSS DSA 100 installation. The advancing contact angle on the glass is 62.8±0.1°, and the receding contact angle is 27.6±0.7°. On a stainless steel plate the advancing contact angle is 90.1±0.1°, and the receding contact angle is 15.7±0.7°. The difference between our advancing and receding angles is our contact angle hysteresis, which characterize the liquid flow in our channel.

Experimental results

Our experiments allow us to distinguish the main regimes of the two-phase flow, which are the jet, bubble, stratified, churn and annular ones; and to draw its regime map, using the Reynolds number of liquid and gas as coordinates, figure 4.

$$Re_L = \frac{U_{SL} \cdot d_h}{\nu_L}; \; Re_G = \frac{U_{SG} \cdot d_h}{\nu_G},$$

IOP Conf. Series: Journal of Physics: Conf. Series **925** (2017) 012034 doi:10.1088/1742-6596/925/1/012034

where $U_{SL} = Q_L/S$ – superficial liquid velocity; $U_{SG} = Q_G/S$ – superficial gas velocity; Q_L – liquid flow rate; Q_G – gas flow rate; S – channel cross-sectional area; d_h – hydraulic diameter; ν_L – liquid kinematic viscosity; ν_G – gas kinematic viscosity.

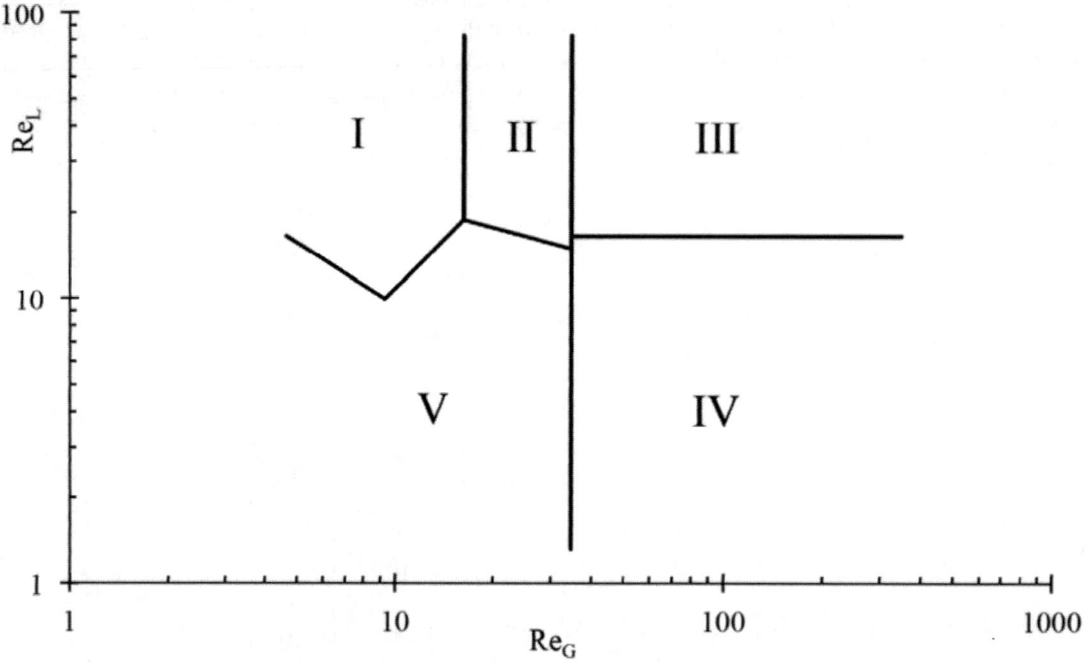

Figure 3. Map of two-phase flow regimes in the microchannel with a cross-section of 0.05×10 mm^2. The flow regimes: (1) bubble; (2) churn; (3) annular; (4) stratified; (5) jet

1.1. Jet regime

Jet regime corresponds to small superficial velocities of liquid and gas. In these conditions, we observed that the gas moves into the central part of the micro channel while most of the liquid moves alongside walls and as jets near the channel's center, as opposed to the jet regime in micro channels with a height of more than 100 μm where the liquid moves only along the channel's sidewalls [10,11]. The major characteristics of the jet regime (as detailed in [11]), which is specific to flat mini and micro channels, are: It can be observed at small superficial velocities of gas and liquid, when the gas flow does not occupy more than half the channel's width. There are no disturbances on the liquid surface. An increase of the liquid superficial velocity makes it occupy a much larger part of the channel (as expected), while the gas flow gets confined to the channel's center.

Two types of jet regimes can exist in microchannels [10]: the stationary and the pulsating jet regimes. The first one complies with the definition given above, where liquid superficial velocity is very low and liquid moves along the sidewalls, the upper wall stays dry and there are no pulsations. The second one corresponds to higher liquid flows where the liquid, still moving along the sidewalls, occupies most of the channel. At a certain moment, when the liquid load on the side reaches a critical amount, the water moves towards the channel's center to form a film on the upper wall before being dragged by the gas flow. After some time, the process is repeated, and the pulsating jet regime is reached.

As opposed to channel height of 300 μm [10], we didn't witness the pulsating jet regime during our experiments. We did observe that an increase in superficial liquid velocity leads to a loss of stability in the two-phase flow, which is in keeping with previous studies on the matter. Eventually, at low superficial gas velocities, keeping increasing the liquid superficial velocity will result in the formation of stable liquid bridges and in the reach of the bubble regime.

1.2. Bubble regime

As stated in the previous section, increasing liquid superficial velocity at low gas superficial velocity will eventually lead to the formation of stable horizontal bridges between the channel's sidewalls, which constitutes the transition between the jet and bubble regimes. In these conditions, the liquid contains many small bubbles of gas and occupies all the available space of the channel. The size and the frequency of the formation of the bubbles increase with the liquid and gas flow rates, but the bubbles' characteristic sizes remain marginal compared to the channel's dimensions.

1.3. Stratified regime

As shown on the regime map, the stratified regime can be observed at low superficial liquid velocities and at high superficial gas velocities. It is characterized by the presence of a liquid film in the bottom of the channel, dragged by the gas flow. As the gas occupies most of the channel's section, the upper wall remains dry. This regime can only be observed in flat microchannels, as in circular ones the liquid film closes to reach the annular regime [4].

1.4. Annular regime

If we come back to the regime described above and increase the superficial liquid velocity, a liquid film forms on the channel's upper wall: this is the transition from the stratified to the annular regime, determined using the schlieren method. In these conditions, moving vertical liquid bridges glide on the liquid films occupying the upper and lower walls of the microchannel. These films contain most of the liquid flow, and form several millimeters downstream the liquid nozzle, as opposed to flows in channels of a height of more than 50 μm where their formation occurs at the nozzle due to frontal instability [10]. The core of the two-phase flow is constituted of the gas flow, which fills most of the available space, partitioned by the moving liquid bridges.

Figure 4 is showing the schlieren photograph of the annular regime at U_{SG} = 11.7 m/s and U_{SL} = 0.233 m/s. Most of the channel is occupied by the core of gas and complex vertical liquid bridges, and the liquid film on the upper wall first forms on the channel's center (at the vertical from the liquid film on the lower wall) before spreading to its sides. The use of the schlieren technique enabled us to distinguish this film from the liquid film on the lower wall of the channel.

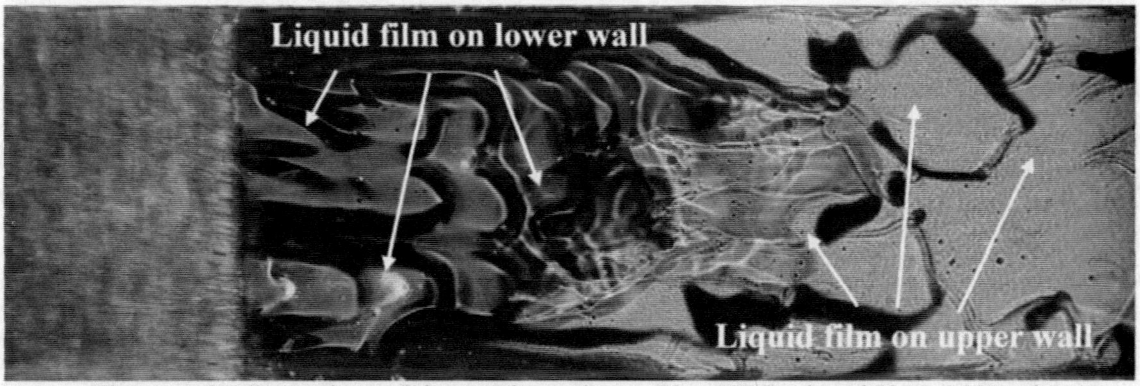

Figure 4. Schlieren photo of the annular regime at U_{SG} = 11.7 m/s and U_{SL} = 0.233 m/s

1.5. Churn regime

This particular regime is observed at superficial liquid velocities superior to 2 m/s and at superficial gas velocities going from 2 to 7 m/s, and is characteristic of vertical channels [12] and of wide horizontal microchannels [13]. It regroups features of both jet and bubble regimes, and is characterized by the presence of fractured horizontal liquid bridges. Its appearance is due to development of instabilities in the jet regime, and in the gas induced increase of the pulsation frequency of the liquid moving along the sidewalls.

The flow structure in this regime is chaotic: at the liquid inlet, there is formation of the liquid film on the channel's lower wall, and in some areas the liquid fills its entire height. A few millimeters downstream the nozzle, we generally observe the formation of the film on the upper wall of the microchannel, and we can witness some ruptures in this entity. However directly at the liquid nozzle, horizontal continuous and/or ruptured liquid bridges are continually formed, and will crumble in the channel over time.

This regime occupies a large area on the map of flow regimes. While the transition from the jet to the churn regime is accompanied by the appearance of numerous continuous and stable horizontal liquid bridges, the transition from the churn to the annular flow regime is accompanied by the disappearance of these bridges.

Conclusions

By comparing regime maps it appears that, in keeping with the previous studies carried by Ronshin et. al. [8] (with the 0.05×20 mm² and 0.05×40 mm² cross sections), decreasing the micro channel's width causes the churn flow regime's area to shrink at the benefit of all other regime flow areas.

Indeed, even though the hydraulic diameter changes insignificantly, we can affirm now that changing the channel's width has a significant effect on the two-phase flow regimes. This is because in wide micro channels, liquid jets forming near the liquid nozzle develop not only along the side walls, but also in the center of the conduct: their interaction results in the appearance of the churn regime. Thus, decreasing the channel's width reduces the number of such jets, which logically brings the constriction of this regime's region.

More specifically, we observed that due to pressure change (coming with the width reduction): The transition between the bubble and the churn regimes moves toward higher superficial velocities of gas. The crossover from the jet to the churn regime moves towards higher liquid superficial velocities. The border between the jet and the stratified areas moves toward lower gas superficial velocity. The transition from the stratified to the annular regime moves toward higher liquid superficial velocities. The transition from the churn to the annular regime happens at lower gas superficial velocities.

Acknowledgments

The work was financially supported by the Ministry of Education and Science of the Russian Federation (Agreement 14.613.21.0067, project identifier RFMEFI61317X0067).

References

[1] Gatapova E Y, Kabov O A 2008 *International Journal of Heat and Mass Transfer* **51** 4797
[2] Chinnov E A, Shatskii E N 2014 *High Temperature* **52** 461–64
[3] Rebrov E V 2010 *Theor. Found. Chem. Eng.* **44** 355–67
[4] Chinnov E A, Ron'shin F V, Kabov O A 2015 *Thermophysics and Aeromechanics* **22** 265–84
[5] Gatapova E Y, Ajaev V S, Kabov O A 2015 *JETP Letters* **101** 160
[6] Cheverda V V, Kabov O A, 2017 *Tech. Phys. Lett.* **43** 293–96
[7] Cheverda V V, Marchuk I V, Karchevsky A L, Orlik E V, Kabov O A 2016 *Thermophysics and Aeromechanics* **23** 415–20
[8] Ronshin F V, Cheverda V V, Chinnov E A., Kabov O A 2016 *Interfacial Phenomena and Heat Transfer* **4** 191–205
[9] Cheverda V, Glushchuk A, Queeckers P, Chikov S B, Kabov O A 2013 *Microgravity sci. technol.* **25** 73–81
[10] Chinnov E A, Ron'shin F V, Kabov O A 2016 *Int. J. Multiphase Flow* **80** 57–68
[11] Chinnov E A, Ron'shin F V, Kabov O A 2015 *Interfacial Phenomena and Heat Transfer* 3 243
[12] Xu J L, Cheng P, Zhao T S 1999 *Int. J. Multiphase Flow* **25** 411–432
[13] Kabov O A, Chinnov E A, Cheverda V 2007 *Microgravity Sci. Technol.* **19** 44–47

IOP Conf. Series: Journal of Physics: Conf. Series **925** (2017) 012035 doi:10.1088/1742-6596/925/1/012035

Study of microwave drying of wet materials based on one-dimensional two-phase model

Vl V Salomatov[1,2], V A Karelin[1,2]

[1] Kutateladze Institute of Thermophysics SB RAS, Kutateladze 2, Novosibirsk, 630090, Russia
[2] Novosibirsk State University, Pirogova 2, Novosibirsk, 630090, Russia

E-mail: salomatov.vv@mail.ru

Abstract. Currently, microwave is one of the most interesting ways to conduct drying of dielectric materials, in particular coal. In this paper, two processes were considered - heating and drying. The temperature field of the coal semi-mass in the heating mode is found analytically strictly with the use of integral transformations. The drying process is formulated as a nonlinear Stephen problem with a moving boundary of the liquid-vapor phase transformation. The temperature distribution, speed and drying time in this mode are determined approximately analytically. Parametric analysis of the influence of the material and boundary conditions on the dynamics of warming up and drying is revealed.

1. Introduction

Wet material drying processes are among the most energy-intensive. At the present stage is a theoretical and practical interest of heat transfer techniques to dried product, based, in particular, to microwave exposure. Important advantage of microwave energy is electromagnetic radiation ability to penetrate into the material to a considerable depth and creating in it the volume distribution of the heat sources without requiring the presence of a heat transfer medium. Microwave exposure is controlled and almost inertialess process that enables instantly change flow of heat. Energy consumption for evaporation during traditional drying types are up to 3.0 kWh / kg [1]. With microwave drying, they are reduced to 1.6-1.8 kWh / kg. The drying time is between 8-20 hours, and under microwave drying under otherwise equal conditions is reduced to 4 hours.

Most large-scale processes are drying coal fuel. Applying electromagnetic microwave radiation will increase the heat of combustion, to reduce the moisture content, ash content and also lower levels of mercury, chlorine, sulfur and other harmful constituents [2].

The combustion of fuel in the presence of the microwave field increases the rate of combustion, combustion becomes self sustainable, improved ignition mode, increase environmental performance and others.

To wide dissemination of microwave technology in coal energy, which in the world today there is a strong interest, interferes mainly developed of weak scientific foundations interconnected, rather complex electro and thermal processes. Use the full benefits of microwave effects on quality of coal preparation and drying as well as eco-friendly combustion is possible only in the presence of advanced mathematical modeling and analysis tools. The study of these problems is the subject of this work.

IOP Conf. Series: Journal of Physics: Conf. Series **925** (2017) 012035 doi:10.1088/1742-6596/925/1/012035

In the article analytically solved the problem of wet coal drying under the influence of a plane electromagnetic wave of the microwave range. The phase transition at the border is formulated in the form of Stephen conditions characterizing the consumption of heat for evaporation and drying, which determines the dynamics of the movement of the front. A mathematical model of microwave drying is considered in two stages.

2. Solutions for the heating stage

Stage 1 - the task of microwave warming. At the first stage we study of microwave warm of wet coal regime, when the absorption of microwave radiation is subject of Bouguer's law. By the time the first phase ends when the maximum temperature within the coal reaches the phase transition temperature "water - steam".

The mathematical formulation of the first phase includes the following key assumptions, which give the possibility of obtaining an analytical solution:

1. Internal power microwave heat absorption follows an exponential function of the coordinate - $q_v(x) = q_{v0}\exp(-\psi x)$.
2. Electrophysical and thermal characteristics of the coal substance conditionally constant.
3. The initial conditions are assumed to be zero.
4. There are no heat flows on the boundary surfaces in a first approximation.

This lead us to the system:

$$\frac{\partial T_2(x,t)}{\partial t} = a_2 \frac{\partial^2 T_2(x,t)}{\partial x^2} + \frac{q_{v_{02}}}{c_2\rho_2} e_2^{-\psi_2 x}, \tag{1}$$

$$0 \le x \le \infty,\ \not> 0;$$

$$T_2(x,0) = 0 \tag{2}$$

$$\frac{\partial T_2(0,t)}{\partial x} = 0 \tag{3}$$

$$\frac{\partial T_2(\infty,t)}{\partial x} = 0. \tag{4}$$

The system of differential equations solved on step 1 with double Laplace transform and cos-Fourier. As a result, we found as the temperature field of the first stage, and the time to achieve the temperature of phase transition "liquid-vapor":

$$T_2(x,t) = \frac{q_{v_{02}}}{\lambda_2\psi_2}\left\{\frac{2}{c_2\rho_2}(a_2t)^{1/2}\left[\frac{1}{\pi}e^{-\frac{x^2}{4a_2t}} - \frac{x}{2(a_2t)^{1/2}}\operatorname{erfc}\left(\frac{x}{2(a_2t)^{1/2}}\right)\right] - \frac{1}{\psi_2}e^{-\psi_2 x} + \Phi_{+}(x,t) + \Phi_{-}(x,t)\right\} \tag{5}$$

where $\Phi_{+(-)}(x,t) = \frac{1}{2\psi_2}\exp\left(\psi_2^2 a_2 t - (+)\psi_2 x\right)\operatorname{erfc}\left[\psi_2(a_2t)^{1/2} - (+)\frac{x}{2(a_2t)^{1/2}}\right]$.

Time of phase transition can be found from $T_2(x,t_{pt}) = T_{pt}$.

Scheme of the problem is shown in figure 1.

IOP Conf. Series: Journal of Physics: Conf. Series **925** (2017) 012035 doi:10.1088/1742-6596/925/1/012035

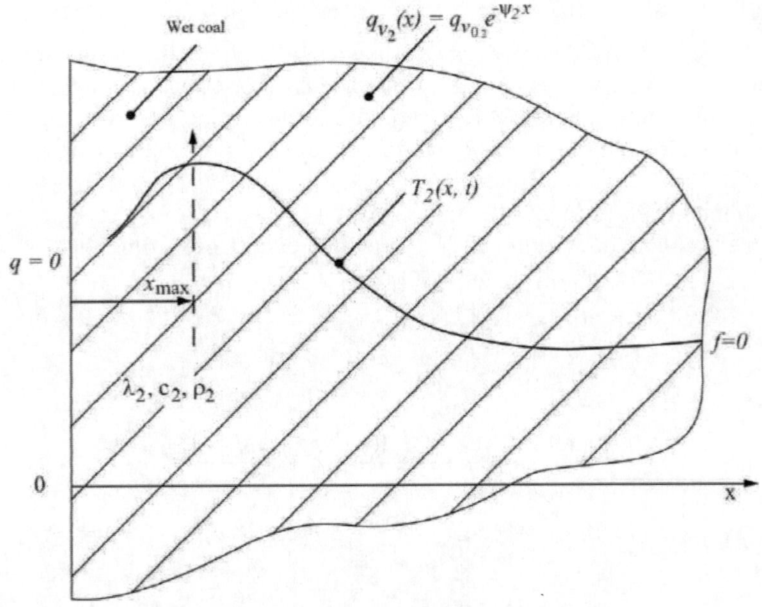

Figure 1. Scheme of the first stage of the task.

3. Solutions for the drying stage

Stage 2 - the problem of microwave drying. The drying process in this article studied as heat transfer medium in the phase-change - known Stephen problem [3]. In our case, we propose a new approach to the microwave drying of coal array of the problem, which is based on a method "heat quasi-stationary approximation" developed by us. It is based on the concept that the movement of the phase transformation front controls part of the quasi-stationary temperature field.

For calculation formulas and the further search for the basic laws of the microwave drying process itself develops as follows. Due to internal heat sources from the microwave absorption of coal mass after reaching the phase transition temperature dries, forming three different areas in their properties in figure 2.

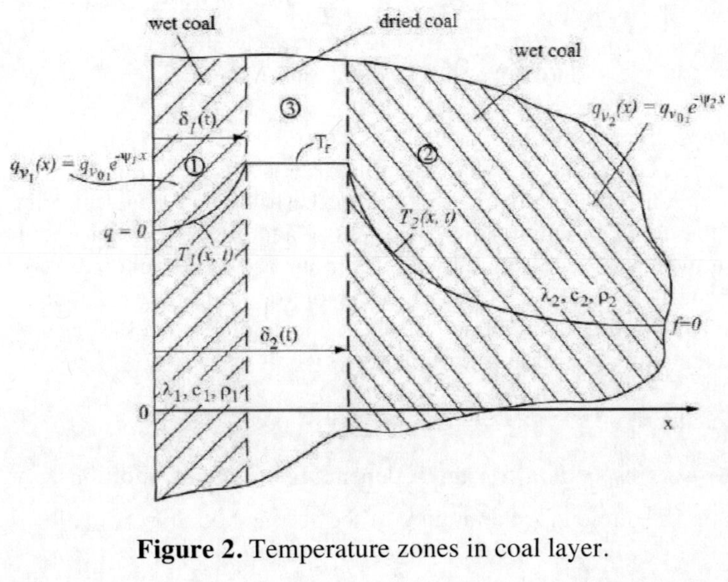

Figure 2. Temperature zones in coal layer.

IOP Conf. Series: Journal of Physics: Conf. Series **925** (2017) 012035 doi:10.1088/1742-6596/925/1/012035

Zones 1 and 2 are partially dried. Zone 2 is taken fully dried. In the future, the heat from the microwave absorption is spent on evaporation of residual moisture and making the phase transition "liquid - vapor" on moving the borders. It is required to determine the dynamics of the reduction of humidity, temperature distribution in zones 1 and 2, the dynamics of the reduction of humidity.

During these assumptions, the system of equations is:
The equation of thermal conductivity for zone 1 (the time count goes from t_{be}):

$$\frac{\partial T_1(x,t)}{\partial t} = a_1 \frac{\partial^2 T_1(x,t)}{\partial x^2} \tag{6}$$

$$t>0 \quad 0<x<\delta_1$$

Heat equation for zone 2:

$$\frac{\partial T_2(x,t)}{\partial t} = a_2 \frac{\partial^2 T_2(x,t)}{\partial x^2} + \frac{q_{v_O}}{c_2\rho_2} e^{-\psi_2 x} \tag{7}$$

$$t>0 \qquad x>\delta_2$$

Initial conditions:

$$T_1(x, t) = \varphi_1(x) \tag{8}$$

$$T_2(x, 0) = \varphi_2(x) \tag{9}$$

Where $\varphi_1(x)$, $\varphi_2(x)$ – known distributions of $T_1(x, t_{be})$, $T_2(x, t_{be})$ in two zones at the time of the onset of evaporation t_{be}.
Border conditions

$$\frac{\partial T_1(0,t)}{\partial x} = 0 \tag{10}$$

$$\frac{\partial T_2(\infty,t)}{\partial x} = 0 \tag{11}$$

Stefan's condition on the moving "liquid-vapor" boundaries

$$-\lambda_1 \frac{\partial T_1(\delta_1,t)}{\partial x} = \rho_1 H_и \frac{d\delta_1}{dt}, \tag{12}$$

$$T_1(\delta_1, t) = T_{ph} \tag{13}$$

$$-\lambda_2 \frac{\partial T_2(\delta_2,t)}{\partial x} = \rho_2 H_и \frac{d\delta_2}{dt}, \tag{14}$$

$$T_2(\delta_2, t) = T_p \tag{15}$$

In the microwave drying step the problem is solved analytically approximated using effective asymptotic procedures. Dynamics of the phase boundary is found within the framework of the quasi-stationary approximation. This approach is based on the following physical principle. Due to lack of comparability of the transfer rate of the heat pulse and movement of the phase boundary, the evaporation front movement is controlled by the limit on part-time temperature field. The fundamental role given to the quasi-stationary asymptotic behavior for large times. Calculations on the derived dependencies found such important characteristics of the studied processes as a warm-up time, drying speed, drying time and other popular options practices.

$$\delta_1(t) = \frac{1}{4P}\left(C_{L1} - C_{L2}\cdot t + 4M^2\cdot t^2\right) \tag{16}$$

Where C_{L1}, C_{L2}, M – corresponding constants, depending on initial conditions.
Similarly, there is find a solution for the motion of the second boundary $\delta_2(t)$

After $\delta_1(t)$ and $\delta_2(t)$ were found – it is possible to find out another parameters:

To calculate the amount of evaporated moisture at the time t, we can use the expression:

$$W_w(t) = \rho_w(\delta_{\max} - \delta_1(t)) + \rho_w(\delta_2(t) - \delta_{\max}) \quad (17)$$

Where $\delta_1(t)$ and $\delta_2(t)$ - coordinates of the motion of the fronts of phase transformation in 1 and 2 zones, respectively.
Thus, knowing the initial and final moisture of the coal, you can calculate the time to reach any humidity and the total drying time.
The drying speed can be found:

$$V = W_w'(t) \quad (18)$$

4. Conclusion

The mathematical model of warming up and drying of coal semi-mass is studied in the article, the results of which can serve as a theoretical basis for the technology of intensive microwave effect on moist material. At the stage of heating by the method of the dual integral Laplace transform and cos-Fourier, the problem is solved analytically strictly. At the stage of microwave drying, the problem is solved approximately analytically. The process of electromagnetic drying is considered as a nonlinear Stephen problem with a "liquid-vapor" phase transition. The motion's dynamics of the phase interface was found within the framework of a quasi-stationary approximation. This approach is based on the following physical principle. Because of the rates incompatibility of transfer of the thermal pulse and the interface's motion, the movement of the evaporation front is controlled by the time-limited part of the temperature field. The fundamental role is given to the quasi-stationary asymptotics for large times. By calculating the derived dependencies, such important characteristics of the studied processes as the warm-up time, drying speed, drying time and other parameters required by practice have been found.

References

[1] Brandt L A 1963 *Investigation of dielectrics at ultrahigh frequencies* (Moscow: Fizmatgiz)
[2] Khaidarova A A 2009 *Influence of microwave energy on brown coal for improvement of its technological characteristics* (Tomsk: Teplofizichesky bases of power technologies, Publishing house TPU) p 109-113
[3] Lykov A V 1968 *The theory of drying* (Moscow: Energia)

IOP Conf. Series: Journal of Physics: Conf. Series **925** (2017) 012036 doi:10.1088/1742-6596/925/1/012036

The drop evaporation on a heated substrate with single wall nanotubes coating

A A Semenov[1,2] and D V Zaitsev[1,2]

[1]Kutateladze Institute of Thermophysics, Siberian Branch of Russian Academy of Sciences, 630090, Novosibirsk, Russia
[2]Novosibirsk State University, 630090, Novosibirsk, Russia

E-mail: semenov.itp@gmail.com

Abstract. Evaporation of a sessile nanoliter water droplet was investigated experimentally at the substrate temperature of 30 ^{0}C. The studies were performed on the two substrates made of copper. One of them had no coating and was polished to the root mean square roughness of 20 nm. Another one was coated with single wall nanotubes (the root mean square roughness is 62 nm). The research has shown that specific rate of evaporation (mass loss per unit of the drop surface area) on the substrate with nanotubes coating is up to 20 percent higher in comparison with the substrate without coating.

1. Introduction

The phenomenon of liquid drop evaporation, which takes place in a variety of technological systems in power engineering, agriculture, medicine, cooling systems, chemical and other industries [1], has been actively investigated during the last few years [2-6]. In most experimental and theoretical studies found in the literature, isothermal evaporation of water droplets has been studied. Publications on nonisothermal evaporation of liquid droplets are very limited.

The main objective of the present work is to study the effect of nanotubes coating on the dynamics and evaporation of the sessile liquid drop in quasistationary conditions when it is heated from the substrate. It is assumed that due to the multifarious influence (variative wettability, thermal conductivity of wall layer, surface structure at the micro- and nanoscale) the use of the nanotubes coating allows to intensify the evaporation process.

In this paper, we study evaporation of a liquid drop into the open atmosphere. The initial drop volume was less than 1 μl. A drop of liquid is placed on a surface with controlled wettability. The drop is heated until quasi-stationary heat transfer between the solid substrate and liquid drop. Distilled deionized nano-filtered water of Milli-Q Company is used as the working fluid. The use of water with relatively high boiling point as a working fluid allows to study in detail the influence of temperature difference between the solid surface and surrounding atmosphere on the rate of evaporation and heat transfer. In this work the temperature difference between the solid and external atmosphere was about 10 °C.

IOP Conf. Series: Journal of Physics: Conf. Series **925** (2017) 012036 doi:10.1088/1742-6596/925/1/012036

2. Experimental setup

The experiments on water drops evaporation in the open atmosphere were made on the experimental stand the scheme of which is presented in Fig. 1. The base of the test section is an aluminum plate of 15 mm thickness with the heater attached to its bottom side. A Peltier element with dimensions 40x40 mm^2 was used as a heater. On the upper side of the aluminum plate, the investigated substrate was fixed. Measurement of the surface temperature of the studied substrate was carried out using a thermocouple, fixed at a distance of 1-2 mm from the contact line of the liquid drop. The temperature under the substrate was measured by three thermocouples of K-type. Thermocouples were placed radially at different distances from the center of the substrate and served to measure the temperature distribution along the substrate radius. The experimental setup was covered with transparent box made of PMMA, which has dimensions of 1000x800x900 mm^3. The relative humidity of air in the box during the experiments was 20-30 %.

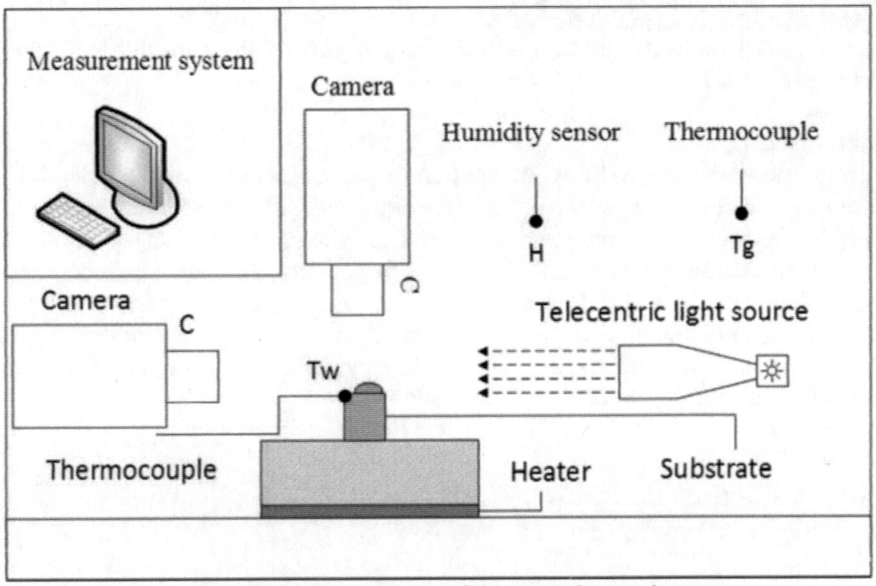

Figure 1. The scheme of the experimental setup.

Two optical methods were used in the work. The first one is the shadow method, consisting of a halogen light source Edmunds Optics MI-150 connected by a fiber optic lightguide with a lens system that produces a beam of parallel light of 50 mm in diameter on one side of the working area, and a PointGray fast speed camera with a 10X Mitutoyo microscope lens, which was connected through Novoflex focusing bellows from the opposite side. The spatial resolution of such a system was 0.5 μm / pixel, and the shooting frequency was up to 500 frames per second. Obtained shadow photos of the liquid drop profile were processed by the KRUSS Drop Shape Analysis software.

The second optical technique used is 5 megapixels Imaging Source camera with 20X Mitutoyo microscope lens, which was placed above the liquid drop and allowed visualization of evaporation process from the top and control the drop symmetry. If the drop lost its symmetry, the experiment was repeated.

Two substrates made of copper were investigated. One of them had no coating and was polished to the root mean square roughness of 20 nm. Another one was coated with single wall nanotubes orientated along the surface. The coated substrate had the root mean square roughness of 62 nm. The morphology of the working surfaces was analyzed using a scanning electron microscope (HITACHI S3400N) and atomic force microscope (Solver Pro NT MDT), Fig. 2.

IOP Conf. Series: Journal of Physics: Conf. Series **925** (2017) 012036 doi:10.1088/1742-6596/925/1/012036

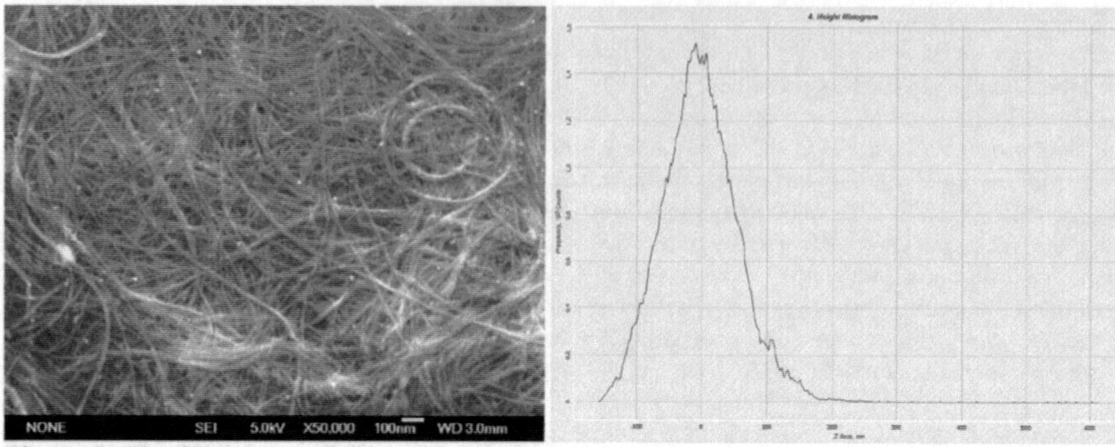

Figure 2. The SEM image (left) and height histogram (right) of the substrate with the single wall nanotubes coating.

3. Experimental results

The experiment began with a preliminary preparation of the investigated substrate, which consisted in a thorough cleaning of the working surface. Then the substrate was placed in a reservoir with distilled water for a day. Thereafter, to remove residual water from the surface the substrate was flushed with clean compressed air designed for cleaning optical components. A liquid drop of pre-set size was placed on the surface, heated to the desired temperature, using a high-precision dispenser Lenpipet. The lens used in the shadow method was focused (depth of field was a few hundreds of μm); and after that the periodic shooting was performed until complete evaporation of the liquid droplet. The upper camera was switched on simultaneously with the shadow method.

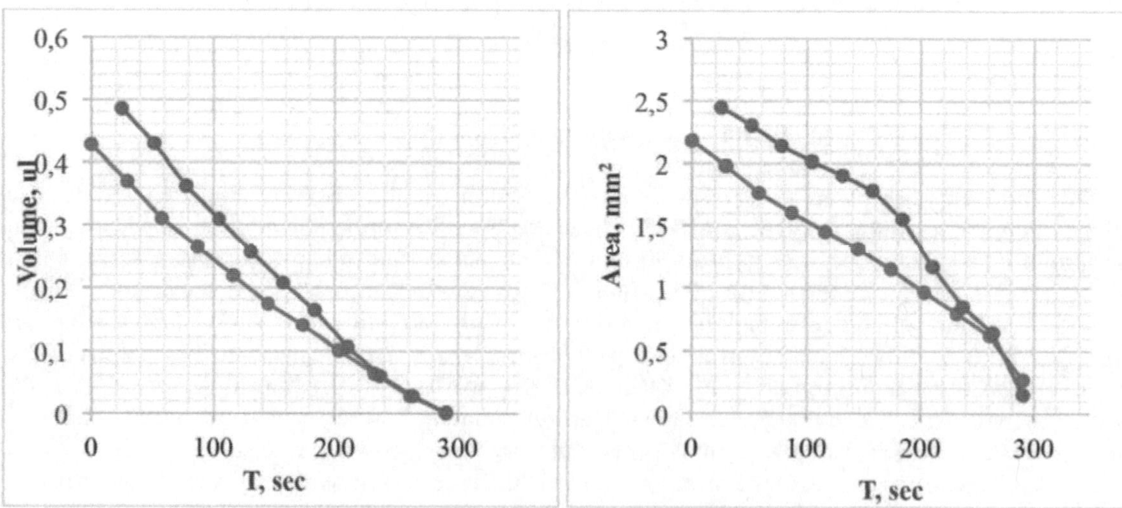

(a) (b)

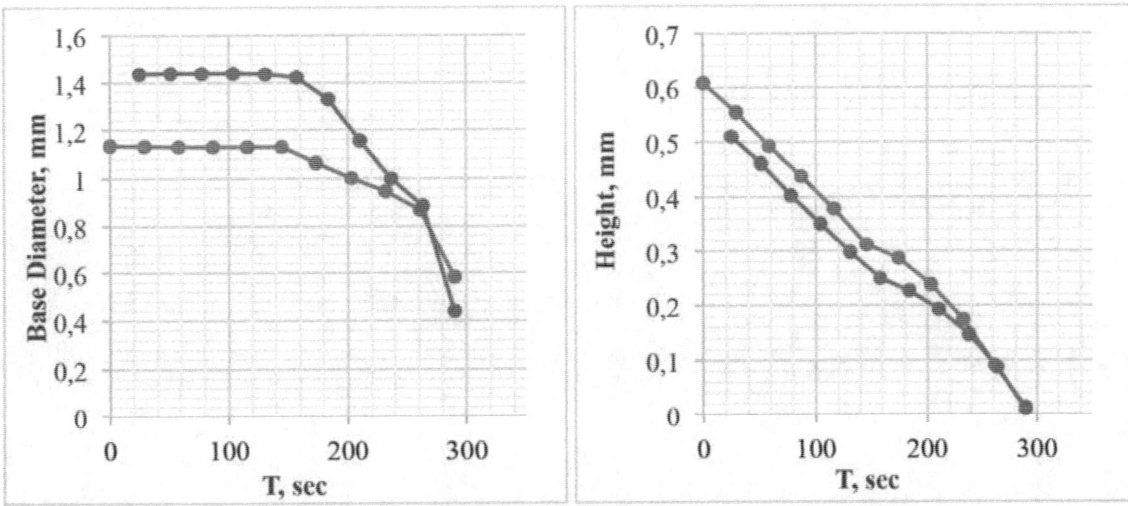

(c) (d)

Figure 3. Dependences of different parameters of droplet vs evaporation time. (a) – the volume of droplet, (b) – the surface area of droplet, (c) – the diameter of droplet base and (d) – the height of droplet. Blue lines and dots are data for polished copper substrate, red – copper substrate with nanotubes coating. Initial temperature of all substrates was T=30 ^{0}C. All data was synchronized for the moment of full droplet evaporation.

Dependences of various parameters of a sessile liquid drop during evaporation vs time are shown in Fig.3 for the substrate temperature of T=30°C. The obtained data are synchronized in time for the moment of complete evaporation of the droplet. It is seen that in spite of the fact that the liquid drop on the coated substrate had slightly higher initial volume, the time of evaporation for this drop is slightly lower, compared with the drop on the substrate without coating.

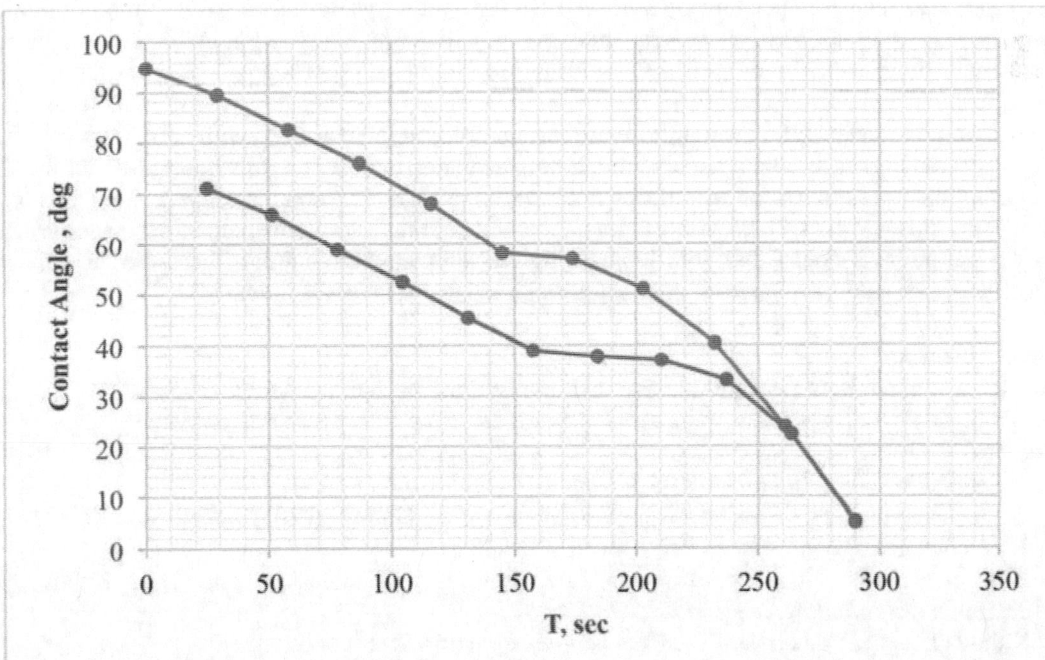

(a)

IOP Conf. Series: Journal of Physics: Conf. Series **925** (2017) 012036 doi:10.1088/1742-6596/925/1/012036

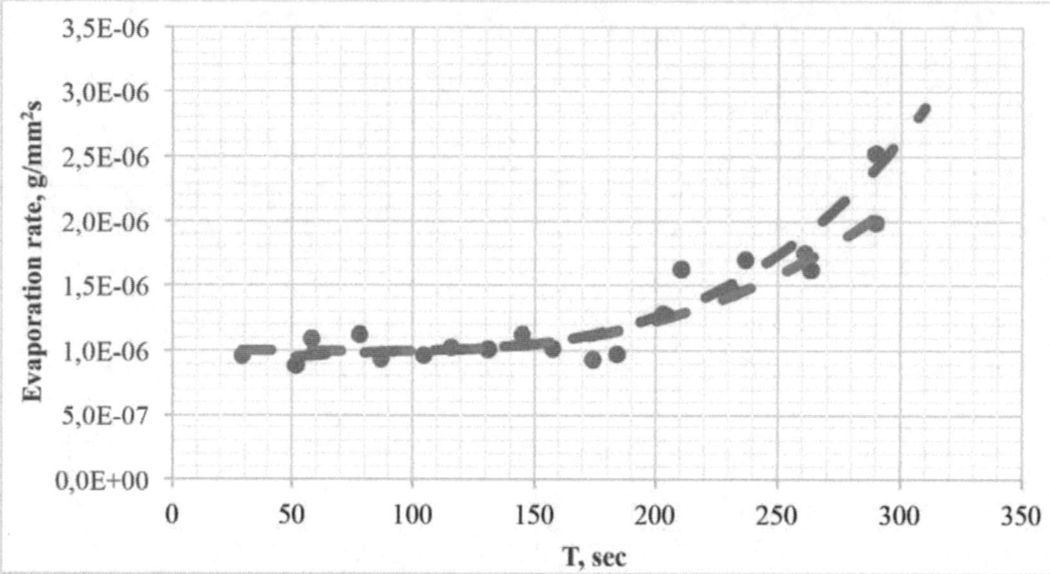

(b)

Figure 4. The dependences of contact angle (a) and specific evaporation rate (b) on time. Blue lines and dots – data obtained with the polished copper substrate; red – copper substrate with nanotubes coating. The dotted lines in figure (b) are the trend lines of the corresponding data. Initial temperature of all substrates was T=30 ^{0}C. All data was synchronized for the moment of full droplet evaporation.

The dependences of the contact angle and specific rate of liquid evaporation on time at different substrates are shown in Fig. 4. The evaporation rate was calculated as a loss in the droplet mass per unit of droplet surface area per unit of time. All the data were synchronized to the time of complete evaporation of liquid droplets. It is seen that the specific rate of evaporation on both substrates increases with time, which is in agreement with the results of [2, 3] where the evaporation of liquid drops with initial volume of about 100 μl was studied. Also, it is seen that at the final stage of the drops evaporation, the specific evaporation rate for the drop on the substrate with nanotubes coating is about 20 percent higher in comparison with the polished substrate. It is worth noting, that at the final stage all the main parameters of both drops (volume, base diameter, contact angle,…) are very close (Figs. 3, 4a). This means that the use of the nanotubes coating allows to intensify the sessile liquid drop evaporation. We expect that with increase of the substrate temperature, the difference between the specific evaporation rate between the coated and uncoated surfaces will increase.

Acknowledgments
The work was funded by RFBR and Novosibirsk region (research project № 17-48-540124).

References

[1] J. Kim 2006 *International Journal of Heat and Fluid Flow* **28** 753
[2] E.Ya. Gatapova, A.A. Semenov, D.V. Zaitsev, O.A. Kabov 2014 *Colloids and Surfaces A: Physicochemical and Engineering Aspects* **441** 776
[3] A.A. Semenov, D.V. Feoktistov, D.V. Zaitsev, G.V. Kuznetsov, and O.A. Kabov 2015 *Thermophysics and Aeromechanics* **22** 771
[4] Kabov O.A. and Zaitsev D.V. 2013 *Doklady Physics* **58** 292
[5] G.V. Kuznetsov, D.V. Feoktistov, E.G. Orlova, K.A. Batishcheva 2016 *Colloid Journal* **78** 335
[6] G. V. Kuznetsov, D. V. Feoktistov, and E. G. Orlova. 2016 *Journal of Engineering Physics and Thermophysics* **89** 317